IAEI

Soares Book on

Grounding

IAEI

Soares Book on
Grounding

Seventh Edition

J. Philip Simmons

International Association of Electrical Inspectors
Richardson, Texas

Notice to the Reader

Publisher does not warrant or guarantee any of the products described herein or perform any independent analysis in connection with any of the product information contained herein. Publisher does not assume, and expressly disclaims, any obligation to obtain and include information referenced in this work.

The reader is expressly warned to consider carefully and adopt all safety precautions that might be indicated by the activities described herein and to avoid all potential hazards. By following the instructions contained herein, the reader willingly assumes all risks in connection with such instructions.

The publisher makes no representations or warranties of any kind, including but not limited to, the warranties of fitness for particular purpose or merchantability, nor are any such representations implied with respect to the material set forth herein, and the publisher takes no responsibility with respect to such material. The publisher shall not be liable for any special, consequential or exemplary damages resulting, in whole or in part, from the reader's use of, or reliance upon, this material.

International Association of Electrical Inspectors
901 Waterfall Way, Suite 602
Richardson, TX 75080-7702
(972) 235-1455 fax: (972) 235-3855

First Edition published 1966
Printed in the United States of America.
03 02 01 00 99   5 4 3 2 1
ISBN: 1-890659-18-5

Library of Congress Cataloging in Publication Data

Simmons, J. Philip
    IAEI Soares book on Grounding, 7th edition
    Includes index.

Library of Congress Catalog Card Number: 96-77947

Contents

About the Author

J. Philip Simmons is the author of the seventh edition of this text and was responsible for updating the work to the 1999 National Electrical Code®. He also was responsible for updating the fourth, fifth and sixth editions of this text and for completely reorganizing and expanding the work.

Mr. Simmons presently serves as education, codes and standards coordinator for the International Association of Electrical Inspectors. As such, he is responsible for developing and updating educational materials. He also serves as a presenter at seminars on electrical codes and safety. Phil served as executive director of IAEI for five years and was international president in 1987, after serving on the board of directors. Other service to the IAEI by Simmons has included secretary-treasurer and chairman of the Puget Sound Chapter and secretary-treasurer of the Northwestern Section.

Simmons is contributing editor of the *IAEI News* and was author of the IAEI books on the Analysis of the 1993 and 1996 National Electrical Code®. He was coauthor of the IAEI book on electrical systems in One- and Two-Family Dwellings and has served as an instructor at many seminars throughout the country. Phil is author of the 1996 and 1993 editions of *Ferm's Fast Finder Index* book. He also is chairman of the Neon Installation Manual committee.

Formerly chief electrical inspector of the state of Washington, Department of Labor and Industries, Mr. Simmons also served the people of the state as a field electrical inspector, electrical plans examiner, health care facility inspector and electrical inspector supervisor.

Mr. Simmons worked in an industrial plant for several years where he learned the electrical trade. He went on to form his own electrical contracting firm which he successfully operated for several years. Simmons is licensed as a journeyman electrician and administrator in the state of Washington. He successfully passed the certification examinations for construction code inspectors and holds a certificate in the One- and Two-Family Electrical Inspector, General Electrical Inspector and Electrical Plan Review classifications.

A student of the NEC® for many years, Mr. Simmons developed and taught Code-related courses for the state of Washington and the IAEI.

Mr. Simmons presently serves on the NFPA, National Electrical Code®, Code Making Panel 5 which is responsible for Article 250, Grounding. He has served on CMP-1 and 17 and as chairman of Code Making Panel 19. He also was a member of the NEC® Technical Correlating Committee. He also serves on the NFPA-73 committee, the Electrical Systems Maintenance Code for One- and Two-Family Dwellings.

Simmons served for six years as a member of the NFPA Standards Council and is past chairman of the NFPA Electrical Section. Other service to NFPA include the Advisory Committee on Adoption and Use of the National Electrical Code® and Related Documents and the Bi-National Correlating Committee for the Harmonization of the Electrical Codes.

Mr. Simmons serves on the Underwriters Laboratories' Electrical Council and is a past UL Trustee. He also is a member of the International Brotherhood of Electrical Workers.

He also is a member of the electrical advisory committee, National Certification Program for Construction Code Inspectors for The Chauncey Group, formerly the Educational Testing Service.

Phil has five children and eleven grandchildren. His education includes a certificate in Public Administration and many developmental and continuing education programs.

Preface

This book is dedicated to the memory of Eustace C. Soares, P.E., one of the most renowned experts in the history of the National Electrical Code® in the area of grounding electrical systems. A wonderful teacher and man of great vision, Eustace foresaw the need for better definitions to clear up the great mystery of grounding of electrical systems.

Eustace Soares' book, *Designing Electrical Distribution Systems for Safety* was originally published in 1966 and was based upon the 1965 edition of the National Electrical Code®(NEC®). Over the years, this book has become a classic.

A great majority of the recommendations contained in the original edition of his book have been accepted as part of Article 250 of the National Electrical Code®. The grounding philosophies represented in the original edition are just as relevant today as they were then. To say that Eustace contributed more than any other man to solving some of the mysteries of grounding of electrical systems would not be an overstatement of fact.

Previous editions have been extensively revised both in format and in information. An effort has been made to bring the work into harmony with the 1999 edition of the National Electrical Code®. This author has attempted to retain the integrity of the technical information for which this work has been well known and at the same time add additional information which may be more recent on the subject of grounding.

It is our intent to revise this work to complement each new National Electrical Code® so this will be an on-going project. If the reader has suggestions for additional pertinent material or comments about how this work could be improved upon, they would be most welcome.

National Electrical Code® and NEC® are registered trademarks of the National Fire Protection Association, Inc., Quincy, MA 02269.

Acknowledgements

This edition of the *IAEI Soares Book on Grounding* has been produced through the combined efforts of a team of talented and highly qualified people. In addition, several firms or organizations have allowed quotations from their works or have provided photographs or illustrations which enhance the book.

The individuals who have made significant efforts during the production of this edition of the book and who deserve special recognition include the following:

- Kathryn Ingley, for putting in many hours editing the text;
- Brady Davis, Elaine Flynn, Joanne Beverly, and Richard Church, who were the production team at the Richardson office and worked on layout and revisions to illustrations. This team put in many hours of work in layout and proofing to ensure the book is of the highest caliber.
- Kirk Massey and Frank Kripaitis, CMBA Advertising, for designing the cover.
- Paul Dobrowsky and Chuck Mello for reviewing the manuscript.
- Travis Lindsey, for providing information on the grounding electrode study.

Several companies furnished text, illustrations, drawings, photographs or granted permission to reprint information from their material for which we are most appreciative. These companies include:

- AEMC Instruments, Chauvin Arnoux, Inc., Boston, MA
- AFC Cable Systems, New Bedford, MA
- AVO International, Biddle Instruments, Dallas, TX
- Cooper Industries, Bussmann Division, St. Louis, MO
- Eagle Electric Company, Long Island City, NY
- Insulated Cable Engineers Association, South Yarmouth, MA
- International Electronic and Electrical Engineers, Piscataway, PA
- National Electrical Manufacturers Association, Rosslyn, VA
- National Fire Protection Association, Quincy, MA
- NFPA Research Foundation, Quincy, MA
- Post Glover Resistors, Inc., Erlanger, KY
- Square D Company, Palatine, IL
- Steel Conduit Section, National Electrical Manufacturers Association, Rosslyn, VA
- Thomas & Betts, Memphis, TN
- Underwriters Laboratories, Northbrook, IL

I wish to thank the above individuals and companies for their assistance and contributions in the preparation and production of this edition of the *IAEI Soares Book on Grounding*.

Phil Simmons

Chapter 1

General

Objectives

After studying this chapter, the reader will be able to understand:

- Fundamentals and purpose of grounding of electrical systems.
- Various definitions relative to grounding.
- Equipment from grounded and ungrounded systems.
- Effect of electric shock hazards on humans and animals.
- Purpose of grounding and bonding.
- Differences between a short circuit and ground fault.
- Basic design requirements for grounding of electrical systems.
- Circuit impedance and other characteristics.

Grounding of electrical systems

Some features of electrical safety are so fundamental they have appeared in some form in every edition of the National Electrical Code®. These include requirements for suitable insulation for conductors and overcurrent protection for circuits.

Another long-time common electrical safety code requirement is grounding of electrical systems and equipment for safety. The grounding of equipment and conductor enclosures, as well as the grounding of one conductor of an electric power and light system, has been practiced in some quarters since the use of electricity began.

At first, there was no uniform standard for grounding. (See the Appendix of this text for information on the early history of grounding including debates that took place on how best to approach the subject.) However, it was not long before it became universal to ground one conductor on all 120-volt lighting circuits. Early code books firmly established the practice of grounding by making it mandatory to ground all such systems where the voltage to ground does not exceed 150 volts. These code books also recommended that alternating current systems be grounded where the voltage to ground did not exceed 300 volts while at the same time stating that higher voltage systems were permitted to be ungrounded. Later, the Code made it mandatory to ground any system having a nominal voltage to ground of not more than 300 volts if the grounded service conductor (usually a neutral) was not insulated.

Grounded

The term "grounded," is defined in the National Electrical Code® (NEC®) in Article 100 as "Connected to earth or to some conducting body that serves in place of the earth."[N] Conducting bodies that serve in place of the earth include conduit, boxes, enclosures, equipment grounding conductors and wiring devices. These are, in fact, an extension of the earth by being electrically connected to the earth by reliable electrical and mechanical means.

Metal frames of buildings that are effectively grounded are acceptable as an extension of the earth so far as being a grounding electrode conductor but are not permitted to be used as an equipment grounding conductor. Frames of vehicles are permitted to serve as a grounding electrode for vehicle-mounted generators and as a return path for low-voltage systems in recreational vehicles.

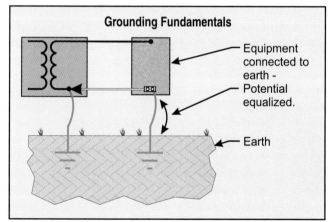

Figure 1-2

The earth as a whole is properly classed as a conductor. For convenience, its electric potential is assumed to be zero. Based upon the composition of the soil or earth, the resistance of segments of the earth can vary widely from one area to another. The earth is composed of many different materials, some of which, especially when dry, are very poor conductors of electricity. Soil temperature, moisture content and chemical composition are factors that have a great influence on soil resistance. As a result of these factors, the capability of the earth to carry electrical current also varies widely. (See

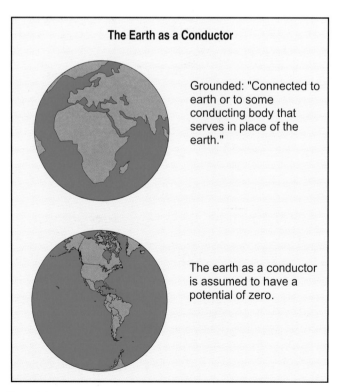

The Earth as a Conductor

Grounded: "Connected to earth or to some conducting body that serves in place of the earth."

The earth as a conductor is assumed to have a potential of zero.

Figure 1-1

Chapter Six of this text for more information on composition of the earth and effectiveness of grounding electrodes.)

A metal object such as a box or other equipment enclosure that is grounded by connecting (bonding) it to the earth by means of a grounding electrode, grounding electrode conductor and/or equipment grounding conductor, is thereby forced theoretically to take the same zero potential as the earth. Slight differences in potential may exist due to differences in impedance of the materials or connections. Any attempt to raise or lower the potential of the grounded object results in current passing over the grounding path until the potential (voltage) of the object and the potential (voltage) of the earth (zero) are equalized. Usually, this potential above ground is caused by a line-to-ground fault. Hence, grounding is a means for ensuring that the grounded object cannot take on a potential differing enough from earth potential to be hazardous. When the grounding conductor is broken, is inadequate in size, or has a poor connection, a hazardous, above ground potential may be present from an abnormal condition.

Resistance and impedance

Though a comprehensive discussion of the subject is beyond the scope of this text, a brief explanation of the terms "resistance" and "impedance" is offered.

For direct current (dc) systems and circuits, the term "resistance" is properly used to describe the opposition to current flow. The term is abbreviated "R." We are accustomed to using the term "ohms" to relate to the resistance of the circuit such as 35 ohms. Ohm's Law can be summarized as follows: in a dc circuit, the current is directly proportional to the voltage and inversely proportional to the resistance. ($I = E/R$) This means, as the voltage is increased, the current will increase through a fixed resistance. As the resistance is reduced, the current will increase if the voltage stays the same. A pressure (voltage) of one volt will cause one ampere to flow through a resistance of one ohm.

For alternating current (ac) systems and circuits, "impedance" is the proper term used to describe the total opposition to current flow. The term is abbreviated "Z." Impedance consists of three components: inductive reactance, capacitive reactance and resistance. You will find that the term "impedance" rather than "resistance" is used most often throughout this text since, for the most

part, ac electrical systems and circuits are being considered. See Chapter 11 for a detailed discussion of the importance of keeping all conductors of the circuit, including the equipment grounding conductor, close together to keep the impedance as low as possible.

Many excellent books on electrical theory cover the subject of impedance, inductive reactance, capacitive reactance and resistance.

Equipment supplied from a grounded system

Where the electrical system is grounded, it is critical to provide a low-impedance path of adequate capacity from all the equipment supplied by the system back to the source of the system. This is required by Section 250-2 and is emphasized in several portions of this text. The first reason is to maintain the metal equipment enclosures as close to earth potential as possible to reduce a shock hazard. The second reason is to ensure the overcurrent protection will operate in the event of a line-to-ground fault.

Where electrical equipment that is supplied by a grounded system is left ungrounded or is poorly grounded, it becomes a silent and often lethal source of electrical shock where a ground-fault occurs.

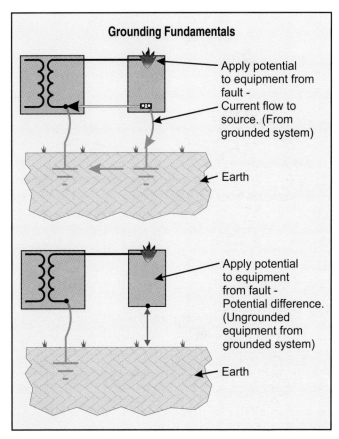

Figure 1-3

Another hazard of equal significance may occur where two pieces of electrical equipment supplied from a grounded system are within reach of a person. If one piece is not grounded or is poorly grounded and becomes energized through a failure of the insulation system (ground-fault), the person making contact with the two (or more) pieces of equipment becomes the circuit (path) for current to pass through as it tries to find its way back to its source. In some cases, the person will receive a mild shock. In other cases, the shock may be fatal. Even though the impedance of the human body may be relatively high, it only takes a few hundred milliamperes (mA) of current flowing through it to be lethal.

Equipment supplied from a ungrounded system

Section 250-21 permits some electrical systems to be operated ungrounded. In this case, the electrical enclosures for the service, feeders, circuits and other equipment are connected to ground but the system itself is not grounded. In many parts of the country, the serving utility will only provide a grounded electrical system.

From a practical standpoint, an ungrounded system exists only in theory or at the distribution transformers hanging on nonmetallic poles before the system is connected to the plant electrical system. Where the ungrounded system conductors are installed in grounded metal raceways or enclosures or connect to motors, the frame of which are grounded, the ungrounded system becomes capacitively coupled to ground. (See Chapter Two of this text for a more detailed discussion of this subject.)

An ungrounded electrical system often becomes grounded through a line-to-ground fault. This essentially creates a grounded system. The other phases of the system then rise to a voltage-to-ground equal to the system voltage. For example, a 480-volt ungrounded delta system is installed in metal conduit and enclosures that are grounded. When checking the voltage to ground, it is found to be about 240 volts. If one of the phases becomes grounded anywhere on the system, the other phases now will have a voltage to ground of approximately 480 volts. Obviously, this becomes a greater shock hazard to personnel who may be servicing the installation as well as adding greater stress on the conductor insulation including transformer and motor windings.

Effect of electricity on humans

Like the bird on the electric wire, the human body is immune to electric shock as long as it is not a part of the electric circuit. The easiest way to avoid danger from electric shock is to keep one's body from becoming a part of the electric circuit. Due to the very common use of electric tools, equipment and appliances, the risk of being exposed to electric shock is multiplied by the numbers of these items the person is exposed to.

When a person becomes a path for electricity, the person will experience an electrical shock. The intensity and damage done to the person by the shock will be determined by the current level (the amount that flows through the person), how long the current flows (the duration), the person's size and the pathway the current takes through the body. Speaking in electrical terms, the person can be thought of as a resistor or impedance in the circuit.

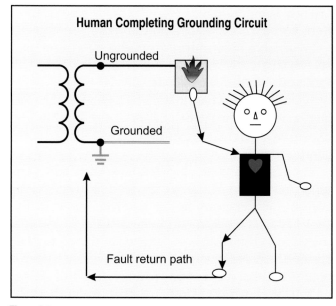

Human Completing Grounding Circuit

Ungrounded

Grounded

Fault return path

Figure 1-4

A person may become a pathway to ground or between conductors in one of two ways: in a series or a parallel circuit. In a series circuit, the person is the only path through which the current will flow in its attempt to return to its source. An example of this is when a person comes into contact with an electrical appliance that is energized through a line-to-ground fault and at the same time touches a grounded appliance like an electric range or a grounded kitchen sink. In general terms, the amount of current that flows through the person's body is determined by Ohm's

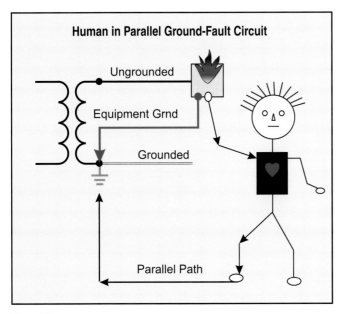

Human in Parallel Ground-Fault Circuit

Ungrounded

Equipment Grnd

Grounded

Parallel Path

Figure 1-5

law by the voltage of the circuit and the resistance offered by the person's body. (See Chapter 14 for a greater discussion of the human body and current flow.)

In a parallel circuit, the human body and another path such as an equipment grounding conductor provide a path for current flow at the same time. The current will divide among the paths based on the impedance of the individual paths. The greater the impedance of the path, the less current will flow in that path.

The ideal situation is where the circuit protective device, usually a fuse or circuit breaker, opens the faulted circuit immediately. This will result in the smallest risk of shock hazard. Another factor is the time it takes for the overcurrent device to clear the fault from the circuit. During this time, the equipment subject to the ground-fault will have a potential above ground depending upon the voltage drop of the equipment grounding circuit.

In some installations, the equipment may be grounded but through a high impedance. This can be the result of an equipment grounding conductor that is too small, is too long for its size, the connection has become loose or it is routed improperly. In this situation, usually not enough current will flow through the equipment grounding conductor to cause the overcurrent protective device to clear the faulted equipment. Part of the fault current will then flow through the person in contact with the energized equipment and part through the equipment grounding conductor path. The current will divide in opposite proportions to the resistance or impedance of the paths. The greatest amount of current will flow through the path providing the lowest resistance or impedance. Remember that even though the human body may have an impedance that is greater than the other path(s), it only takes a small amount of current to cause serious injury or death.

The primary emphasis is to provide a path for ground-fault current that is permanent, is adequately sized and of low-impedance from all electrical equipment back to the source to facilitate the operation of the overcurrent device in a reasonable time. Part of the ground-fault current flows through the person who contacts the faulted equipment and part through the earth or grounded surfaces.

Protection of persons from electric shock is the reason ground-fault circuit-protective devices (GFCIs) have become so popular over the recent years. See Chapter 14 of this text for additional information on the subject.

Burns and other injuries

The most common shock-related injury is a burn. Burns suffered in electrical accidents may be of three types: electrical burns, arc burns and thermal contact burns.

Electrical burns are the result of the electric current flowing through tissues, blood vessels or bone. Tissue damage is caused by the heat generated by the current flow through the body and is often immediately classified as a third degree burn. In many cases, the damage caused to the tissues becomes more apparent in the days or even months following the incident. Severe permanent damage can be caused to internal tissues or organs with little external indication until some time after the electric shock. Burns from electric shock are one of the most serious injuries that can be experienced and should be given immediate medical attention.

Arc or flash burns, on the other hand, are the result of high temperatures produced by an electric arc or explosion near the body. These burns can be of the more minor first-degree type, more severe second-degree or most severe third-degree burns. They should also be attended to promptly and properly.

Finally, thermal contact burns are those normally experienced when the skin comes into contact with hot surfaces of overheated electric conductors, conduits, or other equipment. Additionally, clothing may be ignited in an electrical arc or explosion and a thermal burn will result. All three types of burns may be produced simultaneous.

The proper use of work procedures and personal protective equipment can minimize these injuries.

Electric shock can also cause injuries of an indirect or secondary nature in which involuntary muscle reaction from the electric shock can cause bruises, bone fractures, and even death resulting from collisions or falls. In some cases, injuries caused by electric shock can contribute to delayed fatalities.

In addition to shock and burn hazards, electricity poses other dangers. For example, when an arcing type short circuit occurs, hazards can be created from the resulting arcs. If high-current is involved, these arcs can cause injury or start a fire. Extremely high-energy arcs can damage equipment, causing fragmented metal to fly in all directions or melt steel, copper or aluminum. Even low-energy sparks can cause violent explosions in atmospheres that contain flammable gasses, vapors, or combustible dusts.

Effect of electricity on animals

Some animals are especially sensitive to electricity. For example, past studies claimed that dairy cattle are so sensitive to electricity that a potential of as little as two volts between conductive portions of floors, walls, piping and stanchions caused behavior problems that resulted in loss of production. More recent studies claim the voltage difference to be about four volts. In other cases, severe health problems are attributed to the effects of electricity, which, if not corrected and treated may lead to death of the animal. See Chapter 16 of this text for methods of preventing and correcting these problems in agricultural structures.

Taking the mystery out of grounding

For many years the subject of "Grounding" has been considered one of the most controversial and misunderstood chapters in the National Electrical Code®. There is no real reason why the subject should be treated as a mystery and given so many different interpretations.

Probably the single most important requirement for clearing up the confusion about grounding is to review and clearly understand the definitions of the various elements of the grounding system found in the National Electrical Code®. In addition, these terms should be correctly used

during discussion and instruction on the subject of grounding so everyone will have a common understanding. For example, using the term "ground wire" to mean an "equipment grounding conductor" does no more to help a person understand what specific conductor is referenced than the use of the term "vehicle" when one specifically means a "truck."

Carefully review the terms or words that are defined at the beginning of each chapter in this text for clear understanding. In many illustrations throughout the text, portions of the overall grounding system are emphasized to assist in clearer understanding.

Scope and purpose of grounding and bonding

The "scope of grounding and bonding" and general requirements for grounding and bonding are contained in Section 250-2 of the NEC®. The requirements include:

(a) grounding of electrical systems,

(b) grounding of electrical equipment,

(c) bonding of electrically conductive materials and other equipment, and

(d) performance of the fault-current path.

Grounding of electrical systems. Systems are solidly grounded to limit the voltage to ground during normal operation and to prevent excessive voltages due to lightning, line surges or unintentional contact with higher voltage lines and to stabilize the voltage to ground during normal operation. See Section 250-2(a). Several methods of grounding electrical systems are used depending on National Electrical Code® requirements and system design and function. These methods of

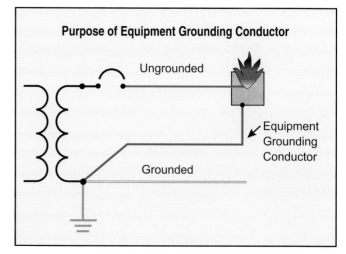

Figure 1-6

grounding electrical systems are covered in detail in Chapter Three of this text.

Grounding of electrical equipment. Conductive materials enclosing electrical conductors or equipment or that are a part of the equipment are grounded to limit the voltage to ground on these materials and bonded to facilitate the operation of overcurrent devices under ground-fault conditions. See Section 250-2(b). Where the electrical system is grounded, the equipment grounding conductor is connected to the grounded system conductor (often a neutral) at the service or the source of a separately derived system. Where the electrical system is not grounded, the electrical equipment is connected to earth at the service to maintain the equipment at or near earth potential and the equipment is bonded together to provide a path for fault current. This occurs where a second ground fault occurs before the first one is cleared.

Bonding of electrically conductive materials and other equipment. Electrically conductive materials, such as metal water piping, metal gas piping and structural steel members, that are likely to become energized, are bonded to provide a low-impedance path for clearing ground faults that otherwise would energize the equipment at a level above earth potential. For systems that are grounded, the equipment is bonded to the grounded system conductor (often a neutral) at the service or source of a separately derived system. Where the electrical system is not grounded, the electrical equipment is connected to earth at the service to maintain the equipment at or near earth potential and the equipment is bonded together to provide a path for fault current. Again, this occurs where a second ground fault occurs before the first one is cleared.

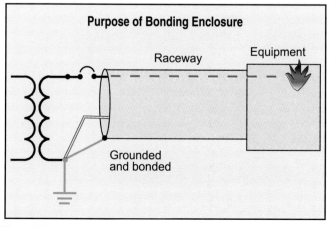

Figure 1-7

Performance of fault-current path. Section 250-2(d) of the NEC®, with a title of "Performance of Fault Current Path," provides requirements on one of the most critical elements of the grounding safety system – the fault-current path. This section requires that the fault-current path:
- be permanent and electrically continuous;
- be capable of safely carrying the maximum fault current likely to be imposed on it; and
- have sufficiently low impedance to facilitate the operation of the overcurrent devices under fault conditions.

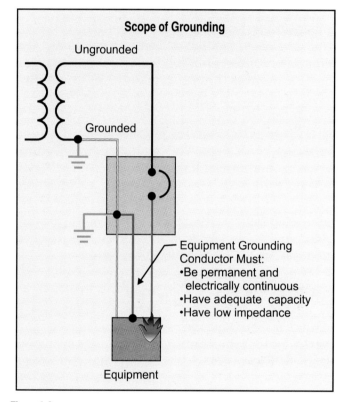

Figure 1-8

Each of these points is important and worthy of discussion. First, a permanent, reliable and electrically continuous grounding system is vital to the overall safety of the electrical system. This includes both a stable voltage reference as well as providing a path for fault current due to abnormal conditions. Intermittent connections are like an unpredictable earthquake, waiting to wreak havoc on the unsuspecting. The grounding system, including all connections, whether a wire, conduit, equipment enclosure, or other element of the path, must be permanent, electrically continuous, and have all joints made up tightly.

If the insulation system of other conductors of the system, like ungrounded (hot) or grounded circuit conductors, fail, it usually is obvious since something such as a piece of equipment or an appliance stops functioning. This is not true of the equipment grounding path. Usually, a failure of the ground-fault path is not known until someone is exposed to an electric shock. Also, the conductors of the equipment grounding conductor system normally only carry current in a fault situation. At this time, the equipment grounding conductor path will usually carry far more current than the ungrounded (hot) or grounded conductor (often a neutral) typically carry.

Providing adequate capacity for the fault current that will flow in the circuit is also of paramount importance. The minimum size system of the grounded conductor, bonding conductor and equipment grounding conductor are given in several locations in the NEC®, many of them in Article 250. Obviously, a grounding conductor that burns off (fuses) due to excessive current flow while it is carrying out its intended purpose is of little value in the safety system. The conductor sizes if equipment grounding conductors given in Table 250-122 are the minimum size. There are cases where the withstand rating of conductors is exceeded by the available fault current and larger conductors are necessary. See Chapter Eleven of this text for additional information on the subject.

The ground-fault path that is permanently connected and of adequate capacity is of little value if it does not have low impedance (measured in ohms). A high impedance path provides resistance to the system thus limiting the amount of current that can flow over the circuit. This allows a voltage above ground to be present on faulted equipment which may then present a shock hazard. ($E = I \times R$) The NEC® stops short of specifying the maximum impedance acceptable for the ground-fault path, except to say that the path must have an impedance that is low enough to facilitate the operation of the overcurrent protective device. See Chapter 11 of this text for additional information on the subject.

The goal of good design of the ground-fault path is to provide a permanent and adequate path of low impedance so enough current will flow in the circuit to cause a circuit breaker to trip or a fuse to open. If the opening of the breaker or fuse on the line side of the faulted circuit does not take place rapidly, thus taking the faulted equipment off the line, the grounding system will have failed to perform its critically important function. This failure may result in greater equipment damage, possibly fires or injury to personnel.

Electrical system design

The ultimate goal of electrical system design is to prevent all faults including ground faults from occurring. This is accomplished by proper design, installation, operation and maintenance of electrical equipment and systems. System conductors that are ungrounded are separated from grounded conductors by insulation. This insulation can be in the form of thermoplastic, thermosetting or other similar insulating medium applied to wires or busbars or by separating conductors having a potential (voltage) between them. This separation is usually accomplished by mounting busbars on insulators. The air then becomes the insulation between phases and a grounded enclosure. However, it is possible to have a failure or breakdown in insulation systems in any electrical installation. Therefore, ground-fault protection must be provided to safely clear line-to-ground faults which may occur.

Section 110-7 of the NEC® requires that, "Completed wiring installations shall be free from short circuits and from grounds other than as required or permitted in Article 250."[N] Though not widely practiced, it is wise to test the insulation of all installations to be certain they are free from unintended short-circuits and/or ground-faults before the wiring is energized. This can be done with some means of continuity testing which may be a battery and a bell and leads (the origin of the term "ring-out"). An ohmmeter or megohm-meter (megger) can also be used. This is more than a simple continuity test. A current is passed through the conductor under test and the amount of leakage through the insulation is measured. Testing of the insulation of the system before it is energized is often required in specifications prepared by architects or engineers for commercial and industrial jobs.

Also, the impedance (rather than the resistance) of the equipment ground-fault return path can be verified by testing instruments. This is especially true for branch circuits where plug-in type testers are available and greatly simplify the testing process.

Designing electrical systems for safety

Electrical systems need to be designed to be certain they are adequate for the loads to be served by them. The NEC® provides that electrical services are to be adequate for the calculated load in accordance with Article 220. See Section 230-42(a). Methods for calculating the minimum capacity for electrical services are found in Article 220 rather than in Article 230. The calculation method for determining the minimum size of the grounded service conductor (often a neutral) is found in Section 220-22. Section 250-24(b)(1) specifies the minimum size grounded service conductor.

The NEC® does not require that electrical systems be designed with any additional capacity for future load expansion. See Section 90-1(b). From a practical standpoint, some additional capacity should be provided for when the system is first installed. Some designers plan the electrical system for at least 25 percent spare capacity.

Electrical systems abused

Electrical systems are intended to provide reliable service for many years. That is generally the case, provided the system is designed for the load to be carried by the system and is installed in a proper manner. Overloading the system is one of the major abuses it must endure. With the expanded use of new or more modern electrical appliances, larger electrical distribution systems are required. It is not uncommon to find temporary power taps, extra wires placed under circuit breaker terminals or splices being made in attics or in crawl spaces to add extra electrical equipment to the electrical system. Often, extra fuse or circuit breaker panels are added to existing systems without taking proper steps to ensure that the conductors and equipment supplying the loads are adequate.

In industrial or commercial installations, additional equipment is often added to existing electrical systems. At times, this is done without consideration for whether the existing system is adequate for the additional load.

In addition, electrical systems may be exposed to overvoltage from lighting or power-line crosses, insulation failure in high-to-low voltage transformers, and to short circuits and ground faults. In some cases, electrical systems are abused until they fail. The opening of a fuse or tripping of a circuit breaker often is the first indication of a failure in the electrical system.

Major problems in electrical systems

The major cause of trouble in an electrical distribution system is insulation failure. The insulation may be air, such as in busways, switchboards and motor control centers, where clearance between uninsulated conductors and grounded metallic electrical equipment is maintained by air space. The clearance is maintained by insulators in the form of rubber, ceramic or thermoplastic or insulation that is applied directly to the conductors. Insulation for wire or bus is usually rubber, thermoplastic or thermosetting material.

Insulation failure may result in two kinds of faults: line-to-line or line-to-ground. The least likely failure is line-to-line or between any two conductors of the system, that is, from one phase conductor to another or from one phase conductor to the neutral or grounded conductor. Experience has shown that most insulation failures (as high as 80 percent) are line-to-ground faults or from one phase conductor to the conductor enclosure or equipment.

While the entire National Electrical Code®, and other safety electrical codes, like the Canadian Electrical Code®, are developed and updated to provide protection against electrical fires and shocks, they are not design specifications. However, by following the rules of these codes, we will have an installation that is essentially free from hazard, but not necessarily efficient, convenient, or adequate for good service or future expansion. See Section 90-1 of the NEC®.

Insulation resistance

Previous editions of the Code (e.g., up to the 1965 NEC®) contained recommended values or results for testing insulation resistance. It was found the values were incomplete and not sufficiently accurate for use in modern installations, and the recommendation was deleted from the Code. However, basic knowledge of and the need for insulation-resistance testing is important.

Measurements of insulation resistance can best be made with a megohm-meter insulation tester. These instruments are available from several manufacturers and vary in cost and features. As measured with such an instrument, insulation resistance is the resistance to the flow of direct current (usually at 500 or 1000 volts for systems of 600 volts or less) through or over the surface of the insulation in electrical equipment. The test is accomplished by energizing the conductor, wire or busbar with the potential and measuring the current that "leaks" through the insulation. By Ohm's law, the insulation resistance is $R = E \div I$. The results are displayed in ohms or megohms, but insulation readings will be in the megohm range.

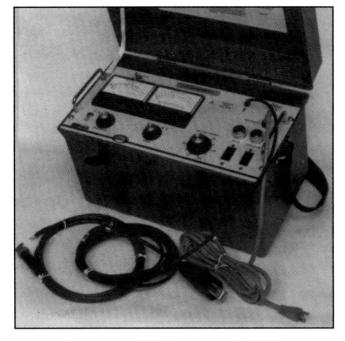

Figure 1-9. *Photo courtesy of Biddle Instruments, Blue Bell, PA.*

The insulation-resistance test is nondestructive, quite different from a high-voltage or breakdown test. It is made with direct current rather than alternating current and is not a measure of dielectric strength as such. However, insulation resistance tests assist greatly in determining when and where not to apply high voltage.

In general, insulation resistance decreases with increased size of a machine or length of cable. This occurs because there is more insulating material in contact with the conductors and the frame, ground, or sheath. The greater volume of insulating material allows more total "leakage" and, therefore, lowers the overall insulation resistance reading.

Insulation-resistance usually increases with higher voltage rating of apparatus because of increased thickness of insulating material. Also different types of insulation, for example air versus thermoplastic, will have different levels of insulation resistance.

Insulation-resistance readings are not only quantitative, but are relative or comparative as well, and since they are influenced by moisture, dirt, and deterioration of the insulation, they are reliable indicators of the presence of those conditions. Cable and conductor installations present a wide variation of conditions from the point of view of the resistance of the insulation. These conditions result from the many kinds of insulating materials used, the voltage rating or insulation thickness,

and the length of the circuit involved in the measurement. Furthermore, such circuits usually extend over great distances and may be subject to wide variations in temperature, which will have an effect on the insulation resistance values obtained. The terminals of cables and conductors will also have an effect on the test values unless they are clean and dry, or guarded.

Records of insulation-resistance should be maintained so failures of conductors or equipment can be predicted. To make valid comparisons or to indicate trends, the recorded values must be corrected to a standard temperature, typically 20°C and differences in the relative humidity must be accounted for. Equipment or conductors can then be repaired or replaced to reduce plant downtime which often is at inconvenient times causing expensive loss of production.

Differentiate between a "short circuit" and a "ground fault"

It is common practice to call all faults or failures in the electrical system "shorts" or "short circuits." That can lead to misunderstanding, and has done so, because we are not using terms properly. A short circuit and a ground fault are different, although they both stem from insulation failure.

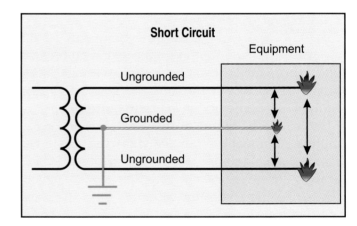

Figure 1-10

To end this confusion and to have a common language for better understanding, we should define the two terms as follows:

Short circuit. "A conducting connection, whether intentional or accidental, between any of the conductors of an electrical system whether it be from line-to-line or from line to the grounded conductor." The grounded conductor often is also the system or circuit neutral. The short circuit can be a solid or "bolted" connection or it may be an arcing fault completing the path through a short air space.

In the case of a "short circuit," the failure may be from one phase conductor to another phase conductor or from one phase conductor to the grounded conductor or neutral. For either condition, the maximum value of fault current is dependent on the available capacity the system can deliver to the point of the fault. The maximum value of short-circuit current from line-to-neutral will vary depending upon the distance from the source to the fault and the impedance of the path. The available short-circuit current is further limited by the impedance of the arc where one is established, plus the impedance of the conductors to the point of short circuit.

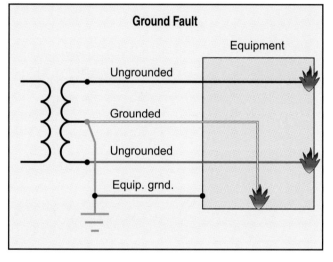

Figure 1-11

Ground fault. "A conducting connection, whether intentional or accidental, between any of the conductors of an electrical system and the grounding conductor or conducting material which encloses the conductors (such as conduit) or any conducting material that is grounded or that may become grounded."

In the case of a "ground fault" there is a fault from a hot or ungrounded phase conductor to the conductor enclosure or from the grounded phase conductor to the conductor enclosure (wire-to-conduit, wire-to-motor frame, etc.). It is not common in practice to refer to a conductor that is intentionally grounded, such as a grounded phase or system conductor or equipment grounding conductor, as being in a ground-fault condition. However, a grounded system conductor such as a neutral that is not permitted to be grounded again past the service can be considered a "ground fault." This

condition is particularly important with equipment ground-fault protection systems where grounding the system grounded conductor downstream from the point of protection will desensitize the protection system.

It is often the case that a fault may easily involve both conditions, short circuit and ground fault. The fault may start as a ground fault or line-to-ground fault and develop into a short circuit or phase-to-phase fault. The fault may also start as a short circuit and expand to a ground fault. For electrical systems of 120-volts-to-ground, the circuit protective devices will usually clear the fault or it will be extinguished when the alternating voltage passes through zero. For 277-volts-to-ground systems, destructive arcing faults can be easily maintained by the system voltage. Equipment ground fault protection systems have been designed to address this problem. In some cases, the use of this equipment is required by the NEC®. See Chapter 15 of this text for additional information on this subject.

NEC® Requirements

Article 250 of the National Electrical Code® covers the subject of "Grounding." Grounding is practiced for the protection of electrical installations, which in turn protects the buildings or structures the electrical systems are installed in. Persons and animals that may come into contact with the electrical system or are in these buildings or structures are also protected if the grounding system is installed and maintained properly. The NEC® does not imply here that grounding is the only method that can be used for the protection of electrical installations, people or animals. Insulation, isolation and guarding are suitable alternatives under certain conditions.

Grounding of specific equipment is covered in several other articles of the National Electrical Code®. An example is health care facilities as covered in Article 517 and swimming pools in Article 680. Refer to the index of this text, to the index of the National Electrical Code®, or in *Ferm's Fast Finder Index* book for specific sections of the Code that apply to the equipment in question.

Circuit impedance and other characteristics

Electrical equipment intended to interrupt current at fault levels must be designed and installed so it has an adequate rating for the nominal circuit voltage and the fault current that is available at its line terminals. See Section 110-9 of the NEC®. Failure to comply with this important requirement can result in disastrous consequences. This includes destruction of the electrical equipment, such as circuit breakers or fuses that are themselves designed to protect the system. Equipment such as time clocks, motor controllers and lighting contactors that are intended to interrupt current at only rated load must be suitable for the nominal circuit voltage and current that must be interrupted. These devices must also withstand the higher fault current levels until an upstream protective device, such as a fuse or circuit breaker, opens.

In addition, Section 110-10 of the NEC® requires that, "The overcurrent protective devices, the total impedance, the component short-circuit current ratings, and other characteristics of the circuit to be protected shall be selected and coordinated to permit the circuit-protective devices used to clear a fault to do so without extensive damage to the electrical components of the circuit. This fault shall be assumed to be either between two or more of the circuit conductors, or between any circuit conductor and the grounding conductor or enclosing metal raceway. Listed products applied in accordance with their listing shall be considered to meet the requirements of this section."[N]

"Busways and associated fittings marked 'Short Circuit Current Rating(s) Maximum RMS Symmetrical Amperes _____ Volts _____' have been investigated for the rating indicated."[U] Electrical system components such as metering equipment, switchboards, panelboards, and motor control centers have short-circuit ratings that must not be exceeded.

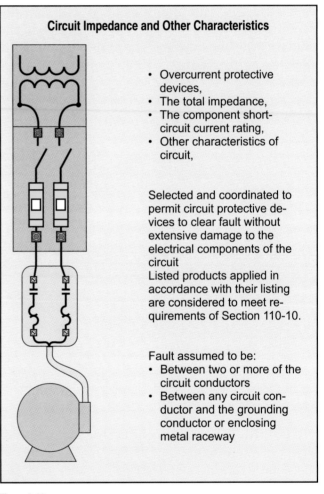

Circuit Impedance and Other Characteristics

- Overcurrent protective devices,
- The total impedance,
- The component short-circuit current rating,
- Other characteristics of circuit,

Selected and coordinated to permit circuit protective devices to clear fault without extensive damage to the electrical components of the circuit
Listed products applied in accordance with their listing are considered to meet requirements of Section 110-10.

Fault assumed to be:
- Between two or more of the circuit conductors
- Between any circuit conductor and the grounding conductor or enclosing metal raceway

Figure 1-13

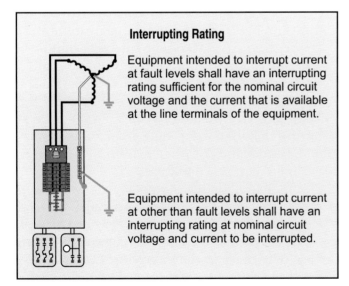

Interrupting Rating

Equipment intended to interrupt current at fault levels shall have an interrupting rating sufficient for the nominal circuit voltage and the current that is available at the line terminals of the equipment.

Equipment intended to interrupt current at other than fault levels shall have an interrupting rating at nominal circuit voltage and current to be interrupted.

Figure 1-12

Listed equipment such as fuses, including the associated switch, and circuit breakers have nominal voltage and short-circuit current ratings. This is the maximum voltage and current the equipment is designed for. In addition, other equipment such as busways may be marked similar to

Overcurrent protective devices such as circuit breakers and fuses should be selected and installed to ensure that the short-circuit current rating of components of the system will not be exceeded should a short-circuit or ground-fault occur. The overcurrent device that is selected must not only safely interrupt the fault current that is available at its line terminals, it also must limit the amount of energy that is let through to downstream equipment so as not to exceed its rating.

Electric utility companies usually provide information on the short-circuit current (often referred to as fault current) that is available at the terminals of their supply transformer or distribution system. The short-circuit current must then be calculated to the point on the electrical system under consideration. Methods for calculating the available fault current at any point on the electrical system can be found in literature from many circuit breaker or fuse manufacturers. See also the Institute of Electrical and Electronic Engineers (IEEE) "Buff Book" (IEEE 242)[1], Underwriters Laboratories product safety standard and "guide card information" for the equipment under consideration[2], as well as in *Ferm's Fast Finder Index*[3] where, in addition to the point-to-point method, many pages of calculations are presented in an easy-to-use table form.

Other components of the electrical system, such as wire and cable, have published withstand ratings, in addition to allowable ampacity ratings, that must be considered for safety. Withstand ratings are not marked on the wire or cable but can be determined from manufacturers data, IEEE Standards or calculated as described in Chapter 11 of this text.

As can be seen upon review of this literature, a conductor can safely carry much more current than allowed in the allowable ampacity table of the NEC® if the time the current flows is reduced. For insulated conductors, the ampacity of the conductor can be thought of as the conductor's long-time or continuous rating and the withstand-rating as

the conductor's short-time rating. This is most important when considering the proper size of equipment grounding conductors to use, as they do not have overcurrent protection on their line side and must carry fault-current levels until an overcurrent device opens. See Chapter 11 for additional information on this subject.

[1] Available from
IEEE Service Center
445 Hoes Lane, P.O. Box 1331
Piscataway, NJ 08855-1331
(800)678-4333

[2] Available from
Publications Stock
Underwriters Laboratories
333 Pfingsten Road
Northbrook, IL 60062-2096
(847)272-8800 Ext. 42612 or 42622

[3] Available from
International Association of Electrical Inspectors
P.O. Box 830848
Richardson, TX 75083-0848
(800) 786-4234,
Fax (972) 235-6858

Chapter One: The questions included here were developed using material included in this chapter. The answers can be found by reviewing the text. It is also important that students make use of the 1999 NEC®, where many answers can be found. See page 279 for answers.

1. A grounded metal object is forced to take the same zero potential as the ____.
   a. electrode
   b. metal raceway
   c. earth
   d. system

2. Electrical systems are solidly grounded in to limit the voltage to ground during normal operation and to prevent excessive voltages because of ____, line surges or unintentional contact with higher voltage lines.
   a. low voltage
   b. overloads
   c. loose connections
   d. lightning

3. The path to ground from circuits, equipment, and metal enclosures must ____.
   a. be permanent and electrically continuous
   b. have capacity to conduct safely any fault current likely to be imposed on it
   c. have sufficiently low impedance to limit the voltage to ground and to facilitate the operation of the circuit protective devices in the circuit
   d. all of the above

4. Insulation is considered to be air where conductors are mounted on insulators such as on poles, in busways or in equipment, or rubber or ____ material or wire or bus.
   a. listed
   b. approved
   c. acceptable
   d. thermoplastic

5. Measurements of insulation resistance can best be made with a ____ insulation tester.
   a. voltage
   b. megohm-meter
   c. low voltage
   d. high voltage

6. Insulation resistance is the resistance to the flow of direct current (usually at ____ or ____ volts for systems of 600 volts or less) through or over the surface of the insulation in electrical equipment megohm range.
   a. 500 or 1000
   b. 600 or 1500
   c. 800 or 1800
   d. 900 or 2000

7. An insulation-resistance test is made with direct-current rather than alternating-current, and is not a measure of ____ strength as such.
   a. conductor
   b. dielectric
   c. terminal
   d. equipment

8. A conducting connection, whether intentional or accidental, between any of the conductors of an electrical system whether it be from line-to-line or to the grounded conductor defines a ____.
   a. phase fault
   b. short circuit
   c. overload
   d. line disorder

9. A conducting connection, whether intentional or accidental, between any of the conductors of an electrical system and the conducting material which encloses the conductors or any conducting material that is grounded or that may become grounded is defined as a ____.
   a. phase fault
   b. overload
   c. system ground
   d. ground fault

To Ground or Not To Ground

Objectives

After studying this chapter, the reader will be able to understand:

- Grounding rules for alternating-current systmes of up to 1000 volts.
- Grounding rules for alternating-current systems 1kV and over.
- Which systems cannot be grounded.
- Which systems can be operated ungrounded.
- Purpose of grounding and bonding.
- Use of ground detection systems for ungrounded systems.
- Factors to consider regarding system grounding.

Electrical systems

When we refer to electrical systems, generally, we are considering the system as a unit of a specific voltage (potential) and often amperage (current or capacity). For example, a common system is 480Y/277-volts, 3-phase, 4-wire at some capacity such as 1,600 amperes. Often, the premises are supplied by an electric utility. These systems are either over 600 volts (often referred to as "primary systems") or 600 volts or less (often referred to as "secondary systems"). In addition, the system is referred to as single-phase or three-phase. Some two-phase systems exist but have fallen out of favor in recent years. Most systems are alternating current at 60 hertz although some direct-current systems are in use, primarily in industrial applications.

In addition, "systems" are produced at various voltages and phases on-site. The most common way to produce an electrical system is through a transformer or generator. In the case of a transformer, other than an autotransformer, one electrical system ends at the primary winding(s) and another system begins at the secondary winding(s). An example is where a plant has a service at 480Y/277-volts and it is necessary to supply receptacle outlets at 120 volts. A single-phase, 480-volt transformer is installed with the winding on the secondary center-tapped to result in 120/240 volts single phase. In this example, the 480-volt system ends at the primary windings and a new 120/240-volt system begins at the secondary winding. In other cases, a 208Y/120-volt system is derived by installing the appropriate transformer(s).Similar examples can be shown for generator-supplied systems, though, for these, we consider the system supplied as the new system.

Grounding electrical systems

Over the years, there have been great debates over the merits of grounding electrical systems versus leaving them ungrounded. This subject is covered at some length in the Appendix of this text where the history of the development of the grounding rules in effect today is covered.

Many of the decisions about whether or not to ground electrical systems are made for us. The National Electrical Code® requires that electrical systems falling within the parameters of Section 250-20 be grounded. Other electrical systems are permitted to be grounded while some systems are not permitted to be grounded due to special conditions. Some systems today, primarily in the industrial or agricultural sector, are operated ungrounded.

Systems that must be grounded

In accordance with Section 250-20, the alternating-current systems that must be grounded are: (a) Alternating-Current Systems of Less than 50 Volts, (b) Alternating-Current Systems of 50 Volts to 1000 Volts, (c) Alternating-Current Systems of 1 kV and Over, and (d) Separately Derived Systems. Each of these systems is discussed in the following sections.

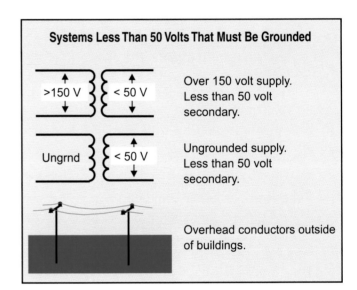

Systems Less Than 50 Volts That Must Be Grounded

>150 V   < 50 V   Over 150 volt supply. Less than 50 volt secondary.

Ungrnd   < 50 V   Ungrounded supply. Less than 50 volt secondary.

Overhead conductors outside of buildings.

Figure 2-1

(a)   Less than 50-volt systems:

The following systems that are less than 50 volts, such as on the secondary of a transformer must be grounded:

    (1) Where supplied by transformers if the supply voltage (primary) to the transformer exceeds 150 volts to ground.

    (2) Where supplied by transformers if the transformer supply system (primary) is ungrounded.

    (3) Where installed as overhead conductors outside of buildings.

(b)   AC systems of 50 to 1000 volts:

Alternating-current systems of 50 to 1000 volts that supply premises wiring and premises wiring systems are required to be grounded under any of the following conditions:

    (1) Where the system can be grounded so the maximum voltage to ground on the ungrounded conductors does not exceed 150 volts. Typical systems include:

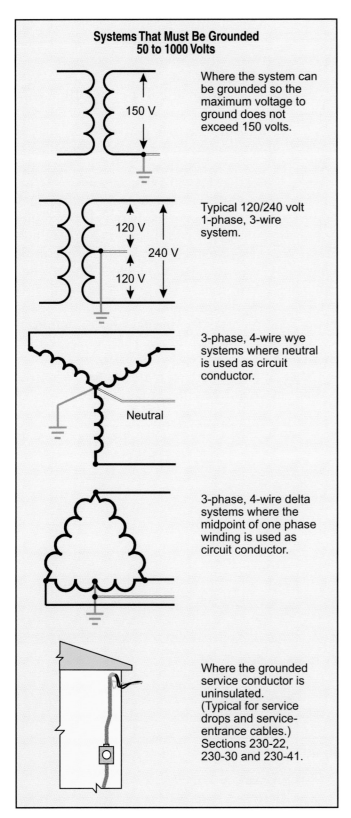

Systems That Must Be Grounded 50 to 1000 Volts

Where the system can be grounded so the maximum voltage to ground does not exceed 150 volts.

150 V

Typical 120/240 volt 1-phase, 3-wire system.

120 V
240 V
120 V

3-phase, 4-wire wye systems where neutral is used as circuit conductor.

Neutral

3-phase, 4-wire delta systems where the midpoint of one phase winding is used as circuit conductor.

Where the grounded service conductor is uninsulated. (Typical for service drops and service-entrance cables.) Sections 230-22, 230-30 and 230-41.

Figure 2-2

- 120 volt, 1-phase,
- 120/240 volt, 1-phase, 3-wire and
- 208Y/120 volt, 3-phase, 4-wire.

(2) Where the system is 3-phase, 4-wire wye connected and where the neutral is used as a circuit conductor. Typical systems include:
- 208Y/120 volt, 3-phase, 4-wire, and
- 480Y/277 volt, 3-phase, 4-wire.

(3) Where the system is 3-phase, 4-wire delta connected in which the midpoint of one phase winding is used as a circuit conductor. Typical systems include:
- 120/240 volt, 3-phase, 4-wire, and
- 240/480 volt, 3-phase, 4-wire.

(4) Where the grounded service conductor is uninsulated as permitted as follows (Note that this requirement appears to have been inadvertently omitted from the 1999 edition of the NEC® and has appeared in previous editions of the Code):
- The grounded conductor of a multi-conductor service drop cable (Section 230-22 Exception),
- The grounded conductor of an underground service lateral (Section 230-30 Exception), and
- The grounded service-entrance conductor (Section 230-41 Exception).

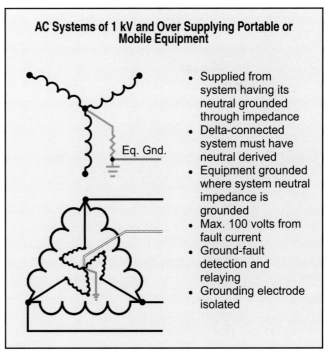

AC Systems of 1 kV and Over Supplying Portable or Mobile Equipment

Eq. Gnd.

- Supplied from system having its neutral grounded through impedance
- Delta-connected system must have neutral derived
- Equipment grounded where system neutral impedance is grounded
- Max. 100 volts from fault current
- Ground-fault detection and relaying
- Grounding electrode isolated

Figure 2-3

(c) AC Systems of 1 kV and Over.

These ac systems must be grounded if they supply mobile or portable equipment as covered in Section 250-188. Systems supplying portable or mobile high-voltage equipment, other than substations installed on a temporary basis, are required to comply with the rules in (a) through (f) below.

(a) Portable or Mobile Equipment. Portable or mobile high-voltage equipment must be supplied from a system having its neutral grounded through an impedance. Where a delta-connected high-voltage system is used to supply portable or mobile equipment, a system neutral shall be derived. This is usually accomplished by means of a zigzag grounding transformer. See Chapter Three of this text for additional information on this subject.

(b) Exposed Noncurrent-Carrying Metal Parts. Exposed noncurrent-carrying metal parts of portable or mobile equipment shall be connected by an equipment grounding conductor to the point at which the system neutral impedance is grounded. This point may be at the service or source of a separately derived system.

(c) Ground-Fault Current. The voltage developed between the portable or mobile equipment frame and ground by the flow of maximum ground-fault current shall not exceed 100 volts. This requirement, no doubt, necessitates an engineering study to determine the voltage drop across the equipment grounding conductor.

(d) Ground-Fault Detection and Relaying. Ground-fault detection and relaying must be provided to automatically de-energize any high-voltage system component that has developed a ground fault. The continuity of the equipment grounding conductor must be continuously monitored so as to de-energize automatically the high-voltage feeder to the portable or mobile equipment upon loss of continuity of the equipment grounding conductor.

(e) Isolation. The grounding electrode to which the portable or mobile equipment system neutral impedance is connected is required to be isolated from and separated in the ground by at least 20 ft. (6.1 m) from any other system or equipment grounding electrode, and there shall be no direct connection between the grounding electrodes, such as buried pipe, fence, etc.

(f) Trailing Cable and Couplers. High-voltage trailing cable and couplers for interconnection of portable or mobile equipment must meet the requirements of Part C of Article 400 for cables and Section 490-55 for couplers.

This type equipment is commonly found in mobile rock crushing plants and batch plants. Other applications are for open pit mining operations. Note that self-propelled mobile surface mining machinery and its attendant electrical trailing cable are not covered by the Code. See Section 90-2(b)(2). Even though exempted from the Code, many requirements for this equipment are incorporated by regulations enforced by the Mine Safety and Health Administration (MSHA).

Other ac systems over 1,000 volts are permitted but are not required to be grounded.

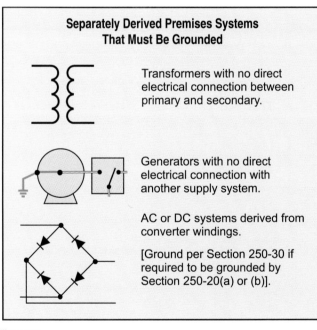

Separately Derived Premises Systems That Must Be Grounded

Transformers with no direct electrical connection between primary and secondary.

Generators with no direct electrical connection with another supply system.

AC or DC systems derived from converter windings.

[Ground per Section 250-30 if required to be grounded by Section 250-20(a) or (b)].

Figure 2-4

(d) Separately Derived Systems.
Electrical systems derived from a battery, a solar photovoltaic system, or from a generator, transformer, or converter windings that have no direct electrical connection, including a solidly connected grounded circuit conductor, to the supply conductors originating in another system, must be grounded if the system that is derived com-

plies with the conditions in (a), (b), or (c) above. The separately derived system then must be grounded as specified in Section 250-30.

Examples of systems that are separately derived include:

(1) Inverters or batteries such as for uninterruptible power systems.

(2) Solar photovoltaic systems.

(3) Transformers with no direct electrical connection between the primary and secondary.

(4) Generator systems that supply power such as for carnivals, rock crushers or batch plants where the neutral is not connected to the utility system.

(5) Generator systems used for emergency, required standby or optional standby power that have all conductors, including a neutral, isolated from the neutral or grounded conductor of another system usually by a transfer switch.

(6) Ac or dc systems derived from inverters or rectifiers.

Figure 2-5 illustrates systems that are not separately derived and thus do not fall under the grounding requirements of Section 250-30. These systems include:

(1) A system supplied by an autotransformer since autotransformers by design have a common conductor.

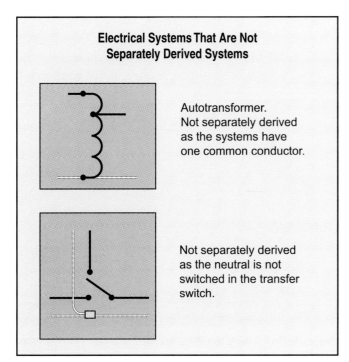

Electrical Systems That Are Not Separately Derived Systems

Autotransformer. Not separately derived as the systems have one common conductor.

Not separately derived as the neutral is not switched in the transfer switch.

Figure 2-5

(2) Systems from a generator that do not have all conductors, including the grounded system (often a neutral) conductor, switched by a transfer switch.

The key to determining whether a generator supplied system is a separately derived system is often to examine electrical connections in the transfer switch. If all the phases and the neutral is switched by the transfer switch, it is a separately derived system and it must be grounded in compliance with Section 250-30. If the neutral is not switched, the system produced is not a separately derived system and the neutral must be grounded at the generator where the system is required to be grounded. See Chapter Twelve of this text for additional information on grounding of separately derived systems.

Systems permitted but not required to be grounded

Five electrical systems are permitted to be but are not required to be grounded. (See Section 250-21.) These alternating-current systems are:

(1) Electric systems used exclusively to supply industrial electric furnaces for melting, refining, tempering, etc.

(2) Separately derived systems used exclusively for rectifiers supplying only adjustable speed industrial drives.

(3) Separately derived systems supplied by transformers that have a primary voltage rating less than 1000 volts if all the following conditions are met:
(a) The system is used for only control circuits.
(b) Only qualified persons will service the installation.
(c) Continuity of power is required.
(d) Ground detectors are installed on the control system.

(4) Isolated power systems as permitted or required in Articles 517 for health care facilities and in Article 668 for electrolytic cells commonly used for the production of aluminum, cadmium, chlorine, copper and other products.

(5) Three-phase ac systems of 480 to 1000 volts that are high-impedance grounded neutral systems. An impedance, usually a resistor, limits the current of the first ground fault to a low value. All the following conditions must be met before high-impedance grounded systems are permitted:

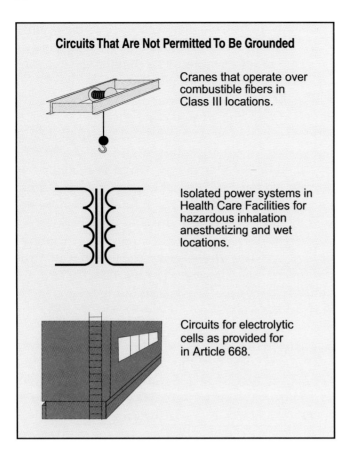

Circuits That Are Not Permitted To Be Grounded

Cranes that operate over combustible fibers in Class III locations.

Isolated power systems in Health Care Facilities for hazardous inhalation anesthetizing and wet locations.

Circuits for electrolytic cells as provided for in Article 668.

Figure 2-6

(a) Only qualified persons will service the system.

(b) Power continuity is required.

(c) Ground detectors are installed.

(d) Line-to-neutral loads are not supplied.

Circuits or systems that are not permitted to be grounded. Only three circuits or systems are not permitted to be grounded by the NEC®. (See Section 250-22) They are as follows:

(1) Circuits for electric cranes that operate over combustible fibers in Class III locations. [See Sections 250-22(1) and 503-13.)] This action reduces the likelihood that sparks from faulted equipment will fall onto combustible fibers causing a fire.

(2) For health care facilities, those isolated power circuits in hazardous (classified) inhalation anesthetizing locations are required to be supplied by an isolation transformer or other ungrounded source [Section 517-61(a)(1)]. In addition, receptacles and fixed equipment in "wet locations" of hospital patient care areas as defined in Section 517-3 must be protected by ground-fault circuit-interrupter devices where interruption of power to equipment under

fault conditions can be tolerated. Where interruption of power under fault conditions cannot be tolerated, protection of these receptacles and fixed equipment are to be supplied from isolated, ungrounded sources by ungrounded electrical systems. [Sections 250-22(2) and 517-20(a)]

3. Circuits for electrolytic cells as provided in Article 668. Equipment located or used within the electrolytic cell line working zone or associated with the cell line dc power circuits are not required to comply with Article 250. [See Sections 250-22(3) and 668-3(c)(3).]

Electrical systems operated ungrounded

Electrical systems that are outside the requirement for grounding in Section 250-20(a), (b), (c) or (d) are often operated ungrounded. At times, the plant owner or engineer will choose to operate electrical systems ungrounded. These systems usually are in industrial or agricultural applications and often are either 240-volt or 480-volt, three-phase, three-wire systems. Some higher voltage systems are also used in heavy-industrial applications. Where ungrounded systems are installed, the engineering decision is often based on an effort to obtain an additional degree of service continuity.

Typical systems that are operated ungrounded include:

• 240 volt, 3-phase, 3-wire, delta connected.
• 480 volt, 3-phase, 3-wire, delta connected.
• 2,300 volt, 3-phase, 3-wire, delta connected.
• 4,600 volt, 3-phase, 3-wire, delta connected.
• 13,800 volt, 3-phase, 3-wire, delta connected.

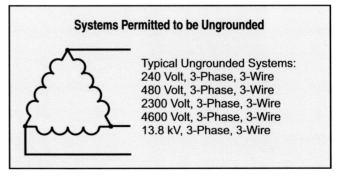

Systems Permitted to be Ungrounded

Typical Ungrounded Systems:
240 Volt, 3-Phase, 3-Wire
480 Volt, 3-Phase, 3-Wire
2300 Volt, 3-Phase, 3-Wire
4600 Volt, 3-Phase, 3-Wire
13.8 kV, 3-Phase, 3-Wire

Figure 2-7

Since the system is ungrounded, the occurrence of the first ground fault (not a short circuit or line-to-line fault) on the system will not cause an overcurrent protective device for the service, feeder or branch circuit to open. This fault does, however, ground the system but in an unspecified and uncontrolled location. In essence, this system becomes a corner grounded delta system. Little, if any, current will flow in this first ground fault. See Chapter Three of this text for additional information on equipment and enclosure grounding of these ungrounded systems.

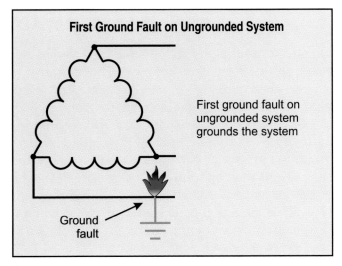

First Ground Fault on Ungrounded System

First ground fault on ungrounded system grounds the system

Ground fault

Figure 2-8

When an ungrounded system with one ground fault experiences a second ground fault, the result is a phase-to-phase fault on the system. This will usually cause one or more overcurrent protective devices to open provided adequate current flows in this path to cause the circuit protective devices to operate. A major concern for this type of system happens where the first and second faults may be located some distance apart. Often, these faults are from line-to-conduit or metallic enclosure such as wireway, pull boxes, busway or motor terminal housings in different parts of the plant. Where this occurs, a relatively high-impedance path for current flow is often established. In some cases, it has been found that a great deal of heat along with arcing and sparking is produced along this fault path due to loose connections or inadequate bonding. Every conduit coupling and locknut connection to enclosures in the fault-current path must be tight to provide an adequate and low-impedance path and to reduce this arcing and sparking.

It is important for safety reasons, as well as for system continuity, that maintenance personnel locate and eliminate ground faults when first identified on ungrounded systems. This should be done as soon as practical and especially before the second ground fault occurs on the system.

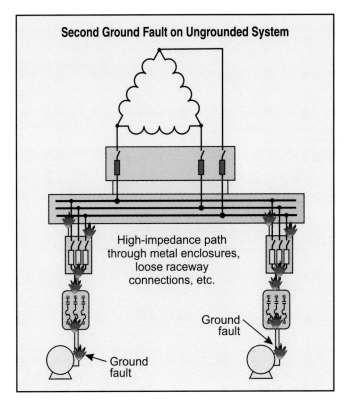

Second Ground Fault on Ungrounded System

High-impedance path through metal enclosures, loose raceway connections, etc.

Ground fault

Ground fault

Ground fault

Figure 2-9

Ground-detector systems

Commercially manufactured ground-detection equipment is available. This equipment, which can be located at the service equipment or in feeder distribution panels, can be set to operate an overcurrent relay or shunt-trip circuit breaker or to operate a visual or audible signaling system to indicate a ground-fault condition.

This monitoring equipment is not required by the electrical code but is a good practice from a plant management perspective. Successful operation of an ungrounded electrical system depends on good system maintenance and prompt elimination of the first ground fault.

Sophisticated equipment is now available to identify the part of the electrical system where the fault is located while the system is energized. This significantly speeds detection and repair.

Ground detector lights

Often, ground detector lights or a neutralizer or potentializer plug are installed to indicate that a ground fault has occurred on the ungrounded system. The 7½ watt indicator lights are connected to the lines through 18,000 ohm resistors. A tap is made to each resistor to give 120 volts to the lamp. The lamp burns until its phase goes to ground, at which time there is no or little potential across the

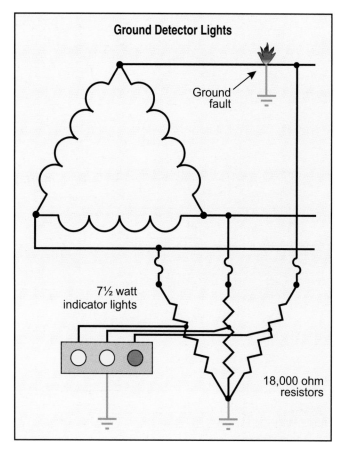

Figure 2-10

installed in metallic enclosures are grounded to varying degrees through the distributed leakage capacitance of the system. Physically, a capacitor exists whenever an insulating material separates two conductors that have a difference of potential between them.

When any conductor is installed in close proximity to grounded metal, there is a capacitance between them that is increased as the distance between the conductors is reduced. In 600-volt systems, the two greatest sources of capacitance to ground are conductors in metal conduit and windings, such as for motors and transformers. In both cases, conductors are separated from grounded metal by fairly thin insulation. The capacitance to ground is known as the leakage capacitance, and the current flowing from the conductors to ground is known as the leakage current or charging current. This capacitance is distributed throughout the electrical system but electrically acts like it is a single, lumped capacitance.

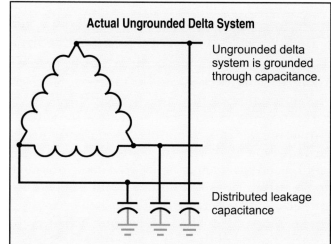

Figure 2-11

lamp, and it stops glowing, thus identifying the faulted phase.

This indication is intended to alert the maintenance personnel to the problem so the ground-fault can be corrected during hours when the plant is not operating. The plant can continue to operate with one-phase grounded, thus preventing costly production down times. In some cases, down-time in production plants can cost thousands of dollars per minute.

A wide variety of "homemade" systems have been installed over the years, some of which are downright dangerous. Some of these have consisted of nothing more than two "barn socket" lampholders with 240-volt lamps that are connected in series from phase-to-ground. Only proven and tested designs for ground detection systems should be used.

Ungrounded system problems

An ungrounded system exists only in theory, in a laboratory or at the electrical distribution transformers hanging on the pole before connection to the plant electrical system. In the real world, ungrounded systems having insulated conductors

Disadvantages of operating systems ungrounded include:

1. Power system overvoltages are not controlled. In some cases, these overvoltages are passed through transformers into the premises wiring system. Some common sources of overvoltages include: lightning, switching surges and contact with a high-voltage system.

2. Transient overvoltages are not controlled, which over time may result in insulation degradation and failure.

3. System voltages above ground are not necessarily balanced or controlled.

4. Destructive arcing burndowns can result if a second fault occurs before the first fault is cleared.

Ungrounded systems have the characteristic that they are subject to relatively severe transient overvoltages. Such overvoltages may be owing to external disturbances as well as to internal faults in the wiring system and easily can reach a value of five to six times normal voltage. An actual case involved a 480-volt ungrounded system. Line-to-ground potentials in excess of 1200 volts were measured on a test meter. The source of the trouble was traced to an intermittent or sputtering (arcing) line-to-ground fault in a motor starting autotransformer. These faults are not uncommon on 480-volt ungrounded systems. During the two-hour period that this arcing fault existed, between 40 and 50 motor windings had failed.

Circuit-switching operations can also be responsible for the creation of transient overvoltages in ungrounded systems. These generally are of short duration and typically only reach two to three times nominal system line-to-ground voltage.

Experience has shown that these overvoltages, that easily reach several times the system voltage, may cause failure of insulation at locations on the system other than at the point of the fault and may result in future system failures. This often occurs at a system weak point such as in a motor or transformer winding.

Locating the ground fault can be troublesome. While it is easy to spot a ground fault on a one-line diagram, locating it in a plant with a complex electrical system can be much more difficult unless sophisticated ground-fault detection equipment has been installed. The first step is to open the feeders one at a time and observe the ground detection indicator. After finding the feeder with the ground fault, branch circuits are disconnected, one at a time, until the offending circuit is located. A significant loss of plant operation time can occur during this process. This is contrasted with a grounded system where only the offending equipment is taken off the line by the circuit protective devices.

Often, overcurrent devices are set above the current level of the fault in ungrounded systems. When arcing faults occur, destructive burn-downs of electrical equipment can result. The fault arc releases a tremendous amount of energy such that conductors and metal enclosures in the vicinity are destroyed.

When the first ground fault occurs on the 480-volt system, the other conductors of the system rise to a level of 480 volts to ground. This presents an additional risk of shock to operation and maintenance staff. This can be contrasted to a 480Y/277-volt grounded wye system where the voltage to ground does not exceed 277 volts while the phase-to-phase voltage is 480 volts, even under ground-fault conditions.

Factors to consider regarding system grounding

Where grounding of the electrical system is optional, the pros and cons of grounding must be carefully weighed by the plant owner or electrical designer so as to make the best decision.

In the long run, greater service continuity may be obtained with grounded systems rather than ungrounded ones. Faults can be isolated to the feeder or circuit affected and cleared without disrupting the entire system. This is obviously a major consideration if the equipment or circuit affected is critical to the plant operation. This has to be balanced against the ungrounded system's tolerance of the first line-to-ground fault but with possible deterioration of conductor insulation from transient overvoltages and possible serious damage caused by a second ground fault on the system.

High-impedance grounded systems

High-impedance grounded systems should be considered where the plant operation cannot tolerate electrical system disruption from the first ground fault. This system has all the advantages of an ungrounded system, so far as operation of the plant or system with one phase faulted to ground is concerned, with none of the disadvantages of an ungrounded system. While the initial cost of the system can be fairly significant, it can pay for itself many times over the installation cost by operational savings in more reliable plant operation. See Chapter Four of this text for additional information on high-impedance grounded neutral systems.

Bolted faults

A common myth is that ground-faults are "bolted" or solidly connected and that a great deal of fault-current will flow to cause the overcurrent device to clear the fault. Bolted faults rarely occur, while sparking, intermittent or arcing faults are quite common. The higher impedance in the arc limits the total current, so standard overcurrent devices may be ineffective.

Arcing faults produce a great deal of heat in the vicinity of the fault and can lead to destructive burn-downs of electrical switchboards and motor control centers. This is the reason the Code requires equipment ground-fault protection systems for 3-phase, 4-wire, wye-connected systems of certain voltage and amperage services and feeders. See Chapter Fifteen of this text for additional information on this subject.

A bolted fault, typically 3-phase bolted faults, must be used for the purposes of determining adequate interrupting ratings of overcurrent protective devices as well as bracing of bus bars, etc. This is generally considered to be the "worst case" condition.

Chapter Two: The questions included here were developed using material included in this chapter. The answers can be found by reviewing the text. It is also important that students make use of the 1999 NEC®, where many answers can be found. See page 279 for answers.

1. Where operating at less than ____ volts, alternating-current systems are required to be grounded where supplied by transformers if the supply voltage to the transformer exceeds ____ volts to ground.
   a. 80 - 100
   b. 70 - 110
   c. 60 - 140
   d. 50 - 150

2. Where operating at less than 50 volts, ac systems are required to be grounded where supplied by transformers if the transformer supply system is ____.
   a. bonded
   b. ungrounded
   c. identified
   d. approved

3. Conductors installed on the outside of buildings where ac systems operate at less than 50 volts are required to be grounded when they are run as ____ conductors.
   a. overhead
   b. underground
   c. optical fiber
   d. Type IGS

4. Alternating-current system grounding is required for 50 to 1000 volt systems supplying premises wiring or premises wiring systems where the system can be grounded so the maximum voltage to ground on the ungrounded conductors does not exceed ____ volts.
   a. 180
   b. 150
   c. 240
   d. 208

5. Alternating-current system grounding is required for 50 to 1000 volt systems supplying premises wiring or premises wiring systems where the system is 3-phase, 4-wire, wye connected in which the neutral is used as a ____ conductor.
   a. bonding
   b. circuit
   c. equipment grounding
   d. isolated

6. Alternating-current system grounding is required for 50 to 1000 volt systems supplying premises wiring, or premises wiring systems where the system operates at 3 phase, 4 wire, delta connected in which the midpoint of one phase is used as a ____.
   a. bonding jumper
   b. equipment grounding conductor
   c. circuit conductor
   d. switch leg

7. Alternating-current system grounding is not required for 50 to 1000 volt electric systems used exclusively to supply industrial electric furnaces for ____.
   a. melting
   b. refining
   c. tempering
   d. all of the above

8. Alternating-current system grounding is not required for 50 to 1000 volt separately derived systems used exclusively for rectifiers supplying only ____.
   a. motor control centers
   b. adjustable speed industrial drives
   c. commercial buildings
   d. class III hazardous (classified) locations

9. Alternating-current system grounding is not required for 50 to 1000 volt separately derived systems supplied by transformers that have a primary voltage rating less than 1000 volts if which of the following conditions is or are met. ____
   a. system is used for only control circuits.
   b. only qualified persons service installation, ground detectors are installed on the control system.
   c. continuity of power is required.
   d. all of the above

10. Alternating-current system grounding is not required for 50 to 1000 volt systems supplying premises wiring, or premises wiring systems where ____ are permitted or required for flammable anesthetizing systems in health care facilities.
   a. grounded power systems
   b. isolated power systems
   c. impedance grounded systems
   d. medium voltage power systems

11. Alternating-current system grounding is not required for 50 to 1000 volt systems supplying premises wiring, or premises wiring systems where three-phase ac systems of 480 to 1000 volts that are high-impedance grounded neutral systems if which one of the following conditions is or are met ____:
    a. only qualified persons will service the system.
    b. power continuity is required.
    c. ground detectors are installed, line-to-neutral loads are not supplied.
    d. all of the above

12. Ac systems of ____ and over must be grounded if they supply mobile or portable equipment as covered in Section 250-118.
    a. 250 volts
    b. 1000 volts
    c. 480 volts
    d. 600 volts

13. Ac systems operating at over ____ volts are permitted (not required) to be grounded where they do not supply mobile or portable equipment.
    a. 1000
    b. 300
    c. 277
    d. 480

14. Equipment operating at over ____ volts is commonly found in mobile rock crushing plants and batch plants. Other applications are for open pit mining operations.
    a. 277
    b. 300
    c. 600
    d. 1000

15. A ____ premises wiring system is one that is derived from a generator, transformer or converter windings that have no direct electrical connection to the supply conductors originating in another system.
    a. identified
    b. open neutral
    c. isolated power
    d. separately derived

16. Separately derived systems must be grounded as specified in Section ____.
    a. 250-24
    b. 250-26
    c. 250-30
    d. 250-32

17. Examples of separately derived systems include ____ with no direct electrical connection between the primary and secondary.
    a. phase converters
    b. transformers
    c. elevator motors
    d. Class 1 systems

18. Examples of separately derived systems include generator systems used for emergency, required standby or optional standby power that have all conductors including a neutral switched by a ____.
    a. double pole
    b. single pole
    c. transfer switch
    d. backfed device

19. Examples of separately derived systems include ac or dc systems derived from ____.
    a. inverters
    b. rectifiers
    c. generators
    d. all of the above

20. Certain systems are permitted to be operated ungrounded and usually are located in industrial or agricultural applications. Typical systems that are operated ungrounded include which of the following ____.
    a. 2,300 volt, three-phase, three-wire, delta connected.
    b. 4,600 volt, three-phase, three wire, delta connected.
    c. 13,800 volt, three phase, three-wire, delta connected.
    d. all of the above

21. Often, ground detector lights or a potentializer plug are installed to indicate that a ground fault has occurred on the ____ system.
    a. ungrounded
    b. grounded
    c. bonded
    d. identified

22. For an ungrounded electrical system, the first ground fault
    a. does nothing
    b. causes a fuse to blow
    c. grounds the system
    d. none of the above

23. For an ungrounded electrical system, the second ground fault
    a. does nothing
    b. causes a fuse to blow
    c. grounds the system
    d. none of the above

24. Disadvantages of operating an ungrounded electrical system include
    a. power system voltages are not controlled
    b. transient overvoltages are not controlled
    c. destructive arcing burndowns can occur on a second ground fault
    d. all of the above

Notes:

Grounding Electrical Systems

Objectives

After studying this chapter, the reader will be able to understand:

- Rules for system conductor to be grounded.
- Proper identification of grounded conductor.
- Methods of grounding electrical systems.
- Delta bank grounding.
- Grounding rules for ungrounded systems.
- Corner grounded delta systems.

Definitions

Ground: "A conducting connection, whether intentional or accidental, between an electrical circuit or equipment and the earth, or to some conducting body that serves in the place of the earth."[N]

Grounded: "Connected to earth or to some conducting body that serves in place of the earth."[N]

Grounded Conductor: "A system or circuit conductor that is intentionally grounded."[N]

The electrical system

As covered in Chapter 2 of this text, Section 250-20 of the NEC® requires that many electrical systems be grounded. It is important that we understand that we are dealing with the electrical system and not the service equipment or disconnecting means at this point.

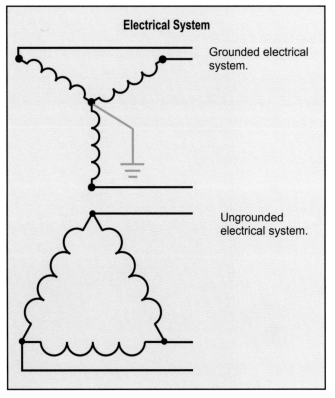

Figure 3-1

Electrical energy is delivered to the customer by the serving utility by either a grounded or ungrounded system or sometimes by both systems. Electrical utilities have tariffs and standards that dictate whether or not they will deliver a system of a certain voltage level and phase configuration as grounded or ungrounded systems. Many power companies insist that all low voltage (600 volts and under) systems be grounded. Others will supply 3-

phase, 480-volt delta connected systems ungrounded, while they insist on furnishing 480Y/277 volt systems in only a grounded wye configuration.

Large industrial plants may purchase power at medium or high voltage levels such as 69,000 volt, 115,000 volt, or 230,000 volt levels and own and

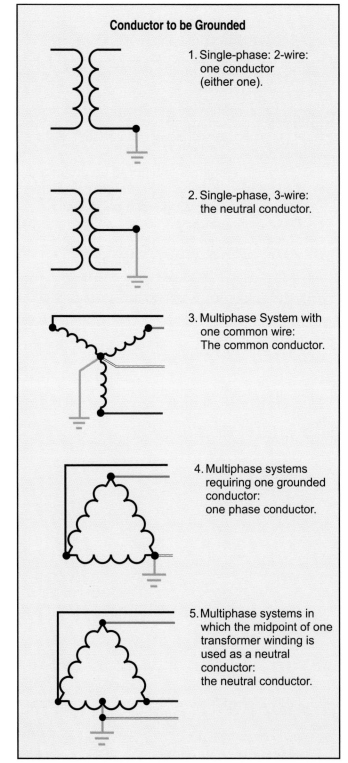

Figure 3-2

maintain their electrical distribution systems. Transformers, capacitor banks, controls, overcurrent devices and relaying systems are then installed at customer-owned switchyards. Power is distributed to utilization points on the premises where transformers are installed as needed for the utilization voltages desired. Generally, electrical systems at the utilization level are grounded at the voltage levels and configurations required by the NEC®.

Conductor to be grounded

Where the electrical system is grounded, Section 250-26 of the NEC® specifies which conductor in an alternating current system shall be grounded. For alternating current premises wiring systems, the conductor that is required to be grounded is as specified in 1 through 5 below.

(1) Single-phase, 2-wire: one conductor (either one).
(2) Single-phase, 3-wire: the neutral conductor.
(3) Multiphase systems having one wire common to all phases: the common conductor.
(4) Multiphase systems requiring one grounded phase: one phase conductor.
(5) Multiphase systems in which one phase is used as in (2) above: the neutral conductor.

Identification of grounded conductor

Grounded conductors are required to be identified by the means specified in Section 200-6 of the NEC®. Rules are provided for identification of grounded conductors of sizes No. 6 or smaller, conductors larger than No. 6, flexible cords, grounded conductors of different systems and grounded conductors of multiconductor cables.

No. 6 and Smaller. Three general means of identification of No. 6 and smaller insulated grounded conductors are provided which include:
- a continuous white outer finish, or
- natural gray outer finish, or
- three continuous white stripes on other than green insulation.

This identification must be along the entire length of the conductor.

For specific conductor assemblies or applications, the following grounded conductor identification is provided regardless of size:

(1) For Type MI cable, the grounded conductor is permitted to be identified by distinctive marking at its termination at the time of installation.
(2) For single-conductor, sunlight-resistant,

outdoor-rated cable for solar photovoltaic systems, the grounded conductor is permitted to be identified at terminations by distinctive white markings. This will usually be made with white adhesive vinyl marking tape.

(3) Fixture wire can be identified as provided in Section 402-8.
(4) For aerial cable, the grounded conductor is permitted to be identified by any of the general means identified above or by a ridge located on the exterior of the cable.

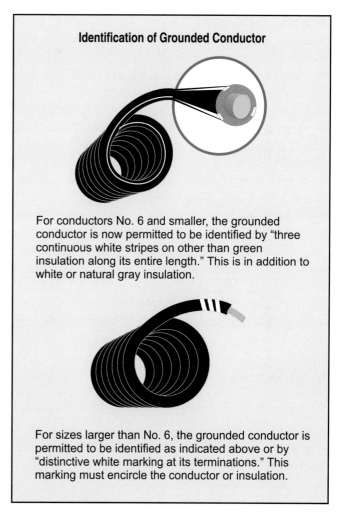

Identification of Grounded Conductor

For conductors No. 6 and smaller, the grounded conductor is now permitted to be identified by "three continuous white stripes on other than green insulation along its entire length." This is in addition to white or natural gray insulation.

For sizes larger than No. 6, the grounded conductor is permitted to be identified as indicated above or by "distinctive white marking at its terminations." This marking must encircle the conductor or insulation.

Figure 3-3

Conductors Larger than No. 6. Conductors larger than No. 6 are permitted to be identified either like conductors No. 6 and smaller, or at the time of installation, by distinctive white marking at each termination. This "distinctive white marking" usually consists of white adhesive vinyl tape or white paint. Where so marked, the marking must encircle the conductor or insulation so it is visible from all sides.

Flexible Cords. The grounded conductor within flexible cords is permitted to be identified by the three methods included for conductors No. 6 and smaller or by the methods included in Section 400-22. These additional methods include: colored braid, tracer in the braid, colored insulation, colored separator, tinned conductors, and surface marking.

Grounded Conductors of Different Systems. In larger electrical installations, it is common to have more than one electrical system installed in the same enclosure such as raceways like conduits and wireways, pull boxes and cables. For example, grounded conductors from both a 480Y/277-volt, three-phase, four-wire system and a 120/240 volt, single-phase system may be in the same raceway or other enclosure. Where this happens, one of the grounded conductors is required to be identified as covered above for conductors smaller or larger than No. 6. Each other system grounded conductor is required to be identified differently by one of the means provided for conductors smaller or larger than No. 6 or by white insulation with a readily distinguishable different colored stripe that is not green and runs along the insulation.

Grounded Conductors of Multiconductor Cables. Where only qualified persons will service and maintain the installation, grounded conductors of multiconductor cables are permitted to be permanently identified at terminations by a distinctive white marking or by other effective means.

Use of Conductors with White Insulation. The previous permission to install conductors with white insulation in a conduit or other raceway and phase-tape them and use them as ungrounded conductors has been removed from the NEC®. See Section 200-7 of the NEC® for additional requirements or restrictions on the use of conductors with white, natural gray or three white stripes on other than green insulation.

Methods of grounding electrical systems

There are a variety of methods used to ground electrical systems. The method chosen will vary, depending upon the system voltage, code requirements, plant owner specifications or engineer's philosophy. The various methods commonly used are shown in Figure 3-4 and are as follows:

(1) **Solidly grounded:** no intentional grounding impedance.

Solidly grounded refers to the connection of the electrical system or of a separately derived system of a generator, power transformer, or grounding

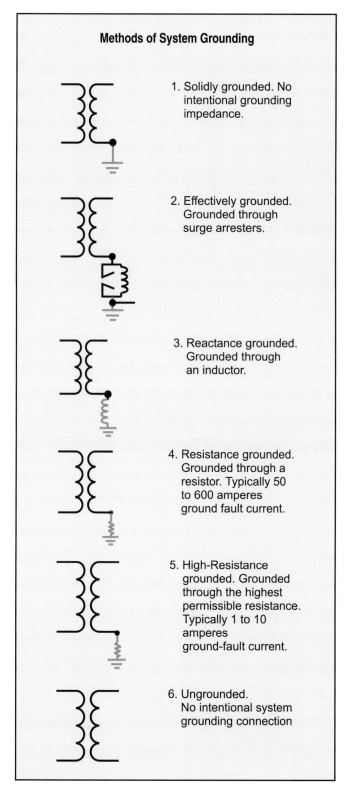

Methods of System Grounding

1. Solidly grounded. No intentional grounding impedance.

2. Effectively grounded. Grounded through surge arresters.

3. Reactance grounded. Grounded through an inductor.

4. Resistance grounded. Grounded through a resistor. Typically 50 to 600 amperes ground fault current.

5. High-Resistance grounded. Grounded through the highest permissible resistance. Typically 1 to 10 amperes ground-fault current.

6. Ungrounded. No intentional system grounding connection

Figure 3-4

transformer directly to the station ground or grounding electrode system without intentionally introducing an impedance.

(2) **Effectively grounded:** to permit the use of reduced rated (80 percent) surge arresters.

Grounding of surge arresters rated near, but not

less than, 80 percent of line-to-line voltage constitutes an effectively grounded system. This will carry with it a line-to-ground circuit current of at least 60 percent of the three-phase short-circuit value.

(3) **Reactance grounded:** the system is grounded through a reactor. This grounding is accomplished by a reactor (grounding transformer) that is a device that introduces impedance to the circuit, the principal element of which is inductive reactance.

(4) **Resistance grounded:** intentional insertion of resistance into the system grounding connection.

For resistance grounding, the system is grounded by connecting the system neutral to ground through a resistor. This grounding is accomplished by a resistor which introduces impedance to the circuit, the principal element of which is resistance.

(5) **High-resistance grounded:** the insertion of nearly the highest permissible resistance into the grounding connection.

This grounding is accomplished by a resistor which introduces impedance to the circuit, the principal element of which is resistance. For high-resistance grounding, the system is grounded by the system neutral through a resistor (may be a grounding transformer) that typically limits the ground-fault current to 10-amperes or less. High-resistance grounding maintains control of transient overvoltages, but may not furnish sufficient current for ground-fault relaying. The protective scheme usually associated with high-resistance grounding is detection and alarm rather than immediate tripout. [See Section 250-36 of the NEC® and Chapter 4 of this text for additional information.] High-resistance grounding is typically applied to 3-phase systems such as 480-volt systems and also higher voltage wye-connected systems, such as 12.47 kV systems.

(6) **Ungrounded:** no intentional system grounding connection.

Ungrounded systems are used in industrial plants and where desired for manufacturing processes to give an additional degree of service continuity. While a ground-fault on one phase of an ungrounded system generally does not cause a service interruption, the occurrence of a second ground fault on a different phase before the first fault is cleared does result in an outage. A ground detection system is recommended, and adequate maintenance and repair will minimize interruptions.

Low-voltage systems, 600 volts and below, that are grounded are almost always solidly or high-impedance grounded; medium-voltage systems are usually either solidly or resistance grounded; and high-voltage systems above 34.5kV are nearly always effectively grounded or ungrounded.

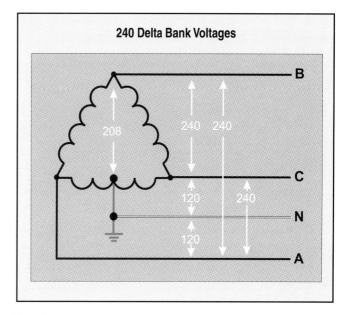

Figure 3-5

Delta bank grounding

Where three transformers that have center taps are connected in a delta bank, only one transformer may have its midpoint grounded. Any attempt to use a second transformer of the delta bank to supply a second 3-wire, single-phase source would require grounding the midpoint of a second transformer. However, since there will be a difference of potential between the midpoints of the different transformers, there would be an abnormal flow of current through both grounded neutrals. This would sooner or later show up in heating or a short circuit, depending on how both neutrals were connected.

With midpoint grounding of one phase in a delta bank, one of the phase conductors will operate at a higher voltage to ground than the other two. In practice this "high leg" is identified as required by Sections 215-8, 230-56 and 384-3(e) by orange color coding or other effective means.

The voltage of the leg with the higher voltage to ground is determined by the following formula: ½ of phase-to-phase voltage x 1.732 = high leg voltage to ground. For example, a 240 volt, 3-phase system center tapped to establish a neutral: ½ of 240 = 120 x 1.732 = 208V high leg voltage to ground.

Grounding of ungrounded systems

In some cases, it is desirable to ground electrical systems that originally were installed ungrounded or are permitted to be but are not required to be grounded. See Chapter Two of this text for a thorough discussion of the subject. There are four methods commonly in use for grounding of ungrounded systems of 600 volts or less. Solid grounding is used in all the methods. They are as follows:

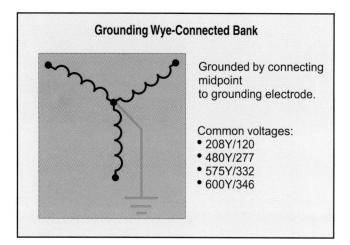

Grounding Wye-Connected Bank

Grounded by connecting midpoint to grounding electrode.

Common voltages:
* 208Y/120
* 480Y/277
* 575Y/332
* 600Y/346

Figure 3-6

(1) Grounding the neutral of wye-connected secondary windings of a transformer bank.

As shown in Figure 3-6, this is the most universal and commonly used method of grounding a system. Standard voltages are 208Y/120, 480Y/277, 575Y/332 and 600Y/346 volts. The first two voltage systems, 208Y/120 and 480Y/277 are in most common use today in the United States with the growth tending in favor of the 480Y/277 volt system owing to the better economies of that system. In general, the primary windings of the transformers serving those systems are delta connected.

(2) Grounding a delta bank with a zigzag grounding transformer.

This method is best adapted to an existing 3-wire, 3-phase delta-connected distribution system which is ungrounded and where it is desired to ground the system to obtain the advantages gained through operating the system grounded. For new systems, the use of transformers with wye-connected secondaries are advisable to obtain a grounded system. The neutral derived by the zigzag transformer can be used as a system current carrying conductor if the zigzag grounding transformers are sized for the maximum unbalanced current. See Section 220-22 for the method for calculating the maximum unbalanced current.

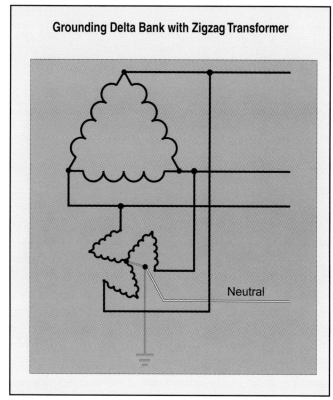

Grounding Delta Bank with Zigzag Transformer

Neutral

Figure 3-7

A zigzag transformer, shown applied to a delta system in Figure 3-7, obtains its name from the manner in which the windings are installed and connected. Windings for each phase are on the same core leg. All windings have the same number of turns but each pair of windings on the same core leg is wound in opposite directions. The impedance of the transformer to 3-phase currents is so high that under normal conditions only a small magnetizing current will flow. If a ground fault develops on one phase of the system, the transformer impedance to ground current is so low that a high ground current will flow. The ground current will divide into three equal parts in the three phases of the grounding transformer.

Such a 3-phase zigzag transformer has no secondary winding. This type of grounding transformer is required to carry rated current for only a short time (the duration of the ground fault). The short time kVA rating of such a grounding transformer may thus be equal to the rating of a regular 3-phase transformer yet be only about 10 percent of the physical size.

Electrically, a zigzag grounding transformer connection appears to superimpose a wye connected system inside the delta connected transformers. This wye-type connection does permit

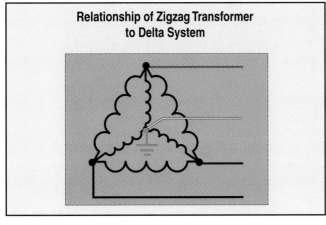

Figure 3-8

the transformer to be used like a wye transformer with phase-to-neutral loads to be utilized. In this case, it is important for the zigzag transformer to be sized for the calculated or connected load.

A wye-delta grounding transformer may also be used to provide a neutral for an existing delta-connected ungrounded system, but the use of the zigzag grounding transformer is more practical and economical.

The grounding transformer may be connected to the system, as shown in Figure 3-7, directly on the power transformer secondaries. Where a grounding transformer is so connected, one grounding transformer is required for each delta-connected power transformer bank. In this manner, the switching out of any one transformer bank will not disturb the secondary system ground. The grounding transformer and the power transformer are considered a single unit, both being protected by the overcurrent protective means provided for the main power transformer.

The grounding transformer also may be connected directly to the main bus through its own overcurrent protection means. In that event, an alarm should be provided to indicate the system is operating ungrounded if the grounding transformer should be disconnected from the line.

Where a grounding transformer is used on low-voltage systems (600 volts and below), it is important that the equipment grounding conductor be connected to the neutral of the grounding transformer in such a way as to provide a low-impedance path for ground-fault currents to flow back to the system. The same grounding electrode should be used for the neutral of the grounding transformer and the equipment

grounding conductor and a common grounding conductor used for both the equipment grounding conductor and the neutral.

The same disconnecting means and overcurrent protection means for the system would be used as described under method (1).

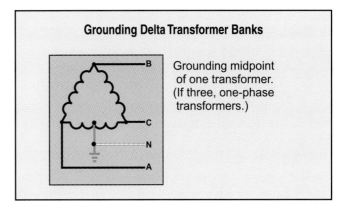

Figure 3-9

(3) Grounding the midpoint of one of the transformers in a delta-connected system.

This method of grounding is commonly used, especially in smaller distribution systems, to provide a three-phase power source and a three-wire single-phase lighting source from the same transformer bank. This then becomes a three-phase, four-wire delta-connected system. At the same time, the advantages of a neutral grounded system are obtained. It is common for the serving utility to use three single-phase transformers connected in a delta configuration for this system. In such a system, the single-phase transformer which supplies the lighting load usually is sized larger than the other two single-phase transformers which supply the power load only.

Since the conductors supplying the 3-phase power will now be from a grounded source, the grounded conductor must be run to the service equipment. There it must be connected to the equipment grounding conductor of the power system and the grounding electrode. This provides a ground-fault current return path of low impedance. Such a connection enhances the safety of the power system. If the power service and the lighting service go to the same building, which is usually the case, the grounding electrode conductor from the power service, as well as the grounding electrode conductor from the lighting service, must be connected to the same grounding electrode. See Chapter Four of this text for additional information on the subject.

(4) Grounding One Corner of a Delta System.

In the past, most 3-phase ungrounded delta distribution systems comprised three single-phase transformers which were connected delta-delta. The main reason was to be able to continue operating if one transformer failed by disconnecting the faulted transformer and operating the bank open-delta, although at reduced capacity.

When the advantages of grounding became very apparent, it first was the practice to ground one corner of the delta. This is shown in Figure 3-10. The grounded leg must be positively identified throughout the system in accordance with Section 200-6. See the information on this subject earlier in this chapter.

Where the grounded conductor is disconnected by the switch or circuit breaker, it is important that a bonding connection be made on the line side of the disconnect or circuit breaker. If this is not done, the electrical equipment will become electri-

cally isolated from the grounded conductor when the switch or circuit breaker is in the open position. In this case, any fault that occurs would cause the enclosure to raise to a hazardous potential above ground equal to the system voltage.

Generally, fuses are not permitted to be installed in the grounded conductor. However, a two-pole fused switch may be used with fuses in the ungrounded phases. Also, a three-pole switch may be used with a solid neutral in the grounded phase. For services, Section 230-90(b) of the NEC® does not allow an overcurrent device, other than a circuit breaker that opens all conductors, to be inserted in the grounded service conductor. Section 240-22 generally prohibits connecting an overcurrent device in series with any conductor that is intentionally grounded. Subsection (1) permits an overcurrent device to be used that opens all conductors of the circuit, including the grounded conductor, and is so designed that no pole can be operated independently. This requirement means that three-pole circuit breakers can be used for this purpose while a three-pole switch with a fuse in series with each phase could not as the switch is not the overcurrent device, the fuse is. Figure 3-9 shows an acceptable use of a fused switch for corner-grounded systems.

Section 430-36 permits a fuse to be inserted in the grounded conductor for the purpose of providing motor overload protection. However, based on the examination of Code requirements discussed in the previous paragraph, this application would be limited to being located downstream from the service equipment.

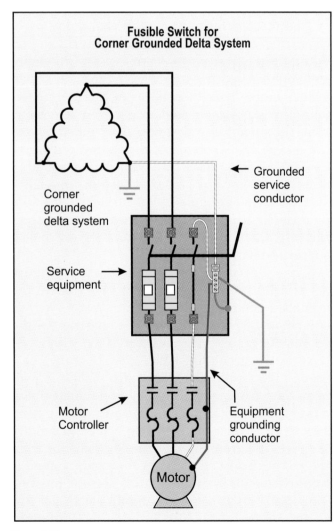

Figure 3-10

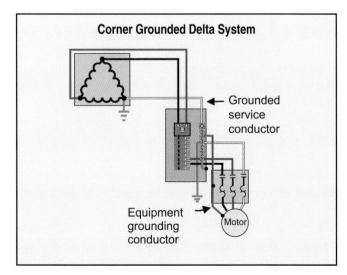

Figure 3-11

Keep in mind that Section 250-24(b) requires that the grounded system conductor be run to each service disconnecting means and be bonded to the service disconnecting means enclosure. This is usually accomplished by connecting it to a terminal bar that is mounted inside the service enclosure. Nothing in the Code permits this bonding connection to be interrupted by a switch or circuit breaker.

Where it is desirable to disconnect the grounded system conductor from the feeder, it is acceptable to route the conductor from the terminal bar through the switch or circuit breaker. It is not uncommon to see three-pole switches with a copper bar or other suitable slug inserted in the grounded leg.

For purposes of installing or grounding a corner grounded delta system, it is helpful to think of it as a single-phase system. This is illustrated in Figures 3-10 and 3-11. The system is grounded at one corner of the delta. Three conductors are taken to the service where the two ungrounded conductors connect to the disconnecting means or circuit breaker. The grounded service conductor is connected to the neutral terminal bar where it is bonded to the enclosure and connected to the grounding electrode conductor.

Where circuit breakers are used as the disconnecting means for corner grounded systems, they must be marked with a voltage rating suitable for the system voltage and will have a "straight voltage rating" such as 240V or 480V. Two-pole circuit breakers that are suitable for a corner grounded delta system are marked "1-phase/3-phase." See Section 240-85.

From the service on throughout the system, the grounded conductor must be insulated from equipment and enclosures. See Section 250-24(a)(5). An equipment grounding conductor is carried with the circuit to ground equipment and enclosures such as disconnecting means, motor controllers, and other noncurrent-carrying equipment that are required to be grounded.

Chapter Three: The questions included here were developed using material included in this chapter. The answers can be found by reviewing the text. It is also important that students make use of the 1999 NEC®, where many answers can be found. See pages 279 for answers.

1. A conducting connection, whether intentional or accidental, between an electrical circuit or equipment and the earth, or to some conducting body that serves in the place of the earth is defined as a ____.
   a. grounded conductor
   b. ground
   c. grounding electrode
   d. bonded conductor

2. Connected to earth or to some conducting body that serves in place of the earth is defined as ____.
   a. grounded
   b. a grounded conductor
   c. being identified
   d. being bonded

3. A system or circuit conductor that is intentionally grounded is defined as one that is ____.
   a. identified
   b. bonded
   c. a grounded conductor
   d. a grounding conductor

4. No. ____ and smaller insulated grounded conductors are required to be identified by a continuous white or natural gray outer finish.
   a. 6
   b. 4
   c. 2
   d. 1

5. For installations that will be serviced by qualified persons only, grounded conductors of ____ cables are permitted to be marked at their terminations.
   a. Type IGS
   b. Type TFFN
   c. multiconductor
   d. single

6. Conductors ____ than No. 6 are permitted to be identified either like conductors No. 6 and smaller, or at the time of installation, by distinctive white marking at their terminations.
   a. of aluminum or smaller
   b. of copper or smaller
   c. smaller
   d. larger

7. Insulated conductors No. 6 or smaller are permitted to be identified by which of the following methods?
   a. a continuous white insulation
   b. a continuous natural gray insulation
   c. three white stripes on other than green insulation
   d. all of the above

8. In practice, a "high leg" is required to be identified by a (an) ____ color coding or other effective means.
   a. green
   b. brown
   c. orange
   d. yellow

9. Grounding of surge arresters rated near, but not less than, ____ percent of line-to-line voltage constitutes an effectively grounded system.
   a. 70
   b. 60
   c. 50
   d. 80

10. For resistance grounding, the system is grounded by connecting the system neutral to ground through a ____.
   a. resistor
   b. terminal
   c. ground rod
   d. generator

11. The insertion of nearly the highest permissible resistance into the grounding connection results in a system that is ____.
   a. extra low resistance grounded
   b. medium resistance grounded
   c. low resistance grounded
   d. high resistance grounded

12. High-resistance grounding maintains control of transient overvoltages, but may not furnish sufficient current for ground-fault ____.
   a. detection
   b. relaying
   c. alarms
   d. conditions

13. Low-voltage systems operating at 600 volts and below are almost always solidly grounded; medium-voltage systems are usually either solidly or resistance grounded; and high-voltage systems above ____ kV are nearly always effectively grounded or ungrounded.
   a. 12.7
   b. 10.4
   c. 15.3
   d. 34.5

14. A wye-delta grounding transformer may also be used to provide a neutral for an existing delta-connected ungrounded system, but the use of the ____ grounding transformer is considered as being more practical and economical.
   a. converter
   b. autotransformer
   c. zigzag
   d. Scott

15. Where a grounding transformer is used on low-voltage systems operating at 600 volts and below, it is important that the equipment grounding conductor be connected to the neutral of the grounding transformer in such a way as to provide a ____ path for ground-fault currents to flow back to the system.
   a. high-impedance
   b. low-impedance
   c. current limiting
   d. straight

16. The same grounding electrode should be used for the ____ of the grounding transformer and the equipment grounding conductor and a common grounding conductor used for both the equipment grounding conductor and the neutral.
   a. unidentified conductor
   b. equipment grounding conductor
   c. bonding jumper
   d. neutral

17. In the past, most 3-phase ungrounded delta distribution systems comprised three single-phase transformers that were connected ____.
   a. wye-delta
   b. delta-wye
   c. delta-delta
   d. delta

18. For purposes of installing or grounding a corner grounded delta system, it is best to think of it as a ____ system.
   a. three phase
   b. single phase
   c. 5-wire
   d. 4-wire

19. Circuit breakers used for corner-grounded delta systems must be marked
   a. with a voltage rating such as 120/240
   b. for at least 600 volt operation
   c. with a straight voltage rating such 240 V or 480 V
   d. none of the above

20. Fuses are permitted to be inserted in a grounded conductor
   a. as desired
   b. for grounded systems only
   c. for ungrounded systems only
   d. for motor overload protection

Notes:

Chapter 4

Grounding Electrical Services

Objectives

After studying this chapter, the reader will be able to understand:

- Important elements for grounding of electrical services.
- Proper location of service grounding connection.
- Rules for low-impedance grounding electrode connections.
- Dwelling unit services and feeders.
- Proper sizing of grounded service conductor.
- Rules for parallel service conductors.
- Rules for multiple services to one building.
- Rules for high-impedance grounded systems.
- Grounding requirements for instrument transformers, relays, etc.
- Hazards of operating an ungrounded service from a grounded system.

General

As discussed in Chapter Two, electrical systems are furnished to the premises by the serving utility as either grounded or ungrounded. At the service disconnecting means, the system is either solidly grounded, left ungrounded or may be resistance or reactance grounded. How the services are treated regarding grounding depends on the type of system installed, design criteria and Code rules.

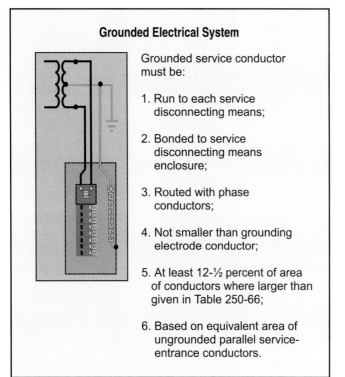

Grounded Electrical System

Grounded service conductor must be:

1. Run to each service disconnecting means;

2. Bonded to service disconnecting means enclosure;

3. Routed with phase conductors;

4. Not smaller than grounding electrode conductor;

5. At least 12-½ percent of area of conductors where larger than given in Table 250-66;

6. Based on equivalent area of ungrounded parallel service-entrance conductors.

Figure 4-1

Grounded electrical systems

One of the most important requirements for grounded systems and services is contained in Section 250-24(b). This section requires that:

(1) Where an ac system operating at less than 1000 volts is grounded at any point, the grounded conductor shall be run to each service disconnecting means.

(2) It must be bonded to each disconnecting means enclosure.

(3) The grounded conductor must be routed with the phase conductors.

(4) It must not be smaller than the required grounding electrode conductor specified in Table 250-66.

(5) For service-entrance phase conductors larger than 1100 kcmil copper or 1750 kcmil aluminum, the grounded conductor shall not be smaller than 12½ percent (0.125) of

the area of the largest service-entrance phase conductor.

(6) Where the service-entrance phase conductors are installed in parallel, the minimum size of the grounded service conductor must

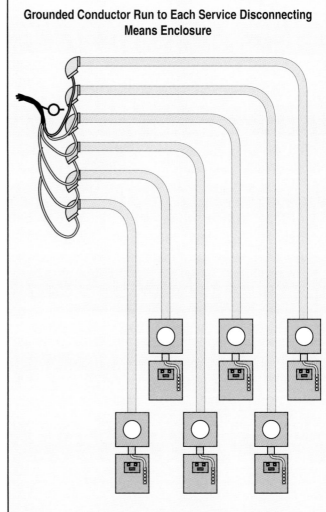

Grounded Conductor Run to Each Service Disconnecting Means Enclosure

Figure 4-2

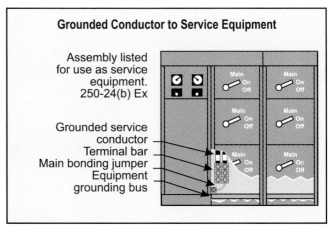

Grounded Conductor to Service Equipment

Assembly listed for use as service equipment. 250-24(b) Ex

Grounded service conductor
Terminal bar
Main bonding jumper
Equipment grounding bus

Main On Off

Figure 4-3

be based on the equivalent area of the ungrounded parallel service conductors.

Three exceptions to the rule for running the grounded service conductor to the service or sizing the conductor are provided.

(1) Where more than one service disconnecting means is located in a single assembly listed for use as service equipment such as a switchboard, the grounded conductor (or conductors in the case of parallel runs) is only required to be run to the assembly, not to each section. It (or they) must be bonded to the assembly enclosure. Since the sections are bolted together to form the assembly, the grounded service conductor is required to be run to only one section. In addition, it is common to have an internal equipment grounding bus connected to each section.

(2) The grounded conductor is not required to be larger than the largest ungrounded service-entrance phase conductor.

(3) See Section 250-36 for high-impedance grounded neutral systems grounding connection requirements.

In addition to providing the return portion of the circuit for unbalanced loads, this grounded service conductor provides the vital low impedance path for ground-fault current to return to the source. The grounded system conductor must be installed to each service disconnecting means and be bonded to the enclosure regardless of whether or not it is needed for the service or is used to supply a load.

Figure 4-4 illustrates a 3-phase power service taken from a grounded system. The grounded conductor is not needed since only 3-phase or phase-to-phase connected loads are supplied. If the grounded conductor is not run to the service, as required in Section 250-24(b), a ground-fault circuit of high impedance is present as shown by the dashed lines. This is a violation of Section 250-2(d) and it becomes virtually impossible to clear a ground fault, thus introducing an unnecessary hazard in the system.

However, the installation complies with Section 250-24(b) if the grounded conductor is run to service. A low-impedance ground-fault path is provided as required by Section 250-2(d). In the latter case, the safety of the service is improved immeasurably. Since the grounded conductor is run only to provide a ground-fault path, the size of the conductor to be run will depend on the size of the phase conductors in the service. This conductor must not be smaller than the grounding electrode conductor given in Table 250-66.

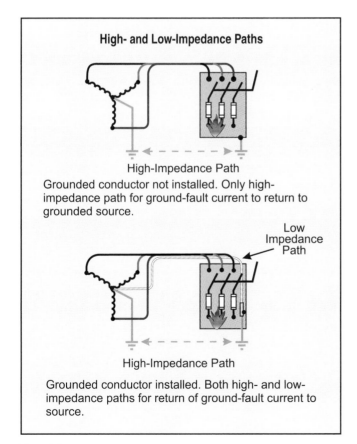

High- and Low-Impedance Paths

High-Impedance Path

Grounded conductor not installed. Only high-impedance path for ground-fault current to return to grounded source.

Low Impedance Path

High-Impedance Path

Grounded conductor installed. Both high- and low-impedance paths for return of ground-fault current to source.

Figure 4-4

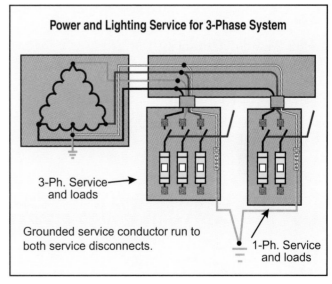

Power and Lighting Service for 3-Phase System

3-Ph. Service and loads

Grounded service conductor run to both service disconnects.

1-Ph. Service and loads

Figure 4-5

Figure 4-5 represents the relationship between the service and equipment ground in a 3-phase, 4-wire delta system with the midpoint of one transformer grounded. Such systems are intended to supply economically both a 3-phase, 3-wire power service and a single-phase, 3-wire lighting service from one transformer bank. The lighting transformer must be big enough to supply all the lighting load plus the three-phase load. The other transformers are sized to carry the 3-phase, 3-wire load. Though two separate panelboards are shown for illustration purposes, all loads can be supplied from a single 3-phase panelboard. Where this is done, the electrician needs to exercise caution in making connections for 120-volt loads. These loads must be connected to only the A and C phases as the B phase will have a voltage to ground of approximately 208 volts which is enough to severely damage equipment designed to operate at 120 volts.

In the case of the 3-phase, 3-wire power service, there is no need for the neutral (or grounded service conductor), as far as meeting voltage or phase requirements on the power service is concerned. This is true where no line-to-neutral loads are supplied. However, it is required to run the grounded conductor of the system to the three-phase power service equipment and to use the main bonding jumper so to satisfactorily clear a ground fault developing on the power supply system.

Location of service grounding connection.

Section 250-24(a) requires that a grounding electrode conductor be used to connect the grounded service conductor to a grounding electrode. The connection to the grounded service conductor must be at an accessible location from the load end of the service drop or service lateral to, and including, the terminal bus at the service disconnecting means. See Section 250-24(a)(1). These locations include current-transformer enclosures, meter enclosures, pull and junction boxes, busways, auxiliary gutters and wireways as well as switchboards, panelboards or motor control centers.

Many inspection authorities and serving utilities will not permit the grounding electrode connection to the system grounded conductor to be within current-transformer cans, meter enclosures or other enclosures that are sealed by the utility. Some utility policies prohibit connection within metering equipment, and many inspection authorities prohibit that connection location as they interpret the connection to be no longer "accessible" as utilities seal the metering equipment to prevent unauthorized access. It is important that serving utilities are consulted to be certain their system grounding policies are complied with.

The most practical and commonly accepted location for the grounding electrode conductor connection to the grounded service conductor is within the service disconnecting means or within a wireway that is not sealed at the service equipment location. The connection is usually made to the neutral bus.

Low-impedance grounding electrode conductor connections

To obtain sufficiently low impedance, the point of connection of the grounding electrode conductor must be within the same metallic enclosure as the phase conductors and the grounded conductor. The most practical point is within the service equipment.

The connection may be at one of the locations shown in Figure 4-6. Locations indicated are within the service panelboard, to the grounded conductor terminal block inside the current transformer cabinet, to the grounded service conductor terminal block inside the service wireway, or within the meter base.

It is also permissible to make the service grounding connections within each service disconnecting means rather than within the wireway. Means and methods of making and sizing these conductors are covered in Chapter Seven.

Where the service is supplied by a transformer located outside the building, an additional ground-

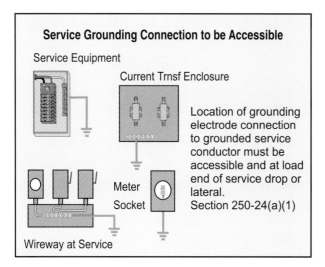

Service Grounding Connection to be Accessible

Service Equipment

Current Trnsf Enclosure

Location of grounding electrode connection to grounded service conductor must be accessible and at load end of service drop or lateral.
Section 250-24(a)(1)

Meter Socket

Wireway at Service

Figure 4-6

ing connection to an electrode must be made outside the building, usually at the transformer. This connection is usually provided by the electric utility.

Section 250-24(a)(5) of the Code prohibits grounding of the grounded circuit conductor (often a neutral) at any point beyond the service. Three exceptions to this rule are provided, several of which will be covered in later chapters of this text. An additional grounding connection beyond the service is permitted: (1) for separately derived systems, (2) where more than one building or structure on the same premises is supplied by the same service and, (3) for existing circuits for electric ranges and dryers. The use of the words "grounded circuit conductors" is intended to cover all such conductors whether they are feeders or branch circuits.

Sizing of grounded service conductor

The size of the grounded service conductor (often the neutral) varies with the load as calculated in accordance with Section 220-22. This becomes the minimum size grounded service conductor unless modified by Section 250-24(b)(1).

First of all, a load calculation should be performed in accordance with Section 220-22. Then, the minimum size of conductor must be compared with Table 250-66. Section 250-24(b)(1) requires the grounded conductor to be no smaller than the grounding electrode conductor as determined in Table 250-66. If the load calculation permits a conductor smaller than provided for in Table 250-66, the conductor must be increased in size to be not smaller than provided for in the table. If the load calculation requires a conductor larger than required by Table 250-66, the larger conductor must be used.

For example, if a 400 ampere service is to be installed from a grounded system, 500 kcmil copper conductors are selected for the ungrounded service conductors. By referring to Table 250-66, we see that the grounded service conductor can be no smaller than 1/0 copper or 3/0 aluminum. This conductor

Table 250-66 Grounding Electrode Conductor for AC Systems			
Size of Largest Service-Entrance Conductor or Equivalent Area for Parallel Conductors[1]		Size of Grounding Electrode Conductor	
Copper	Aluminum or Copper-Clad Aluminum	Copper	Aluminum or Copper-Clad Aluminum[2]
2 or smaller	1/0 or smaller	8	6
1 or 1/0	2/0 or 3/0	6	4
2/0 or 3/0	4/0 or 250 kcmil	4	2
Over 3/0 thru 350 kcmil	Over 250 kcmil thru 500 kcmil	2	1/0
Over 350 kcmil thru 600 kcmil	Over 500 kcmil thru 900 kcmil	1/0	3/0
Over 600 kcmil thru 1100 kcmil	Over 900 kcmil thru 1750 kcmil	2/0	4/0
Over 1100 kcmil	Over 1750 kcmil	3/0	250 kcmil

Notes:
1. Where multiple sets of service-entrance conductors are used as permitted in Section 230-40, Exception No. 2, the equivalent size of the largest service-entrance conductor shall be determined by the largest sum of the areas of the corresponding conductors of each set.
2. Where there are no service-entrance conductors, the grounding electrode conductor size shall be determined by the equivalent size of the largest service-entrance conductor required for the load to be served.
[1]This table also applies to the derived conductors of separately derived ac systems.
[2]See installation restrictions in Section 250-64(a).

(FPN): See Section 250-24(b) for size of ac system conductor brought to service equipment.

must be routed with the phase conductors and bonded to the service disconnecting means enclosure.

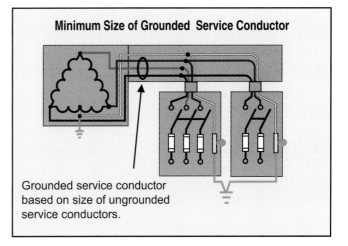

Minimum Size of Grounded Service Conductor

Grounded service conductor based on size of ungrounded service conductors.

Figure 4-7

This is the minimum size grounded service conductor permitted and must be installed as noted even though there is no neutral load on the system or the calculated neutral load would permit a smaller conductor. Follow a similar procedure for other service installations.

It is important to note that the minimum size of the grounded service conductor that must be run to the service disconnecting means is based on the size of the ungrounded (hot) service-entrance conductors and not on the rating of the circuit breaker or fuse that is installed in the service. It is helpful to remember that Table 250-66 generally applies up to the service, and Table 250-122 applies beyond the service.

Dwelling unit services and feeders

Special rules or provisions for sizing service entrance conductors and feeders are provided in Section 310-15(b)(6) and Table 310-15(b)(6). These rules apply only to 120/240 volt, 3-wire, single-phase dwelling services and feeders. By following the conditions of the section and table, the specified size of service-entrance or feeder conductors shown in the table are permitted to be used based upon the service or feeder rating.

The grounded conductor (often a neutral) is permitted to be smaller than the ungrounded (hot) conductors provided the rules of Sections 215-2, 220-22 and 230-42 are met.

Section 215-2 provides that the feeder neutral must be adequate for the load, must be a minimum size for certain loads and does not have to be larger than the service-entrance conductors that supply them.

Section 220-22 provides the method for calculating the feeder neutral load. The basic requirement is that the neutral must carry the maximum unbalanced load from the ungrounded conductors.

Section 230-42 generally requires that service-entrance conductors be of sufficient size to carry the loads as computed by Article 220. For grounded conductors in Section 230-42(c), the conductor cannot be smaller than required by Section 250-24(b).

Note, that for dwelling unit services and feeders, the previous permission to size the grounded conductor not more than two sizes smaller than the ungrounded conductors given in Section 310-15(b)(6) (old Note 3 to Table 310-16) has been removed from the Code. As a result, the grounded system conductor now has to be sized according to the above rules.

Parallel service conductors

Where parallel service conductors as permitted by Section 310-4 are installed, the minimum size grounded service conductor is determined by multiplying the circular mil area of the ungrounded conductors installed by the number of conductors installed in parallel.

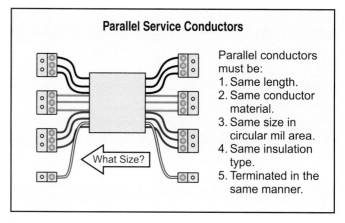

Parallel Service Conductors

What Size?

Parallel conductors must be:
1. Same length.
2. Same conductor material.
3. Same size in circular mil area.
4. Same insulation type.
5. Terminated in the same manner.

Figure 4-8

Example No. 1.

Given: Three, No. 4/0 copper conductors per phase are installed in parallel (electrically connected at each end to form a single conductor). Before multiplying the conductor size (4/0), it must be converted from the American Wire Gauge (AWG) 4/0 designation to the circular mil area.

- Refer to Chapter 9, Table 8 to determine the circular mil area of the No. 4/0 conductors. There we find the area to be 211,600 circular mils.
- Three times 211,600 = 634,800 circular mils.
- By referring to Table 250-66, we find the minimum size grounded service conductor is 2/0 copper or 4/0 aluminum.

This example assumes that all of the conductors are installed in the same raceway. This may not be practical due to the requirement that the derating or ampacity adjustment factors of Table 310-15(b)(2)(a) (old Note 8(a) to the Allowable Ampacity Tables) be applied. This obviously has the effect of requiring larger conductors to be installed than if the individual sets of service entrance conductors were installed in individual conduits or raceways. Where conductors are installed in sheet-metal wireways, derating may be avoided if the number of conductors does not exceed 30 and the conductors do not fill more than 20 percent of the square-inch area. See Section 362-5 of the NEC® for additional information.

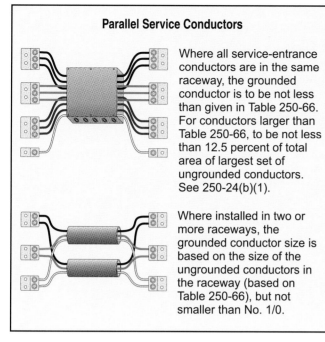

Parallel Service Conductors

Where all service-entrance conductors are in the same raceway, the grounded conductor is to be not less than given in Table 250-66. For conductors larger than Table 250-66, to be not less than 12.5 percent of total area of largest set of ungrounded conductors. See 250-24(b)(1).

Where installed in two or more raceways, the grounded conductor size is based on the size of the ungrounded conductors in the raceway (based on Table 250-66), but not smaller than No. 1/0.

Figure 4-9

Where installed in separate metal raceways, Section 300-3(b) requires that the neutral conductor must be installed in each raceway. In this case, Section 310-4 requires that the paralleled conductors be not smaller than No. 1/0, so the minimum size grounded conductor permitted in parallel is No. 1/0.

Example No. 2.

If six No. 4/0 copper conductors are installed parallel, the minimum size grounded service conductor is determined as follows:

As explained above, the area of No. 4/0 conductors (211,600 circular mils) in Chapter 9, Table 8 is used.

- Six times 211,600 = 1,269,600 circular mils. Since the total conductor area exceeds the 1100 kcmils for copper conductors given in Table 250-66, the rule in Section 250-24(b)(1) must be followed. There we find a requirement that the grounded service conductor be not smaller than 12½ percent (0.125) of the equivalent area for parallel conductors.

- 1,269,600 times 0.125 = 158,700 circular mils.

- By again referring to Chapter 9, Table 8, the conductor that is the next size larger than 158,700 circular mils is a No. 3/0 conductor which has a circular mil area of 167,800.

This example also assumes that all the service

entrance conductors are installed in the same raceway, which, due to derating requirements, may not be too practical.

Again, where installed in separate metal raceways, a grounded service conductor must be installed in each raceway and must not be smaller than No. 1/0 AWG. The grounded service conductor must also comply with Section 230-42(a). This section requires that the conductor be adequate to carry the load as determined by Article 220.

Section 220-22 requires that the neutral be sized for the maximum unbalance of the load. Examples of neutral conductor load calculations are found in Appendix D of the National Electrical Code®. In addition, where the length of run of the grounded conductor from the transformer to the service equipment is long, the size of grounded conductor should be increased.

Underground parallel service conductors

As illustrated in Figure 4-10, for only underground installations in nonmetallic raceways, it is permitted by Section 300-5(i) Exception No. 2 to install all the conductors of each phase in the same raceway. This is also permitted in Section 300-3(b)(1) Exception. To comply with these sections, all the ungrounded conductors of phase A are installed in one raceway, phase B in another, C in the third and the grounded service conductors in another. This method is often chosen to allow phase conductors to more readily line up with bus terminations in bottom-fed switchboards. As can be seen in the illustration, this reduces the "rats nest" in the bottom of these enclosures caused by

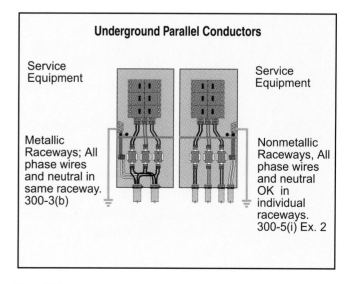

Underground Parallel Conductors

Service Equipment

Metallic Raceways; All phase wires and neutral in same raceway. 300-3(b)

Service Equipment

Nonmetallic Raceways, All phase wires and neutral OK in individual raceways. 300-5(i) Ex. 2

Figure 4-10

many conductors criss-crossing each other for termination. Another advantage is it is much easier to comply with the requirement that parallel conductors be the same length.

When this type of installation is made, care must be taken to eliminate the inductive heating of metal enclosures with magnetic properties by cutting a slot in the metal between the conduit entries or arranging for the manufacturer of the equipment to install a nonmetallic plate in the bottom of the equipment where the conduits terminate. [See Sections 300-20(a) and (b).] Usually, a slot cut with a single hacksaw blade is adequate to provide the desired relief. Another option is to terminate these conduits above the floor in the compartment of an open-bottom, floor-standing switchboard.

See Section 384-10 where the conduit or raceways, including their end fittings, are not permitted to rise more than 3 in. above the bottom of the enclosure.

Multiple services to one building

Section 230-2 permits several services to one building under one of several conditions given. Each service that is supplied from a grounded system must be provided with a grounded service-entrance conductor. The size of the ungrounded service-entrance conductor for each service determines the minimum size of grounded service con-

ductor for that service. Each service is considered individually for sizing the grounded service conductor to it.

For example, a building has a 400 ampere 480-volt 3-phase and a 100 ampere 120/240-volt service. The size of grounded service conductor is determined as follows:

400 ampere service
- 750 kcmil THW aluminum ungrounded service conductors.
- Table 250-66 = 1/0 copper or 3/0 aluminum grounded service conductor.

100 ampere service.
- No. 2 copper ungrounded service conductors.
- Table 250-66 = No. 8 copper or No. 6 aluminum grounded service conductor.

Again, we must emphasize that this method determines the minimum size of grounded service conductor to comply with Section 250-24(b). A larger conductor may be required to carry the maximum unbalanced load on the neutral conductor as determined by Section 220-22.

High-impedance grounded systems

Industrial plants having a continuous industrial process often have a need for uninterrupted electrical power and systems. It is quite common to see these plants located near a power company substation and to have more than one high voltage feeder to improve system reliability.

Another step that is commonly taken to improve system reliability is to install high-impedance grounded systems rather than solidly grounded systems. Advantages include improved reliability, the ability to have ground-fault relaying and fault indicating alarms, as well as fewer problems to the system from transient overvoltages.

Four conditions must be met before the Code will permit high-resistance grounded systems to be installed. They are as follows:
1. Qualified persons must be available to service and maintain the system.
2. Continuity of power is required.
3. Ground detectors must be installed to indicate an insulation failure.
4. Line-to-neutral loads are not served.
   Specific rules are provided in Section 250-36 for installing these systems. The grounding impedance, usually a resistor, is installed between the transformer supplied

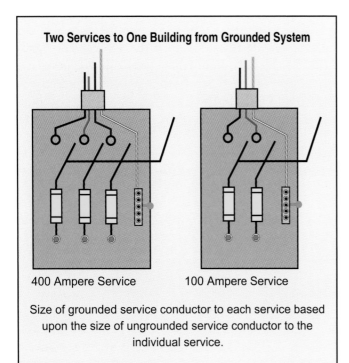

Two Services to One Building from Grounded System

400 Ampere Service   100 Ampere Service

Size of grounded service conductor to each service based upon the size of ungrounded service conductor to the individual service.

Figure 4-11

High Impedance Grounded Systems

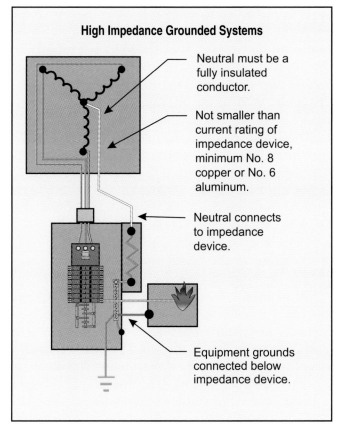

Neutral must be a fully insulated conductor.

Not smaller than current rating of impedance device, minimum No. 8 copper or No. 6 aluminum.

Neutral connects to impedance device.

Equipment grounds connected below impedance device.

Figure 4-12

Equipment for High Resistance Grounded Neutral System

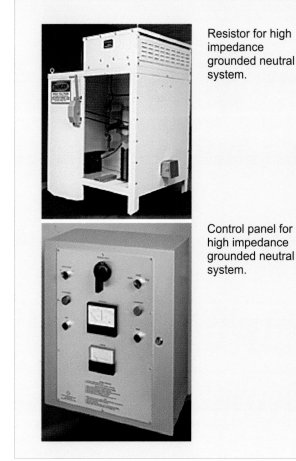

Resistor for high impedance grounded neutral system.

Control panel for high impedance grounded neutral system.

Figure 4-13. *Photo courtesy of Post Glover*

grounded service-entrance conductor and the grounding electrode. Usually, the impedance device is sized to a value just greater than the capacitive charging current of the system. For 480-volt systems, this is usually about 10 amperes.

A fully insulated grounded service-entrance conductor must be run to the impedance device. This conductor must have an ampacity not less than the maximum current rating of the grounding impedance. The minimum size grounded service conductor cannot be smaller than No. 8 copper or No. 6 aluminum or copper-clad aluminum. Since the grounding impedance will limit the fault current to a fairly low value, usually 10 amperes or less, the minimum size of neutral conductor is primarily related to mechanical strength, not to its current-carrying capabilities.

It is not required that the grounded service conductor be routed with the ungrounded service-entrance conductors since it will carry very little current in the event of a fault. [See Section 250-36(d)] At the service, it is connected to the impedance device which may be located outside the service enclosure to dissipate heat. An unspliced equipment bonding conductor is installed from the

load side of the impedance device to the service enclosure where a terminal bar is installed for connection of equipment grounding conductors. [See Section 250-36(e)] The grounding electrode conductor is permitted to be connected to any point from the grounded side of the grounding impedance to the equipment grounding connection at the service equipment or the first system disconnecting means. [See Section 250-36(f)]

This system is designed to limit the fault current on the first ground fault that might occur on the system. As can be seen in Figure 4-12, the impedance device is in series with the first ground fault. The electrical system will continue to function normally with the first ground fault present on the system. The ground-detection system will indicate the presence of the faulted condition by either a visual or audible signal or both. This is intended to alert maintenance personnel of the ground-fault condition so corrective action can be taken, hopefully during a period the plant is not planned to be in operation. One difference from

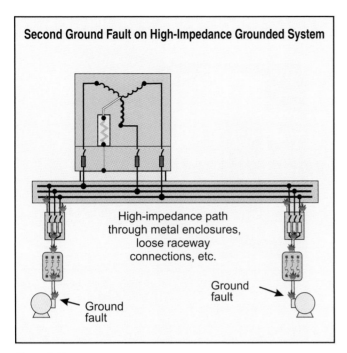

Second Ground Fault on High-Impedance Grounded System

High-impedance path through metal enclosures, loose raceway connections, etc.

Ground fault

Ground fault

Ground fault

Figure 4-14

ground-detection on an ungrounded system is that the faulted circuit can be identified without shutting the plant down.

A second ground fault (illustrated in Figure 4-14) that occurs before the first fault is cleared will be a line-to-line or phase-to-phase fault that would be cleared by the service or feeder overcurrent device which will result in a power outage. This can, and at times has, involved two pieces of equipment in separate parts of the plant supplied by different feeders. In this situation, a great deal of current can flow in the circuit and can cause extensive damage to electrical equipment.

Instrument transformers, relays, etc.

Part J of Article 250 deals with grounding requirements for instrument transformers, relays, etc. The Code requires the grounding of secondary circuits of current and potential instrument transformers where the primary windings are connected to circuits of 300 volts or more to ground and, where on switchboards, shall be grounded irrespective of voltage. (See Section 250-170) If a switchboard has no live parts or wiring exposed or accessible to other than qualified persons, such circuits need not be grounded where the primary windings are connected to circuits of 1000 volts or less.

Instrument transformer cases or frames are required to be grounded where accessible to other than qualified persons. Such cases or frames of current transformers need not be grounded where the primaries are not over 150 volts to ground and the current transformers are used exclusively to supply current to meters. (See Section 250-172.)

The cases of instruments, meters and relays, where operated with windings or working parts at 1000 volts or less, are required to be grounded. If such instrument cases and relays are not on switchboards and are accessible to other than qualified persons, the cases or other exposed metal parts shall be grounded if they operate at 300 volts or more to ground. [See Section 250-174(a).]

For dead-front switchboards having no live parts on the front of the panel, such instrument cases shall be grounded. [See Section 250-174(b).]

For live-front switchboards which have exposed live parts on the front of the panel, such instrument cases are not required to be grounded. However, the Code recommends the use of an insulating mat for the protection of the operator where the voltage to ground exceeds 150 volts. [See Section 250-174(c).]

If the operating voltage exceeds 1000 volts, the cases of instruments shall not be grounded but shall be isolated by elevation or protected by suitable barriers, grounded metal or insulating covers or guards. In the case of electrostatic ground detectors, the internal ground segments of the instruments are to be connected to the instrument case and grounded. The ground detector is to be isolated by elevation. (See Section 250-176.)

Size of instrument grounding conductor

The Code specifies that the grounding conductor for circuits of instrument transformers and for instrument cases shall not be smaller than No. 12 copper or No. 10 aluminum. However, if cases are mounted directly on grounded metal surfaces of enclosures, they shall be considered to be grounded. (See Section 250-178.)

Ungrounded systems

Ungrounded systems are subject to relatively severe transient overvoltages which can reach several times normal voltage to ground. See Chapter Two of this text for more information on this subject. Such abnormal voltages become potential hazards and often cause insulation failure and equipment breakdowns. If a system has one conductor grounded and is thus a grounded system, the value of such transient overvoltages as they develop is greatly reduced.

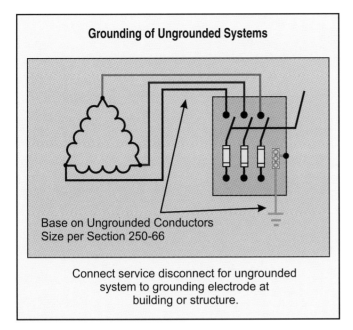

Grounding of Ungrounded Systems

Base on Ungrounded Conductors
Size per Section 250-66

Connect service disconnect for ungrounded system to grounding electrode at building or structure.

Figure 4-15

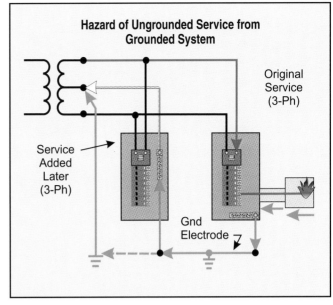

Hazard of Ungrounded Service from Grounded System

Original Service (3-Ph)

Service Added Later (3-Ph)

Gnd Electrode

Figure 4-16

An ungrounded system must have its conductor and equipment enclosures connected to a grounding electrode system at the building or structure served. This keeps such enclosures as near to ground potential as possible and reduces shock hazards to a minimum. These service equipment enclosures are grounded by connecting them to a grounding electrode system per Section 250-24(d). Grounding electrode systems are treated extensively in a later chapter of this text.

Hazard of operating a service ungrounded from a grounded system

Figure 4-16 illustrates the hazard of operating an ungrounded service from a grounded system. The original ungrounded service on the right was installed some time before the service on the left. The first and original service was supplied by an ungrounded utility system. The service and feeder shown supplying equipment were protected by large overcurrent devices.

Some time later, the service on the left which included a grounded service conductor was installed. At that time, the serving utility grounded the transformer bank and extended a grounded system conductor to the new service weatherhead. However, the grounded service conductor was not extended nor connected to the older, existing service. The older, existing service, which supplied only 3-phase equipment loads continued to supply power to a portion of the building, though without a grounded conductor connection.

The two services were connected with properly-sized grounding electrode conductors to a common grounding electrode system which included a metal water piping system.

Some time later a ground fault occurred in the equipment being supplied from the original ungrounded service equipment. Since the system supplying the equipment was now grounded, current attempted to find its way back to the source of power to complete the circuit. Current flowed over the grounding electrode conductor at the ungrounded service, to the grounding electrode, through the grounding electrode to the grounded service and returned to the transformer bank over the grounded service conductor. In this and similar situations, the current will seek a path back to the system through all available paths and will divide among the paths based upon the impedance of the paths that are available.

A great deal of current flowed over the grounding electrode conductors that were not designed or installed for that amount of current which caused the conductors to get extremely hot. The resulting fire burned a great deal of the industrial plant to the ground.

The action necessary to prevent the fire was to run a properly-sized grounded system conductor to the existing service when the new service from a grounded supply was installed. That conductor should have been extended to the service disconnecting means and bonded to it. The grounded service conductor would have provided a low-impedance path back to the transformer bank and would have allowed the overcurrent devices to clear the ground fault.

Methods of clearing ground faults and short circuits are discussed in detail in Chapter 11.

Chapter Four: The questions included here were developed using material included in this chapter. The answers can be found by reviewing the text. It is also important that students make use of the 1999 NEC®, where many answers can be found. See page 279 for answers.

1. The grounded conductor is required to be run to each ac service disconnecting means where they operate at less than ____ volts and are grounded at any point.
   a. 2000
   b. 1500
   c. 1000
   d. 1200

2. Where an ac system operating at less than ____ volts is grounded at any point, the grounded conductor is required to be bonded to each disconnecting means enclosure.
   a. 1500
   b. 1200
   c. 2000
   d. 1000

3. Where an ac system operating at less than ____ volts is grounded at any point, the grounded conductor is required to be routed with the phase conductors.
   a. 1000
   b. 1200
   c. 1300
   d. 2000

4. Where an ac system operating at less than 1000 volts is grounded at any point the grounded conductor cannot be smaller than the required _____ conductor specified in Table 250-66.
   a. service-entrance
   b. grounding-electrode
   c. equipment grounding
   d. ungrounded service

5. For service-entrance phase conductors larger than 1100 kcmil copper or 1750 kcmil aluminum, the grounded conductor cannot be smaller than ____ percent of the area of the largest service-entrance phase conductor.
   a. 12½
   b. 14½
   c. 13¾
   d. 12

6. Where the service-entrance phase conductors are installed in ____, the size of the grounded service conductor is required to be based on the equivalent area of the ungrounded service conductors.
   a. a raceway
   b. a trench
   c. parallel
   d. a cable tray

7. Where more than one service disconnecting means is located in an assembly listed for use as service equipment, only one grounded conductor is required to be run to the assembly. It is required to ____ to the assembly enclosure.
   a. be sized at No. 8 and bonded
   b. not be bonded
   c. be bonded
   d. be sized at No. 6 and bonded

8. Some utility policies prohibit connection within metering equipment, and many inspection authorities prohibit that connection location because they interpret the connection to be no longer ____.
   a. accessible
   b. identified
   c. serviceable
   d. visible

9. The grounding electrode conductor connection to the grounded service conductor is most often made within the service disconnecting means and is connected to the ____ bus.
   a. unidentified
   b. isolated ground
   c. floating neutral
   d. neutral

10. Where the transformer supplying the service is located outside the building an additional grounding connection to an electrode is required to be made outside the building, usually at the ____.
   a. transformer
   b. pole
   c. meter
   d. grid

11. Where copper conductors sized at 500 kcmil are selected for the ungrounded service conductors supplying a 400 ampere service, and are installed from a grounded system, the size of the copper grounding electrode conductor cannot be less than No. ____.
   a. 8
   b. 6
   c. 4
   d. 1/0

12. Section 310-15(b)(6) requires or permits the grounded service conductor for dwelling services and feeders to be:
 a. two sizes smaller than the ungrounded conductors.
 b. used for 60-ampere feeders.
 c. sized by calculation from Section 220-22.
 d. Type TW conductors.

13. Industrial plants having a continuous industrial process often have a need for ____ electrical power and systems.
 a. isolated
 b. larger
 c. five-wire, two-phase
 d. uninterrupted

14. Certain conditions must be met before the NEC® will permit high-resistance grounded systems to be installed. They include ____.
 a. qualified persons must be available to service and maintain the system.
 b. continuity of power is required, line-to-neutral loads are not served.
 c. ground detectors must be installed to indicate an insulation failure.
 d. all of the above.

15. Secondary circuits of current and potential instrument transformers where the primary windings are connected to circuits of ____ volts or more to ground and, where on switchboards, are required to be grounded irrespective of voltage.
 a. 300
 b. 100
 c. 200
 d. 150

16. Where a switchboard has no live parts or wiring exposed or accessible to other than qualified persons, such instrument transformer secondary circuits are not required to be grounded where the primary windings are connected to circuits of ____ volts or less.
 a. 480
 b. 1000
 c. 277
 d. 150

17. Instrument transformer cases or frames are required to be grounded where accessible to other than qualified persons. Such cases or frames of current transformers are not required to be grounded where the primaries are not over ____ volts to ground, and the current transformers are used exclusively to supply current to meters.
 a. 100
 b. 120
 c. 130
 d. 150

Chapter 5

Service and Main Bonding Jumpers

Objectives

After studying this chapter, the reader will be able to understand:

- Definitions of bonding and bonding jumpers.
- Functions of the main bonding jumper.
- Sizing of main and equipment bonding jumpers.
- Methods for bonding at service equipment.
- Use of neutral for bonding on line side of service.
- Requirements for grounding and bonding of remote metering.

Definitions:

Bonding (Bonded): "The permanent joining of metallic parts to form an electrically conductive path that will ensure electrical continuity and the capacity to conduct safely any current likely to be imposed." [N]

Bonding jumper, Main: "The connection between the grounded circuit conductor and the equipment grounding conductor at the service." [N]

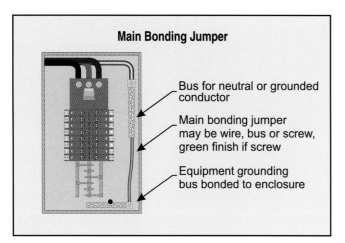

Main Bonding Jumper

Bus for neutral or grounded conductor

Main bonding jumper may be wire, bus or screw, green finish if screw

Equipment grounding bus bonded to enclosure

Figure 5-1

Main bonding jumper

The main bonding jumper is one of the most critical elements in the safety grounding system. This conductor is the link between the grounded service conductor, the equipment grounding conductor and in some cases, the grounding electrode conductor. The primary purpose of the main bonding jumper is to carry the ground-fault current from the service enclosure as well as from the equipment grounding system that is returning to the source. In addition, where the grounding electrode conductor is connected directly to the grounded service conductor bus, the main bonding jumper ensures that the equipment grounding bus is at the same potential as the earth.

For a grounded system, Section 250-28 requires that an unspliced main bonding jumper be used to connect the equipment grounding conductor(s) and the service-disconnect enclosure to the grounded conductor of the electrical system. The connection is required to be made within the enclosure for each service disconnect.

An example of this is where two or more service disconnecting means in individual enclosures are grouped at one location. This type of installation often is made with a wireway or a short section of busway installed downstream from the metering

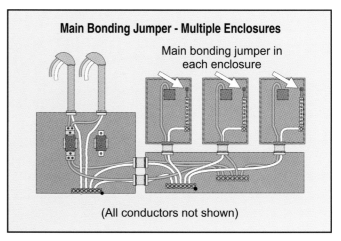

Main Bonding Jumper - Multiple Enclosures

Main bonding jumper in each enclosure

(All conductors not shown)

Figure 5-2

equipment. In other cases, a wireway or short section of busway is installed ahead of metering and is supplied by a service lateral or service-entrance conductors. Sets of service-entrance conductors supply each of the service disconnecting means. Service disconnecting means are installed from the wireway or auxiliary gutter. (If there are nipples between the disconnecting means and the metal or nonmetallic trough, the trough meets the definition of a wireway from Article 362 rather than an auxiliary gutter from Article 374.) Section 250-28 requires a main bonding jumper be installed in each service disconnect enclosure. As previously mentioned, Section 250-24(b) requires that the grounded service conductor be brought to each service disconnecting means and be bonded to the enclosure. The main bonding jumper is the means to accomplish this requirement.

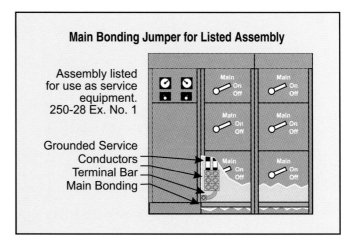

Main Bonding Jumper for Listed Assembly

Assembly listed for use as service equipment. 250-28 Ex. No. 1

Grounded Service Conductors Terminal Bar Main Bonding

Figure 5-3

The rules are a little different where more than one service disconnecting means is in a common enclosure. This equipment usually consists of listed switchboards, panelboards or motor control centers. Where more than one service disconnecting means is located in an assembly listed for use as service equipment, Section 250-28 Exception No. 1 permits the grounded service conductors to be run to a single grounded conductor bus in the enclosure and then be bonded to the assembly enclosure. This means that only one main bonding jumper connection is required to be installed from the common grounded conductor bus to the assembly enclosure. The sections of the assembly are bonded together by means of an equipment grounding conductor bus or by being bolted together.

Exception No. 2 to Section 250-28(b) permits alternate means for bonding of high-impedance grounded neutral systems. See Chapter Four of the *IAEI Soares Book on Grounding* for methods and requirements for grounding high-impedance grounded neutral systems. Also see NEC® Sections 250-36 and 250-186 for the specific requirements and allowances.

The main bonding jumper is permitted to consist of a wire, bus, screw or other suitable conductor. It must be fabricated of copper or other corrosion-resistant material. Aluminum alloys are permitted where the environment is acceptable. In addition, where the main bonding jumper consists of a screw, it must have a green finish that is visible with the screw installed. This green finish assists in identifying the bonding-jumper screw from the other screws that are on or near the neutral bus. See Sections 250-28(a) and (b).

Functions of Main Bonding Jumper

The main bonding jumper performs three major functions:

1. Connecting the grounded service conductor to the equipment grounding bus or conductor and the service enclosure.

2. Providing the low-impedance path for the return of ground-fault currents to the grounded service conductor. The main bonding jumper completes the ground-fault return circuit from the equipment through the service to the source as is illustrated in Figure 5-4.

3. Connecting the grounded service conductor to the grounding electrode conductor. Under certain conditions given in Section 250-24(a)(4), it is permitted to connect the grounding electrode conductor to the equipment grounding terminal bar rather than to the terminal bar for the grounded service conductor. This scheme is common on larger switch-

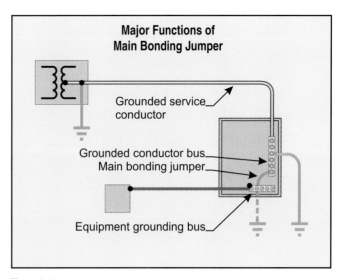

Figure 5-4

board services and is necessary for proper operation of certain types of equipment ground fault protection systems. See Chapter 15 of the *IAEI Soares Book on Grounding* for additional information on this subject.

Size of main bonding jumper in listed enclosures

Where listed service equipment consisting of a switchboard, panelboard or motor control center is installed, the main bonding jumper that is provided with the equipment is rated for the size of conductors that would normally be used for the service. The method for sizing of the main bonding jumper in listed service equipment is found in Underwriters Laboratories Safety Standard for the equipment under consideration and is verified by the listing agency. Therefore, if a main bonding jumper that is a bus bar, strap, conductor, or screw is furnished by the manufacturer as part of the listed equipment, it may be used without calculating its adequacy. Section 384-3(c) requires the equipment manufacturer to provide the main bonding jumper.

Size of main bonding jumper at single service-disconnect or enclosure

Since the main bonding jumper must carry the full ground-fault current of the system back to the grounded service conductor (which may be a neutral), its size must relate to the rating of the service conductors which supply the service. The minimum size of the main bonding jumper is found in Table 250-66 as required by Section 250-28(d). This relationship is based on the conductor's abil-

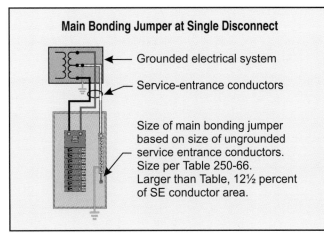

Main Bonding Jumper at Single Disconnect

← Grounded electrical system

← Service-entrance conductors

Size of main bonding jumper based on size of ungrounded service entrance conductors. Size per Table 250-66. Larger than Table, 12½ percent of SE conductor area.

Figure 5-5

ity to carry the expected amount of fault current for the period of time needed for the overcurrent device to open and stop the flow of current.

For example, where 250 kcmil aluminum service-entrance conductors are installed, the main bonding jumper is found to be No. 4 copper or No. 2 aluminum by reference to Table 250-66.

The size of the main bonding jumper does not directly relate to the rating of the service overcurrent device. Do not attempt to use Table 250-122 for this purpose. Table 250-122 gives the minimum size of equipment grounding conductors for feeders and circuits on the load side of the service.

Sizing of main bonding jumper for parallel service conductors

Where service conductors are installed in parallel, (connected together at each end to form a larger conductor) the total circular mil area of the conductors connected in parallel for one phase are added together to determine the minimum size main bonding jumper required. See Section 250-28(d). For example, where three 250 kcmil conductors are connected in parallel per phase, they are treated as a single 750 kcmil conductor. By reference to Table 250-66 the main bonding jumper, if aluminum service-entrance conductors are used, is 1/0 copper or 3/0 aluminum.

Where the service-entrance conductors are larger than the maximum given in Table 250-66, Section 250-28(d) requires the main bonding jumper to be not less than 12½ percent (0.125) of the area of the largest phase conductors.

This is illustrated by the following example:

Three 500 kcmil copper conductors are installed in parallel as service-entrance conductors.

3 x 500 kcmil = 1500 kcmil.

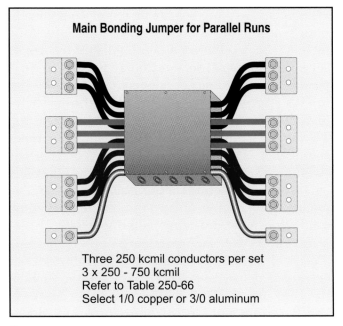

Main Bonding Jumper for Parallel Runs

Three 250 kcmil conductors per set
3 x 250 - 750 kcmil
Refer to Table 250-66
Select 1/0 copper or 3/0 aluminum

Figure 5-6

1500 x .125 = 187,500 circular mils.

Since a 187,500 circular mil conductor is not a standard size, we next refer to Chapter 9, Table 8 to find the area of conductors.

The next conductor exceeding 187,500 circular mils is a No. 4/0 AWG conductor which has an area of 211,600 circular mils. It is always necessary to go to the next larger size conductor since the 12½ percent size is the minimum size permitted.

Follow a similar procedure for determining the minimum size main bonding jumper required for other sizes of parallel service-entrance conductors.

Bonding of service conductor enclosures

Special rules are provided for bonding enclosures on the line side of the service disconnecting means. This is due to the fact that this equipment does not have overcurrent protection on its line side such as feeders and branch circuits have. Fault current of sufficient magnitude must flow during a short period of time to allow the fuse on the line side of the utility transformer to open. The level of fault current and particularly the duration the current may flow could be much larger than would flow in a feeder or branch circuit as there is not an overcurrent device in series with the conductor.

The basic rule is that all metallic enclosures that contain a service conductor must be bonded together. The bonding ensures that none of the equipment enclosures can become isolated electrically and become a shock hazard should a line-to-ground fault occur. The bonding also provides a low impedance path for fault current to flow in so the fuse or circuit breaker on the line side of the electric utility transformer will open.

Sizing of equipment bonding jumper on line (supply) side of service

Equipment bonding jumpers on the line side of the service and main bonding jumper must be sized to comply with Table 250-66. This is required by Section 250-102(c). For example, where 250 kcmil copper conductors are installed as service-entrance conductors, Table 250-66 requires a No. 2 copper or 1/0 aluminum bonding jumper.

Where the sum of the circular mil area of the service-entrance phase conductors exceeds 1100 kcmil copper or 1750 kcmil aluminum, the equipment bonding conductor must be not less than 12½ percent (0.125) of the area of the ungrounded phase conductors.

This is determined as follows:
Five x 250 kcmil = 1250 kcmil.
1250 kcmil x .125 = 156,250 circular mils.

The next larger conductor found in Chapter 9, Table 8 is 3/0 with an area of 167,800 circular mils.

In this case, a 3/0 copper equipment bonding conductor must be connected from the grounded service conductor or equipment grounding bus to each metal raceway in series (daisy-chain fashion from one raceway to another).

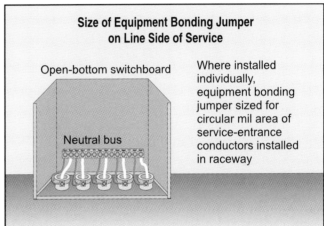

Size of Equipment Bonding Jumper on Line Side of Service

Open-bottom switchboard

Neutral bus

Where installed individually, equipment bonding jumper sized for circular mil area of service-entrance conductors installed in raceway

Figure 5-8

A more practical method of performing the bonding for services supplied by multiple raceways may be to connect an individual bonding conductor between each raceway and the grounded service conductor terminal bar or equipment grounding bus. This is permitted by Section 250-102(c). This will usually result in a smaller equipment bonding conductor which is easier to install.

Again, using the example above and referring to Table 250-66, the minimum size equipment bonding conductor for the individual raceways containing 250 kcmil copper service-entrance conductors is No. 2 copper or 1/0 aluminum. A properly sized equipment bonding jumper is installed from the terminal bar for the grounded service conductor or from the equipment grounding terminal bar to each conduit individually.

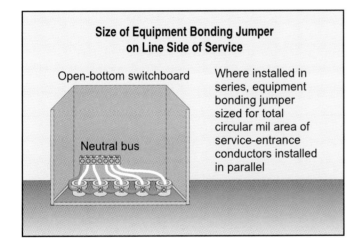

Size of Equipment Bonding Jumper on Line Side of Service

Open-bottom switchboard

Neutral bus

Where installed in series, equipment bonding jumper sized for total circular mil area of service-entrance conductors installed in parallel

Figure 5-7

Sizing of equipment bonding jumper for parallel conductors

Two methods are provided for bonding service raceways that are installed in parallel. The first method is to add the circular mill area of the service-entrance conductors per phase together and treat them as a single conductor. The bonding jumper size is determined from Table 250-66 and is connected to each conduit bonding bushing in a "daisy-chain fashion." This method often results in an equipment bonding jumper that is quite large and difficult to work with.

For example, if five 250 kcmil copper conductors are installed in parallel for a phase, the equipment bonding jumper for bonding the metal raceways must not be smaller than 3/0 copper.

Different conductor material

Section 250-28(d) provides instructions on sizing the main bonding jumper or equipment bonding jumper on the supply side of the service where different conductor materials are used for the service-entrance conductors and the bonding jumper. The procedure involves assuming the phase conductors are of the same material (copper or aluminum) as the bonding jumper and that they have an equivalent ampacity to the conductors that are installed. This is illustrated as follows:

Assume aluminum phase conductors and a copper bonding jumper are installed.

Three 750 kcmil Type THW aluminum conductors are installed.

From Table 310-16, 385 amperes x 3 = 1155 amperes. The smallest type THW copper conductor that has an equivalent rating is 600 kcmil with an ampacity of 420.

Next, determine the total circular mil area of the copper conductors.

Three x 600 kcmil = 1800 kcmil.

1800 kcmil x .125 = 225 kcmil.

The next standard size is 250 kcmil copper which is the minimum size bonding jumper permitted to bond equipment at or ahead of the service equipment in this example.

Bonding service equipment enclosures

The Code requires that electrical continuity of service equipment and enclosures that contain service conductors be established and maintained by bonding. The items required to be bonded together are stated as follows in Section 250-92(a):

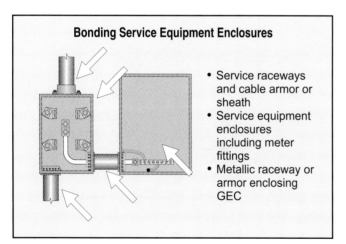

Bonding Service Equipment Enclosures

- Service raceways and cable armor or sheath
- Service equipment enclosures including meter fittings
- Metallic raceway or armor enclosing GEC

Figure 5-9

(1) The service raceways, cable-trays, cablebus framework or service cable armor or sheath.

(2) All service equipment enclosures containing service conductors, including meter fittings, boxes or the like, interposed in the service raceway or armor.

(3) Any metallic raceway or armor which encloses the grounding electrode conductor. (This subject is covered in detail in Chapter 7 of this text.)

An exception to this requirement for bonding at service equipment is mentioned in Section 250-92(a)(1). It refers to Section 250-84 which has rules on underground service cables that are metallically connected to the underground service conduit. The Code points out that if a service cable contains a metal armor, and if the service cable also contains an uninsulated grounded service conductor which is in continuous electrical contact with its metallic armor, then the metal covering of the cable is considered to be adequately grounded.

Use of neutral for bonding on line side of service

Section 250-94(1) permits the use of the grounded service conductor (may be the neutral) for grounding and bonding equipment on the line side of the service disconnecting means. This is also permitted by Section 250-142(a)(1). (Two other applications of this bonding are explored in later chapters of the *IAEI Soares Book on Grounding*.) Often, connecting the grounded service conductor to equipment such as meter bases, current transformer enclosures, wireways and auxiliary gutters is the most practical method of bonding these enclosures.

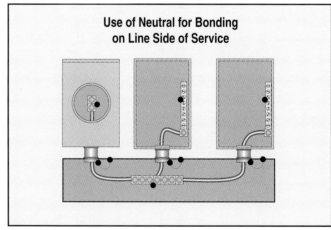

Use of Neutral for Bonding on Line Side of Service

Figure 5-10

Usually, self-contained meter sockets and meter-main combination equipment are produced with the grounded conductor terminals or bus (often a neutral) bonded directly to the enclosure. The enclosure is then effectively bonded by the connection of the grounded circuit conductor to these terminals. No additional bonding conductor connection to the meter enclosure is required. Current from a ground fault to the meter or meter-main enclosure will return to the source by the grounded service conductor (may be a neutral) and, hopefully, will allow enough current to flow in the circuit to operate the overcurrent protection on the line side of the utility or other transformer.

In addition, meter enclosures installed on the load side of the service disconnecting means are permitted to be grounded (bonded) to the grounded service conductor provided that:

(a) Service ground-fault protection is not installed; and

(b) The meter enclosures are located near the service disconnecting means. (No distance is used to clarify what is meant by the word "near."), and

(c) The size of the grounded circuit conductor is not smaller than the size specified in Table 250-122 for equipment grounding conductors. See Section 250-142(b) Exception No. 2.

Means of bonding at service equipment

The methods for bonding at service equipment are outlined in Section 250-94. These requirements for bonding are more restrictive at services than downstream from the service. The reason this is so important is service equipment and enclosures may be called upon to carry heavy fault currents in the event of a line-to-ground fault. The service conductors in these enclosures have only short-circuit protection provided by the overcurrent device on the line side of the utility trans-

former. Only overload protection is provided at the load end of the service conductor by the overcurrent device. This is one of the reasons the Code limits the length of service conductors inside of a building.

Bonding of these enclosures is to be done by one or more of the following methods from Section 250-94:

(1) Bonding to the grounded service conductor through the use of exothermic welding, listed pressure connectors such as lugs, listed clamps, or other listed means. These connections cannot depend solely upon solder.

(2) Threaded couplings and threaded bosses in a rigid or intermediate metal conduit system where the joints are made up wrench-tight. Threaded bosses include hubs that are either formed as a part of the enclosure or are supplied as an accessory and installed according to the manufacturer's instructions.

(3) Threadless couplings and connectors are permitted where they are made up tight for rigid and intermediate metal conduit and electrical metallic tubing and metal-clad cables.

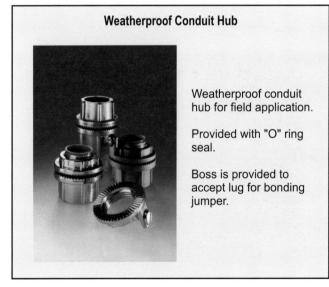

Weatherproof Conduit Hub

Weatherproof conduit hub for field application.

Provided with "O" ring seal.

Boss is provided to accept lug for bonding jumper.

Figure 5-12. Photo courtesy of Crouse-Hinds, Myers Hubs

(4) Other approved devices such as bonding-type locknuts and bushings.

Bonding jumpers are required to be used around concentric or eccentric knockouts that are punched or otherwise formed so as to impair an adequate electrical path for ground-fault current. It is important to recognize that concentric and eccentric knockouts in enclosures such as panelboards, wireways and auxiliary gutters have not been

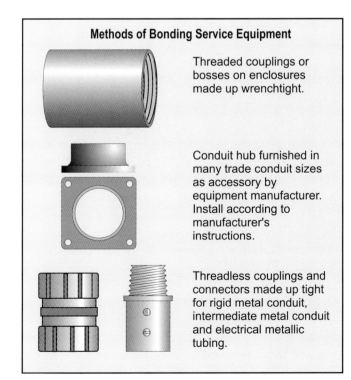

Methods of Bonding Service Equipment

Threaded couplings or bosses on enclosures made up wrenchtight.

Conduit hub furnished in many trade conduit sizes as accessory by equipment manufacturer. Install according to manufacturer's instructions.

Threadless couplings and connectors made up tight for rigid metal conduit, intermediate metal conduit and electrical metallic tubing.

Figure 5-11

Bonding Fittings

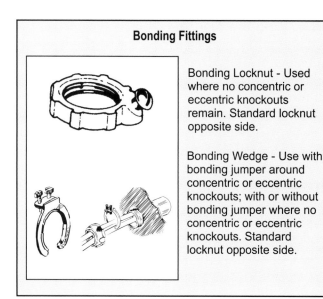

Bonding Locknut - Used where no concentric or eccentric knockouts remain. Standard locknut opposite side.

Bonding Wedge - Use with bonding jumper around concentric or eccentric knockouts; with or without bonding jumper where no concentric or eccentric knockouts. Standard locknut opposite side.

Figure 5-13. *Photo courtesy of Thomas & Betts*

investigated for their ability to carry fault current. Where any of these knockout rings remain at the conduit connection to the enclosure, they must always be bonded around to ensure an adequate fault-current path.

The Code states here that "Standard locknuts or bushings shall not be the sole means for the bonding required by this section." This statement does

Conduit Bonding Bushing

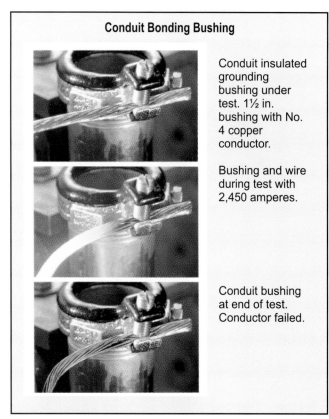

Conduit insulated grounding bushing under test. 1½ in. bushing with No. 4 copper conductor.

Bushing and wire during test with 2,450 amperes.

Conduit bushing at end of test. Conductor failed.

Figure 5-14 *Photo courtesy of Thomas & Betts*

not intend to prevent the use of "standard" lock-nuts and bushings, it is just that they cannot be relied upon as the sole means for the bonding that is required by this section. "Standard" locknuts are commonly used outside the enclosure on conduit that is bonded with a bonding bushing or bonding locknut inside the enclosure. Standard locknuts are used to make a good, reliable mechanical connection as required by Section 300-10.

Parallel Bonding Conductors

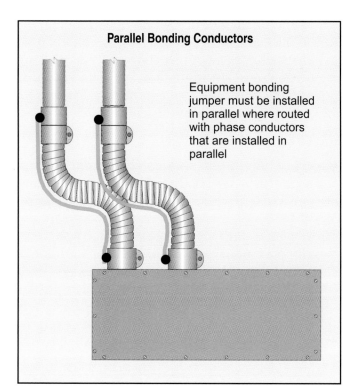

Equipment bonding jumper must be installed in parallel where routed with phase conductors that are installed in parallel

Figure 5-15

Parallel bonding conductors

Section 250-102(c) requires that where service-entrance conductors are paralleled in two or more raceways or cables and the equipment bonding jumper is routed with the raceways or cables, the equipment bonding jumper must be run in parallel.

In this case again, the size of the bonding jumper for each raceway is based upon the size of the service-entrance conductor in the raceway by referring to Table 250-66.

Grounding and bonding of remote metering

As mentioned before, Section 250-92(a) requires all equipment containing service conductors to be bonded together and to the grounded service conductor. This includes remote (from the service equipment) meter cabinets and meter sockets.

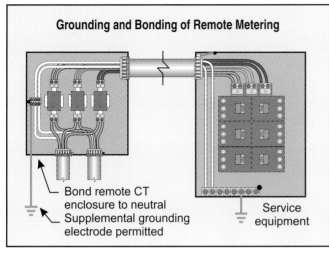

Grounding and Bonding of Remote Metering

Bond remote CT enclosure to neutral

Supplemental grounding electrode permitted

Service equipment

Figure 5-16

Grounding and bonding of equipment such as meters, current transformer cabinets and raceways to the grounded service conductor at locations on the line side of and remote from the service disconnecting means increases safety.

This equipment should never be grounded only to a grounding electrode such as a ground rod. Figures 5-16 and 5-17 show why. If a ground-fault occurred at this line-side equipment, and it is not bonded as required, the only means for clearing a

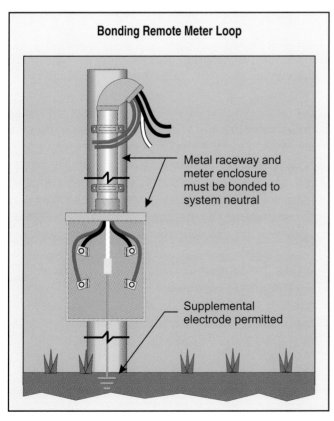

Bonding Remote Meter Loop

Metal raceway and meter enclosure must be bonded to system neutral

Supplemental electrode permitted

Figure 5-17

ground fault would be through the grounding electrodes and earth. Given the relatively high impedance and low current-carrying capacity of this path through the earth and high resistance of grounding electrodes such as rods, little current will flow in this path. This leaves the equipment enclosure(s) at a dangerous voltage above ground potential just waiting to shock or possibly electrocute a person or animal that may contact it. The voltage drop across this portion of the circuit can easily be calculated by using Ohms Law. (Resistance times the current gives the voltage.) There are many records of livestock being electrocuted while contacting electrical equipment that was improperly grounded. Sections 250-2 and 250-54 require that the earth not be used as the sole equipment grounding conductor or fault-current path.

The most practical method for grounding and bonding this line-side equipment is to bond the grounded service conductor to it. As can also be seen in Figures 5-16 and 5-17, a ground fault to the equipment will have a low impedance path back to the source through the grounded service conductor. This will allow a large current to flow in the circuit to cause the overcurrent protection on the line side of the transformer to clear the fault.

Supplementary grounding electrodes

In accordance with Section 250-54, it is permissible to install a grounding electrode at the remote meter location shown in Figures 5-16 and 5-17 to supplement the grounded service conductor. This Code section refers specifically to grounding electrodes supplementing the equipment grounding conductors. Some electric utilities require a grounding electrode at meter equipment installed remote from service equipment such as on poles. The Code in Section 230-66 makes it clear that individual meter socket enclosures are not to be considered service equipment. The same is true for metering equipment installed in remote current-transformer enclosures. As mentioned earlier, it is critically important that these meter enclosures be properly bonded as they contain service conductors.

This additional grounding electrode will attempt to keep the equipment at the earth potential that exists at the meter location. In addition, the electrodes at the remote meter and at the service location are bonded together by the grounded service conductor installed between the metering and service equipment. This brings the installation into compliance with Section 250-58 which requires a common grounding electrode or where two or more electrodes are installed, they must be bonded together.

As previously stated, these grounding electrodes should never be used as the only means for grounding or bonding these enclosures or to carry fault current.

More extensive discussion of this subject is found in Chapter Six of the *IAEI Soares Book on Grounding*.

Bonding of multiple service disconnecting means

Installation of multiple services as permitted by Section 230-2(a) through (d) and installations of services that have multiple disconnecting means can take several forms. Additional services are permitted by Section 230-2 for:

(a) Fire pumps, emergency, legally required, standby, optional standby or parallel power production systems.

(b) By special permission, for multiple occupancy buildings where there is no available space for service equipment that is accessible to all occupants, or, for a single building or structure that is large enough to make two or more services necessary.

(c) Capacity requirements; where the service capacity requirements exceed 2,000 amperes at 600 volts or less, where load requirements of a single-phase installation is greater than the serving utility normally provides through a single service, or by special permission (related to capacity requirements).

(d) Different characteristics of the services such as different voltages, frequencies, or phases, or for different uses, such as for different rate schedules.

The basic rule for sizing of the equipment bonding jumper for bonding these various configurations is found in Section 250-102(c). This section requires that the bonding jumpers on the line side of each service the main bonding jumper be sized from Table 250-66. Also, the size of the bonding jumper for each raceway is based on the size of service-entrance conductors in each raceway. As discussed earlier, conductors larger than given in Table 250-66 are required for larger services. Since different sizes of service-entrance conductors may be installed at various locations, the minimum size of the equipment bonding conductor and main bonding jumper is based on the size of the service-entrance conductors at each location.

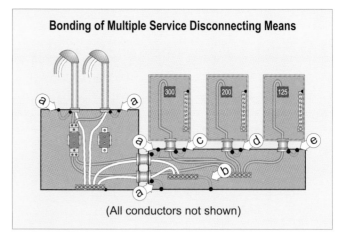

Bonding of Multiple Service Disconnecting Means

(All conductors not shown)

Figure 5-18

For example, the appropriate size of bonding jumper for the installation in Figure 5-18 with the assumed size of conductors is as follows: (all sizes copper)

Service-Entrance Conductor	Bonding Jumper
a. 500 kcmil in service mast and nipple	1/0
b. 1000 kcmil in wireway	2/0
c. 300 kcmil to 300 ampere service	No. 2
d. 3/0 to 200 ampere service	No. 4
e. No. 2 to 125 ampere service	No. 8

A practical method for bonding the current transformer enclosure and wireway (sometimes referred to as a "hot gutter") is to connect the grounded service conductor directly to the current transformer enclosure or wireway. This may be done by bolting a multi-barrel lug directly to the wireway and connecting the neutral or grounded service conductors to the lug. Be sure to remove any nonconductive paint or other coating that might insulate the connector from the enclosure.

As previously discussed, the grounded service conductor must also be extended to each service disconnecting means and be bonded to the enclosure.

Chapter Five: The questions included here were developed using material included in this chapter. The answers can be found by reviewing the text. It is also important that students make use of the 1999 NEC®, where many answers may be found. See page 279 for answers.

1. The permanent joining of metallic parts to form an electrically conductive path that will assure electrical continuity and the capacity to conduct safely any current likely to be imposed is defined as ____.
   a. grounding
   b. bonding
   c. welded
   d. grounded

2. The connection between the grounded circuit conductor and the equipment grounding conductor at the service is defined as the ____.
   a. main bonding jumper
   b. grounding electrode conductor
   c. equipment bonding jumper
   d. neutral conductor

3. The main bonding jumper is permitted to consist of a ____ or other suitable conductor.
   a. wire
   b. bus
   c. screw
   d. all of the above

4. Where the main bonding jumper consists of a ____ only, it is required to have a green finish that is visible with the ____ installed.
   a. bus
   b. screw
   c. wire
   d. jumper

5. Where ____ kcmil aluminum service-entrance conductors are installed, the main bonding jumper is required to be No. 4 copper or No. 2 aluminum.
   a. 1/0
   b. 2/0
   c. 3/0
   d. 250

6. Where the service-entrance conductors are larger than the maximum sizes given in Table 250-94, the main bonding jumper cannot be less than ____ percent of the area of the largest phase conductor.
   a. $9^1/_2$
   b. $10^1/_2$
   c. $11^1/_2$
   d. $12^1/_2$

7. Where conductors are used in parallel, and the sum of the circular mil area exceeds ____ kcmil copper or ____ kcmil aluminum, the equipment bonding conductor must be not less than $12^1/_2$ percent of the area of the ungrounded phase conductors.
   a. 1100 - 1750
   b. 1000 - 1650
   c. 1050 - 1500
   d. 1075 - 1400

8. The Code requires that electrical continuity of the grounding circuit of service equipment and enclosures that contain service conductors be obtained by ____.
   a. grounding
   b. welding
   c. approval
   d. bonding

9. The grounded circuit conductor (may be neutral) is permitted to be used for grounding and bonding on the ____ side of the service disconnecting means.
   a. load
   b. line
   c. subpanel
   d. control center

10. Grounding and bonding of equipment such as meters, current transformer cabinets and raceways to the grounded service conductor at locations on the line side of and remote from the service disconnecting means increases ____.
   a. voltage
   b. current
   c. safety
   d. cost

11. The size of the bonding jumper for equipment containing service-entrance conductors must be sized according to
   a. Table 250-66
   b. Table 250-122
   c. Table 310-16.
   d. none of the above

12. Which of the following are NOT permitted to be the sole means for bonding enclosures for service-entrance conductors
   a. bonding bushings
   b. bonding locknuts
   c. threadless couplings
   d. standard locknuts

Notes:

Chapter 6

Grounding Electrodes

Objectives

After studying this chapter, the reader will be able to understand:

- Definition and general requirements for grounding electrodes.
- Grounding electrode system to be used.
- Sizing bonding jumper for grounding electrode system.
- Description and installation of made electrodes.
- Common grounding electrode.
- Objectionable current flow and resistance of grounding electrodes.

Definition

Grounding electrode. A conducting element used to connect electrical systems and/or equipment to the earth.

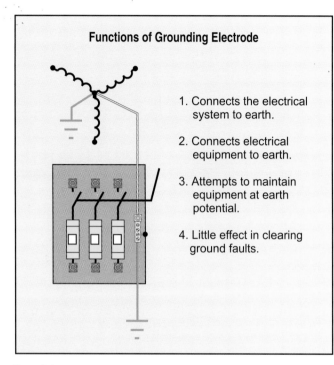

Functions of Grounding Electrode

1. Connects the electrical system to earth.

2. Connects electrical equipment to earth.

3. Attempts to maintain equipment at earth potential.

4. Little effect in clearing ground faults.

Figure 6-1

General

For many applications, grounding electrodes provide the essential function of connecting the electrical system to the earth. The earth is considered to be at zero potential. In some cases, the grounding electrode serves to ground the electrical system. In other instances, the electrode is used to connect noncurrent carrying metallic portions of electrical equipment to the earth. In both situations, the primary purpose of the grounding electrode is to maintain the electrical equipment at the earth potential present at the grounding electrode.

Another essential function of the grounding electrode is to dissipate over-voltages into the earth. These over-voltages can be caused by high-voltage conductors being accidentally connected to the lower-voltage system such as by a failure in a transformer or by an overhead conductor dropping on the lower-voltage conductor. Over-voltages can also be caused from lightning.

In Section 250-24(c), we find a requirement to connect the equipment grounding conductors, the service-equipment enclosures, and where the system is grounded, the grounded service conductor to a grounding electrode. The conductor used to make this connection is the grounding electrode conductor.

Grounding electrode system

The NEC® in Section 250-50 requires that, where available on the premises at each building or structure served, all grounding electrodes including "made" electrodes be bonded together to form the grounding electrode system. This includes metal underground water pipes, metal frames of buildings, concrete-encased electrodes, and ground rings. The general requirement is that a bonding jumper must be installed between the grounding electrodes to bond them together. A grounding electrode conductor is run from the service enclo-

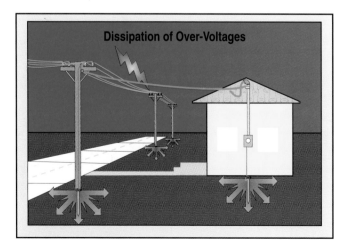

Dissipation of Over-Voltages

Figure 6-2

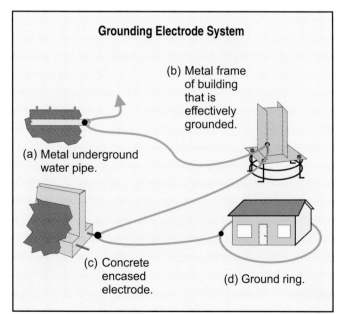

Grounding Electrode System

(b) Metal frame of building that is effectively grounded.

(a) Metal underground water pipe.

(c) Concrete encased electrode.

(d) Ground ring.

Figure 6-3

sure to one of the grounding electrodes that are bonded together. The NEC® also provides for the option of running a grounding electrode conductor to each grounding electrode individually.

Where the interior metal water pipe is used as a part of the grounding electrode system or as a conductor to bond other electrodes together to create the "grounding electrode system," Section 250-50 requires that all bonding take place within the first 5 feet from the point the water pipe enters the building. This section does not require that the interior water pipe be used for the purpose of interconnecting other electrodes to form the grounding electrode system. Any of the other electrodes, such as the metal frame of the building, concrete encased electrode or ground ring, can be used for the purpose of interconnecting the other grounding electrodes. Where these other electrodes are used for this purpose, no restrictions are placed on where the connections are permitted to be made or how far inside the building they are permitted. Section 250-68(a) requires grounding electrode conductor connections to grounding electrodes to be accessible except for connections to a buried, driven or concrete encased electrode.

Grounding electrodes required to be used

All of the identified grounding electrodes are required to be used where "available on the premises at each building or structure served." The Code does not define what is meant by "available" nor does it require that the electrodes be made available where they are not. For example, if the building has encased concrete reinforcing rods when the electrical system is installed, it is not required that the rods be exposed for connection. On the other hand, the concrete reinforcing rods must always be used when "available." Several electrical inspection agencies require that a concrete-encased grounding electrode be connected to the system before approval of the service for utility connection is granted. The grounding electrodes are not listed in an order of preference nor is it optional to choose which ones to use.

Electrodes that must be used, in addition to any "made" electrodes that exist or are installed at the building or structure served, where "available" are as follows:

1. Metal Underground Water Pipe. Defined in Section 250-50(a) as "A metal underground water pipe that is in direct contact with the earth for 10 feet or more including any metal well casing

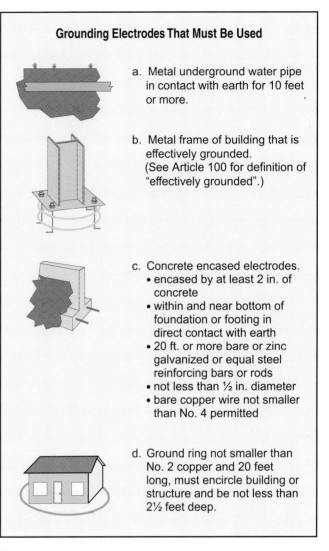

Grounding Electrodes That Must Be Used

a. Metal underground water pipe in contact with earth for 10 feet or more.

b. Metal frame of building that is effectively grounded. (See Article 100 for definition of "effectively grounded".)

c. Concrete encased electrodes.
• encased by at least 2 in. of concrete
• within and near bottom of foundation or footing in direct contact with earth
• 20 ft. or more bare or zinc galvanized or equal steel reinforcing bars or rods
• not less than ½ in. diameter
• bare copper wire not smaller than No. 4 permitted

d. Ground ring not smaller than No. 2 copper and 20 feet long, must encircle building or structure and be not less than 2½ feet deep.

Figure 6-4

that is effectively bonded to the pipe." There is no minimum or maximum pipe size given. Types of metal, such as steel, iron, cast iron, stainless steel or even aluminum are not distinguished. Different types of water pipes such as for potable water, fire protection sprinkler systems, irrigation piping, etc., are also not defined. As a result, all of these metal underground water pipes must be used where "available at each building or structure served."

Continuity of the grounding path of the water pipe grounding electrode or the bonding of interior piping systems cannot depend on water meters or on filtering devices or similar equipment. See Section 250-50(a)(1). Where a water meter or filtering equipment is in this metal water piping system, a bonding jumper must be installed around the equipment to maintain continuity even if the water meter or filter is removed.

2. Metal Frame of the Building. Section 250-50(b) requires the metal frame of the building be used as a grounding electrode where it is effectively grounded. "Grounded effectively" is defined in Article 100 and means that the metal frame of the building is, "Intentionally connected to earth through a ground connection or connections of sufficiently low impedance and having sufficient current-carrying capacity to prevent the buildup of voltages that may result in undue hazards to connected equipment or to persons."[N]

To be an effective grounding electrode, the metal frame of the building must have a sufficiently low-impedance contact with the earth to pass current when called upon to do so and to maintain the electrical system at or near the electrical potential of the surrounding earth. The building steel can be connected to the earth by bolted or welded connection to reinforcing steel in foundations or footings that are in turn encased in concrete. Also, building structural steel may itself be encased in concrete that is in contact with the earth. In both of these cases, the concrete that encases the building steel or reinforcing steel must be in direct contact with the earth.

Certain back-fills such as gravel or vapor barriers may render the building steel an ineffective electrode. Building steel that is connected to concrete footings or foundations by only "J" bolts are not considered "effectively grounded" unless these "J" bolts are in turn connected to structural members such as reinforcing steel. The reinforcing steel needs to be near the base of the footing or foundation.

The structural steel should be tested with an earth resistance tester if in doubt about its resistance to ground and adequacy as a grounding electrode.

3. Concrete-Encased Electrodes. Section 250-50(c) defines this grounding electrode as one or more steel reinforcing bars or rods that are not less than 20 feet in length and ½ inch in diameter or 20 feet or more of bare copper conductor not smaller than No. 4. These electrodes must be located within or near the bottom of the foundation or footing and be encased by at least 2 inches of concrete. A single 20 ft. length of reinforcing bar is not required. Reinforcing bars are permitted to be bonded together by the usual steel tie wires or other effective means like welding. Where subjected to high currents such as lightning strikes, welding is preferred.

Reinforcing rods must be of bare, zinc galvanized or other electrically conductive coated steel material. Obviously, insulated reinforcement rods would not perform properly as a grounding electrode. Some complaints have been made that lightning surges, that are dissipated through this electrode, break out chunks of concrete where the surge exits the footing.

This grounding electrode is commonly referred to as the "Ufer ground" after H.G. Ufer who spent many years documenting its effectiveness. See the Appendix of this text for additional information on the development of the concrete-encased electrode.

Several electrical inspection agencies require that a concrete-encased electrode be installed or connected to the service prior to authorizing electrical service due to its effectiveness in most any climatic and soil condition.

4. Ground Ring. Section 250-50(d) recognizes a copper conductor, not smaller than No. 2 and at least 20 feet long, as a ground ring grounding electrode. The conductor must "encircle" the building or structure and be buried not less than 2½ feet deep. Ground rings often are installed at telecommunication central offices, radio and cellular telephone sites. Where available on the premises served, ground rings must be used as one or more of the grounding electrodes making up the grounding electrode system.

Supplemental electrode

Section 250-50(a)(2) requires that where the only grounding electrode available and connected at the building or structure served is a metal underground water pipe, it be supplemented by another grounding electrode. Electrodes suitable to supplement the metal underground water pipe

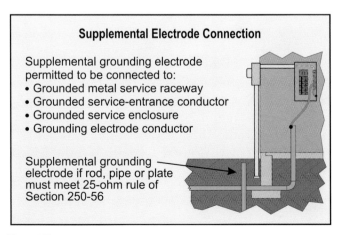

Figure 6-5

include: the metal frame of the building, a concrete-encased electrode, ground ring, other local metal underground systems or structures, rod and pipe electrodes, or plate electrodes. This supplemental grounding electrode is required since, often, metal underground water pipes are replaced by plastic water services or the system continuity is interrupted by nonmetallic couplings or repairs. The effectiveness of the water pipe grounding electrode would thus be lost.

Specific locations are provided where the supplemental grounding electrode is permitted to be connected. Where an underground metal water pipe is the only grounding electrode, the supplemental grounding electrode is permitted to be connected to only the grounding electrode conductor, the grounded service-entrance conductor, the grounded service raceway or to any grounded service enclosure. An exception to this requirement permits the bonding connection to the interior metal water piping in a qualifying industrial or commercial plant to be made at any location if the entire length of interior metal water pipe that is being used as a conductor is exposed.

Often, changes, repairs or modifications are made to the metallic water piping systems with nonmetallic pipe or fittings or dielectric unions. In this case, it is possible to inadvertently isolate portions of the grounding system from the grounding electrode conductor. This is another in several steps that has been taken over recent years to reduce the emphasis and reliance on the metal water piping system for grounding of electrical systems.

With a change to the 1999 NEC®, where the supplemental grounding electrode is of the rod, pipe or plate type, it is now required to meet the 25-ohm-to-ground rule in Section 250-56. This means that the supplemental grounding electrode must have a resistance of not more than 25 ohms or a second supplemental grounding electrode must be used. This has the effect of the system being served by only the supplemental grounding electrodes in case the underground metal water pipe grounding electrode is interrupted for any reason.

Size of bonding jumper for grounding electrode system

The bonding jumper used to bond the grounding electrodes together to form the grounding electrode system must be sized in accordance with Section 250-66 based on the size of the ungrounded

service-entrance conductor. The conductor that connects the grounding electrodes together is a bonding conductor and not a grounding electrode conductor. The bonding conductors are not required to be installed in one continuous length as grounding electrode conductors are. Also, the exceptions for sizing the grounding electrode conductor in Section 250-66 apply for the sizing of the bonding jumpers.

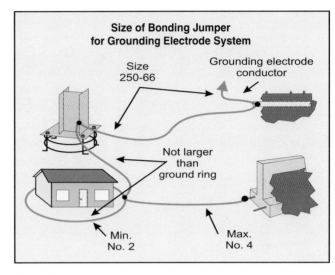

Figure 6-6

For example, if the service-entrance conductor is 500 kcmil copper, the minimum size of bonding jumper is determined by reference to Section 250-66 and Table 250-66, including the rules in Sections 250-66(a), (b) and (c) are as follows:

- To metal underground water pipe and metal frame of a building; No. 2 copper or No. 1/0 aluminum conductor. (From Table 250-66.)
- To "made" electrodes as in Section 250-52(c) or (d) such as pipes, rods or plates; that portion of the bonding conductor that is the sole connection to the made electrode; No. 6 copper or No. 4 aluminum. The term "sole connection" means that the bonding conductor is not connected to the made electrode being considered and then another grounding electrode is connected to it. See Section 250-66(a).
- To a concrete encased electrode as in Section 250-50(c); that portion of the bonding conductor that is the sole connection to the concrete encased electrode; No. 4 copper conductor. See Section 250-66(b).
- To a ground ring as in Section 250-50(d); that portion of the bonding conductor that

is the sole connection to the ground ring is not required to be larger than the ground ring conductor. See Section 250-66(c).

Note that aluminum is not permitted to be installed as a grounding electrode conductor where in direct contact with masonry or the earth or where subject to corrosive conditions. Where used outside, aluminum or copper-clad aluminum grounding conductors are not permitted within 18 inches of the earth. See Section 250-64(a).

No sequence for installing the bonding jumper or jumpers is given. However, the minimum wire size required to the various grounding electrodes must be observed. In addition, the point where the grounding electrode connects to the grounding electrode system must provide for the largest required grounding electrode conductor. For example, it would be a violation to connect a No. 4 bonding conductor from a concrete encased grounding electrode a building steel grounding electrode which would require a 3/0 grounding electrode conductor. The installation would be acceptable if the 3/0 copper grounding electrode conductor connects to the building steel and a No. 4 copper bonding jumper extends to the concrete encased electrode. In addition, the unspliced grounding electrode conductor is permitted to run from the service equipment to any convenient grounding electrode.

Section 250-50. The minimum size of each grounding electrode conductor to the individual grounding electrode is shown in Figure 6-5. Note that a grounding electrode conductor is permitted to "supply" or "serve" any number of grounding electrodes but must be sized for the largest grounding electrode conductor required. For example, a bonding conductor is permitted to be run to a concrete encased electrode and then to the underground metal water pipe. The bonding jumper must be sized for the largest grounding electrode conductor required for the grounding electrode or electrodes served.

Made electrodes

Where the electrodes described in Section 250-50 are not available at the service location, a grounding electrode must be "made" or installed. The made electrode as provided for in Section 250-52 may be local metal underground systems or structures, driven pipes, or rods or buried plates conforming to the following requirements:

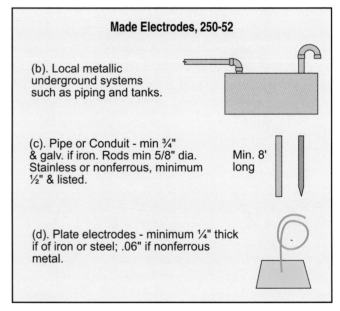

Made Electrodes, 250-52

(b). Local metallic underground systems such as piping and tanks.

(c). Pipe or Conduit - min ¾" & galv. if iron. Rods min 5/8" dia. Stainless or nonferrous, minimum ½" & listed.

Min. 8' long

(d). Plate electrodes - minimum ¼" thick if of iron or steel; .06" if nonferrous metal.

Figure 6-8

(b) Local systems. Local metallic underground systems as piping, tanks, etc. These objects must have the metal in direct contact with the earth. Protective coatings may render them ineffective as a grounding electrode.

(c)(1) Pipe electrodes. Pipe or conduit electrodes shall be not less than 8 feet in length nor smaller than ¾-inch trade size and if of iron or steel, shall be galvanized or metal-coated for corrosion protection.

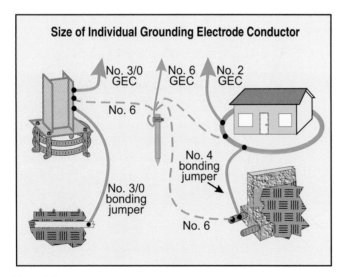

Size of Individual Grounding Electrode Conductor

No. 3/0 GEC

No. 6 GEC

No. 2 GEC

No. 6

No. 4 bonding jumper

No. 3/0 bonding jumper

No. 6

Figure 6-7

Alternately, individual grounding electrode conductors are permitted to be installed from the service equipment to one or more grounding electrodes rather than the electrodes being bonded together in a circular or "daisy-chain" manner. See

(c)(1) Rod electrodes. Electrodes of steel or iron shall be at least 5/8 inch diameter. Rods of nonferrous metal or stainless steel that are less than 5/8 inch in diameter shall be listed and be at least ½-inch in diameter.

(d) Plate electrodes. Electrodes shall have at least 2 square feet of surface in contact with exterior soil. If of iron or steel, the plate shall be at least ¼-inch thick. If of nonferrous metal, they shall be at least 0.06 inch thick.

Note that underground metal gas piping systems are not permitted to be used as a grounding electrode. This does not eliminate the requirement that interior metal gas piping systems be bonded. See Chapter 8 of this text for additional information on bonding of metal piping systems.

Installation of made electrodes

Where practicable, made electrodes must be installed below permanent moisture level. This is a key ingredient in establishing an effective made electrode. They also are required to be free from nonconductive coatings such as paint and enamel.

Rod and pipe electrodes must be installed so at least 8 feet is in contact with the soil. They must be driven vertically unless rock bottom is encountered. If rock bottom is encountered which prevents the rod from being driven 8 feet vertically, the rod is permitted to be installed at an oblique angle of not more than 45 degrees from vertical or it can be buried in a trench that is at least 2½ feet deep. See Section 250-52(c)(3).

The upper end of the rod must be flush with or below ground level unless the above ground end of the rod and the grounding electrode attachment are protected from physical damage. This, of course,

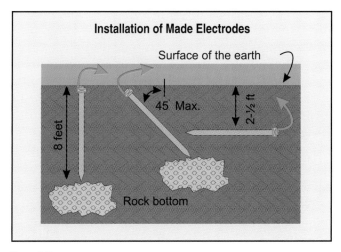

Installation of Made Electrodes

Figure 6-9

requires that a ground rod longer than 8 feet be used if any of the rod is exposed above ground level. For an eight-foot ground rod or pipe, the ground clamp must be listed for direct earth burial as the electrode must be driven to its full length.

Plate electrodes are required to be buried not less than 2 1/2 feet in the soil.

Section 250-10 requires that ground clamps or other fittings be approved (acceptable to the authority having jurisdiction) for general use without protection or be protected from physical damage by metal, wood or equivalent protective covering.

Common grounding electrode

Section 250-58 of the NEC® requires that a common grounding electrode be used for all alternating-current system grounding in or at a building. In addition, where more than one service supplies a building, the common grounding electrode must be used for all services. This section recognizes that where two or more grounding electrodes are bonded together, they are considered to be one electrode.

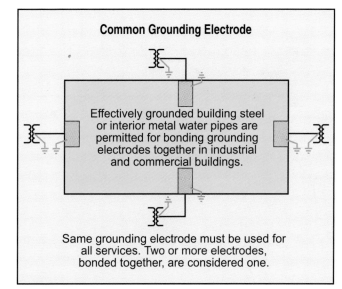

Common Grounding Electrode

Effectively grounded building steel or interior metal water pipes are permitted for bonding grounding electrodes together in industrial and commercial buildings.

Same grounding electrode must be used for all services. Two or more electrodes, bonded together, are considered one.

Figure 6-10

Interestingly, no distance between electrodes is given beyond which the electrodes do not have to be bonded together. Buildings of "large area" are permitted by Section 230-2(b)(2) to have more than one service. However, nothing in the Code defines the dimensions of a "large building." Some inspection authorities use voltage drop of major feeders for guidance in determining when a building is one of "large area." Where feeder conductors would have to be increased in size unreasonably to maintain voltage regulation, one or more additional services are permitted.

Section 250-58 requires the grounding electrodes for the multiple services be bonded together no matter how far apart they are in the same building. This is important so there is not more than one earth potential impressed on equipment in or on the building. Section 250-50 requires the bonding jumper(s) used for this purpose to be sized from Table 250-66 and be installed in accordance with Sections 250-64(a), (b) and (e). See Chapter Seven of this text for additional information on installation of grounding electrode conductors. It is permitted to use the steel frame of a building that is effectively grounded or a concrete encased grounding electrode to bond grounding electrodes from other services together.

Section 250-58 also requires that a common grounding electrode be used to ground conductor enclosures and equipment in or on the building and that the same grounding electrode be used to ground the system. This does not mean that one cannot use more than one grounding electrode. But, if more than one is used, then all the grounding electrodes must be bonded together to form a common grounding electrode. Where multiple grounding electrodes are bonded together as cited above, such multiple grounding electrodes become, in effect, a common grounding electrode system.

Earth return prohibited

No mention is made in the Code to providing a low-resistance, low-impedance, common grounding electrode for clearing ground faults. Reference to Figure 6-10 will show that the grounding electrode is in the earth return circuit. Even if the grounding electrode resistance to earth was very low, it would have little affect on the clearing of a ground fault, because the reactance of the earth and the soil resistance in the return circuit is very high. Where a parallel path exists through the earth and through a grounded service conductor, about 95 percent or more of the ground fault current will return to the source over the grounded service conductor. A low-resistance, common grounding electrode is valuable, however, in holding equipment close to earth potential. It simply is not effective in clearing a line-to-ground fault.

Section 250-2(d) and 250-54 make it clear that grounding electrodes are not permitted to be used instead of equipment grounding conductors. The

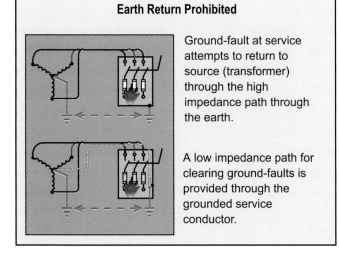

Earth Return Prohibited

Ground-fault at service attempts to return to source (transformer) through the high impedance path through the earth.

A low impedance path for clearing ground-faults is provided through the grounded service conductor.

Figure 6-11

earth is not to be used as the sole or only equipment grounding conductor. However, grounding electrodes are permitted to supplement equipment grounding conductors.

If a ground fault should develop as shown in the upper drawing in Figure 6-10 where two separate grounding electrodes are used, the fault current flow will be through the service conductor then through the impedance of the ground fault, the grounding electrode conductor, the grounding electrode, the path through the earth to the grounding electrode at the transformer and finally through the grounding conductor to complete the circuit to the transformer. It would be a rare case where that circuit resistance would add up to less than 12 ohms (while the impedance would be higher). At best, therefore, the fault current would not reach a value high enough to operate a 15-ampere overcurrent device on a 120 volt-to-ground circuit. (120 ÷ 12 = 10 amperes).

Considering resistance only, the circuit shown has two grounding electrodes in series. Compared to the much lower resistance parallel path of the grounded circuit conductor, a resistance ratio between the two parallel paths is about 50 times for a 100 ampere service, to well over 100 times for the larger services. When impedance of the two paths is considered, the ratio will be higher. Thus, almost all the current from a line-to-ground fault will return to the transformer over the grounded service conductor.

Under normal operating conditions some unbalanced current will flow in the neutral. Some unbalanced neutral current will thus flow through the earth, but it will be small in comparison to that which will flow through the grounded service conductor.

Any belief that the circuit to the grounding electrode can be depended on to clear a ground fault is clearly erroneous no matter how large a grounding electrode conductor is used or how good a grounding electrode is. However, when the high-impedance earth path is short-circuited by installing the grounded circuit conductor as shown in the lower drawing in Figure 6-11, a low-impedance path is established as required in Section 250-2(d). This will allow a large current to flow over the equipment grounding and service grounded conductor to allow the branch-circuit, feeder or service overcurrent device to clear the fault and thus provide the safety contemplated by the Code.

Resistance of grounding electrodes

There is no requirement in Article 250 that the grounding electrode system required by Section 250-50 (consisting of metal underground water pipe, metal frame of the building, concrete-encased electrode or ground ring) meet any maximum resistance to ground. No doubt it is felt that the grounding electrode system will have a resistance to ground of 25 ohms or less.

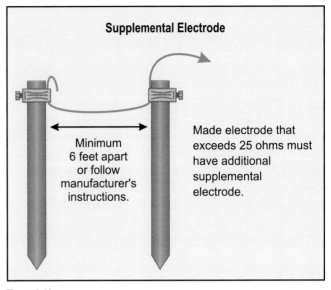

Figure 6-12

Supplemental Electrode

Minimum 6 feet apart or follow manufacturer's instructions.

Made electrode that exceeds 25 ohms must have additional supplemental electrode.

Maximum Resistance of Grounding Electrodes

- **No maximum for grounding electrode system.**
- **Maximum 25 ohms for "made" electrodes.**
- **Where "made" electrodes exceed 25 ohms - supplement with one additional electrode.**

The rules change for "made" electrodes. The Code states, in Section 250-56, that where a single rod, pipe, or plate electrode does not achieve a resistance to ground of 25 ohms or less, it shall be supplemented by one additional electrode. This means that where driven ground rods are utilized, two ground rods would be the maximum required under any condition. There is no requirement that additional made electrodes such as ground rods or plates be installed until the 25 ohm-to-ground resistance is obtained.

In general, metallic underground water piping

systems, metallic well piping systems, metal frame buildings and similar grounding electrodes may be expected to provide a ground resistance of not over 3 ohms and in some cases as low as 1 ohm.

However, from a practical standpoint, no grounding electrode, no matter how low its resistance, can ever be depended upon to clear a ground fault on any distribution system of less than 1,000 volts.

If a system is effectively grounded as pointed out in the Code under Section 250-2(d), a path of low impedance (not through the grounding electrode) must be provided to facilitate the operation of the overcurrent devices in the circuit. See Chapter 11 of this text "Clearing Ground-fault Circuits on Distribution Systems."

The lowest practical resistance of a grounding electrode is desirable and will better limit the voltage to ground when a ground fault occurs. It is more important to provide a low-impedance path to clear a fault promptly, for a voltage to ground can only occur during the period of time that a fault exists. Clearing a ground fault promptly thus will enhance safety.

Even though the grounding electrode has low resistance, it is a part of a high-impedance circuit and plays virtually no part in the clearing of a fault on a low-voltage distribution system. This is due in part to being a higher resistance path through the earth than through the grounded service conductor. In addition, the remote path through the grounding electrode and earth is a high-impedance path compared to the circuit where the grounded service conductor is installed and routed with the ungrounded phase conductors.

Objectionable currents

The Code in Section 250-6 recognizes that conditions may exist which may cause an objectionable flow of current over grounding conductors such as the grounding electrode conductor, other than temporary currents that may be set up under accidental conditions. We should recognize that grounding conductors are not intended to carry current under normal operating conditions. They are installed for and are intended to carry current to perform some safety function.

The Code does not define what is meant by "objectionable" currents. Clearly, any current over a grounding electrode conductor that would prevent it from maintaining the equipment at the earth potential would be objectionable. Since every conductor has resistance, current flow through the conductor will produce a voltage drop across it. Any voltage drop on a grounding conductor that would create a shock hazard certainly would also not be acceptable.

Section 250-6(b) permits the following corrective actions to be taken where there is an "objectionable" flow of current over grounding conductors:

1.   If due to multiple grounds, one or more, but not all, of such grounds may be discontinued,
2.   The location of the grounding connection may be changed,
3.   The continuity of the grounding conductor or conductive path between grounding connections may be suitably interrupted, or
4.   Other means satisfactory to the authority enforcing the Code may be taken to limit the current over the grounding conductors.

The Code points out that temporary currents that result from accidental conditions such as ground-fault currents, that occur only while the grounding conductors are performing their intended protective functions, are not considered the "objectionable" currents covered in these sections.

Section 250-6(d) points out that currents that introduce noise or data errors in electronic equipment are not considered to be objectionable currents. Electronic data processing equipment is not permitted to be operated ungrounded or by connection to only its own grounding electrode.

Soil Resistivity[1]
Why measure soil resistivity?

Soil resistivity measurements have a threefold purpose. First, such data are used to make subsurface geophysical surveys as an aid in identifying ore locations, depth to bedrock and other geological phenomena. Second, resistivity has a direct impact on the degree of corrosion in underground pipelines. A decrease in resistivity relates to an increase in corrosion activity and therefore dictates the protective treatment to be used. Third, soil resistivity directly affects the design of a grounding system, and it is to that task that this discussion is directed. When designing an extensive grounding system, it is advisable to locate the area of lowest soil resistivity in order to achieve the most economical grounding installation.

Effects of soil resistivity on ground electrode resistance

Soil resistivity is the key factor that determines what the resistance of a grounding electrode will be, and to what depth it must be driven to obtain low ground resistance. The resistivity of the soil varies widely throughout the world and changes seasonally. Soil resistivity is determined largely by its content of electrolytes, which consist of moisture, minerals and dissolved salts. A dry soil has high resistivity if it contains no soluble salts (Figure 6-13).

Soil	Resistivity (approx.), -cm		
	Min.	Average	Max.
Ashes, cinders, brine, waste	590	2.370	7,000
Clay, shale, gumbo, loam	340	4,060	16,300
Same, with varying proportions of sand and gravel	1,020	15,800	135,000
Gravel, sand, stones with little clay or loam	59,000	94,000	458,000

Figure 6-13

Factors affecting soil resistivity

Two samples of soil, when thoroughly dried, may in fact become very good insulators having a resistivity in excess of 109 ohm-centimeters. The

Moisture content, % by weight	Resistivity, -cm	
	Top soil	Sandy loam
0	>10^9	>10^9
2.5	250,000	150,000
5	165,000	43,000
10	53,000	18,500
15	19,000	10,500
20	12,000	6,300
30	6,400	4,200

Figure 6-14

resistivity of the soil sample is seen to change quite rapidly until approximately 20% or greater moisture content is reached (Figure 6-14).

The resistivity of the soil is also influenced by temperature. Figure 6-15 shows the variation of the resistivity of sandy loam, containing 15.2% moisture, with temperature changes from 20° to -15°C. In this temperature range the resistivity is seen to vary from 7200 to 330,000 ohm-centimeters.

Temperature		Resistivity,
C	F	Ohm-cm
20	68	7,200
10	50	9,900
0	32 (water)	13,800
0	32(ice)	30,000
-5	23	79,000
-15	14	330,000

Figure 6-15

Because soil resistivity directly relates to moisture content and temperature, it is reasonable to assume that the resistance of any grounding system will vary throughout the different seasons of the year. Such variations are shown in Figure 6-16. Since both temperature and moisture content become more stable at greater distances below the surface of the earth, it follows that a grounding system — to be most effective at all times — should be constructed with the ground rod driven down a considerable distance below the surface of the earth. Best results are obtained if the ground rod reaches the water table.

In some locations, the resistivity of the earth is

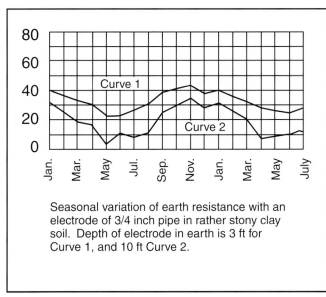

Seasonal variation of earth resistance with an electrode of 3/4 inch pipe in rather stony clay soil. Depth of electrode in earth is 3 ft for Curve 1, and 10 ft Curve 2.

Figure 6-16

so high that low-resistance grounding can be obtained only at considerable expense and with an elaborate grounding system. In such situations, it may be economical to use a ground rod system of limited size and to reduce the ground resistivity by periodically increasing the soluble chemical content of the soil. Figure 6-17 shows the substantial reduction in resistivity of sandy loam brought about by an increase in chemical salt content.

THE EFFECT OF SALT* CONTENT ON THE RESISTIVITY OF SOIL	
(Sandy loam, Moisture content, 15% by weight, Temperature, 17°C)	
Added Salt (% by weight of moisture)	Resistivity (Ohm-centimeters)
0	10,700
0.1	1,800
1.0	460
5	190
10	130
20	100

Figure 6-17

Chemically treated soil is also subject to considerable variation of resistivity with temperature changes, as shown in Figure 6-18. If salt treatment is employed, it is necessary to use ground rods which will resist chemical corrosion.

THE EFFECT OF TEMPERATURE ON THE RESISTIVITY OF SOIL CONTAINING SALT*	
(Sandy loam, 20% moisture. Salt 5% of weight of moisture)	
Temperature (Degrees C)	Resistivity (Ohm-centimeters)
20	110
10	142
0	190
-5	312
-13	1,440

Figure 6-18

*Such as copper sulfate, sodium carbonate, and others. Salts must be EPA or local ordinance approved prior to use.

Soil Resistivity Measurements (4-Point Measurement)

Resistivity measurements are of two types; the 2-point and the 4-point method. The 2-point method is simply the resistance measured between two points. For most applications the most accurate method is the 4-point method. The 4-point method (Figures 6-19 and 6-20), as the name implies, requires the insertion of four equally spaced and in-line electrodes into the test area. A known current from a constant current generator is passed between the outer electrodes. The potential drop (a function of the resistance) is then measured across the two inner electrodes. Most earth resistance meters read directly in ohms.

$$\rho = \frac{4\pi AR}{1 + \dfrac{2A}{\sqrt{(A^2 + 4B^2)}} - \dfrac{2A}{\sqrt{(4A^2 + 4B^2)}}}$$

Where: A = distance between the electrodes in centimeters

B = electrode depth in centimeters

If A > 20 B, the formula becomes:

$\rho = 2\pi$ AR (with A in cm)
$\rho = 191.5$ AR (with A in feet)
$\rho =$ Soil resistivity (ohm-cm)

This value is the average resistivity of the ground at a depth equivalent to the distance "A" between two electrodes.

Soil resistivity measurements

Given a sizable tract of land in which to determine the optimum soil resistivity some intuition is in order. Assuming that the objective is low resistivity, preference should be given to an area containing moist loam as opposed to a dry sandy area. Consideration must also be given to the depth at which resistivity is required.

Example

After inspection, the area investigated has been narrowed down to a plot of ground approximately 75 square feet (7 m²). Assume that you need to deter-

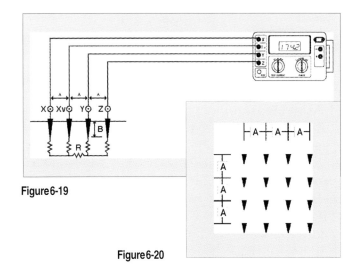

Figure 6-19

Figure 6-20

mine the resistivity at a depth of 15 feet (450 cm). The distance "A" between the electrodes must then be equivalent to the depth at which average resistivity is to be determined (15 ft, or 450 cm). Using the more simplified Wenner formula ($\rho = 2\pi$ AR), the electrode depth must then be 1/20th of the electrode spacing or 8-7/8" (22.5 cm).

Lay out the electrodes in a grid pattern and connect to the meter as shown in Figures 6-19 and 6-20. Proceed as follows:

- Remove the shorting link between X and Xv (C1, P1)
- Connect all four auxiliary rods (Figure 6-19)

For example, if the reading is R = 15

ρ (resistivity) = 2π A x R
A (distance between electrodes) = 450 cm
ρ = 6.28 x 15 x 450 = 42,390 Ω-cm

Ground electrodes

The term "ground" is defined as a conducting connection by which a circuit or equipment is connected to the earth. The connection is used to establish and maintain as closely as possible the potential of the earth on the circuit or equipment connected to it. A "ground" consists of a grounding conductor, a bonding connector, its grounding electrode(s), and the soil in contact with the electrode.

Grounds have several protection applications. For natural phenomena, such as lightning, grounds are used to discharge the system of current before personnel can be injured or system components damaged. For foreign potentials due to faults in electric power systems with ground returns, grounds help ensure rapid operation of the protection relays by providing low resistance fault current paths. This provides for the removal of the foreign potential as quickly as possible. The ground should drain the foreign potential before personnel are injured and the power or communications system is damaged.

Ideally, to maintain a reference potential for instrument safety, protect against static electricity, and limit the system to frame voltage for operator safety, a ground resistance should be zero ohms. In reality, as we describe further in the text, this value cannot be obtained.

Last, but not least, low ground resistance is essential to meet NEC, OSHA and other electrical safety standards.

Figure 6-21 illustrates a grounding rod. The resistance of the electrode has the following components:

(A)   the resistance of the metal and that of the connection to it.

(B)   the contact resistance of the surrounding earth to the electrode.

(C)   the resistance in the surrounding earth to current flow or earth resistivity which is often the most significant factor.

More specifically:

(A)   Grounding electrodes are usually made of a very conductive metal (copper or copper clad) with adequate cross sections so that the overall resistance is negligible.

(B)   The National Institute of Standards and Technology has demonstrated that the resistance between the electrode and the surrounding earth is negligible if the electrode is free of paint, grease or other coating, and if the earth is firmly packed.

(C)   The only component remaining is the resistance of the surrounding earth. The electrode can be thought of as being surrounded by concentric shells of earth or soil, all of the same thickness. The closer the shell to the electrode, the smaller its surface; hence, the greater its resistance. The farther away the shells are from the electrode, the greater the surface of the shell; hence, the lower the resistance. Eventually, adding shells at a distance from the grounding electrode will no longer noticeably affect the overall earth resistance surrounding the electrode. The distance at which this effect occurs is referred to as the effective resistance area and is directly dependent on the depth of the grounding electrode.

In theory, the ground resistance may be derived from the general formula:

$$R = \frac{\rho\,L}{A} \quad \text{Resistance} = \text{Resistivity} \ \times \ \frac{\text{Length}}{\text{Area}}$$

This formula illustrates why the shells of concentric earth decrease in resistance the farther they are from the ground rod:

$$R = \text{Resistivity of Soil} \ \times \ \frac{\text{Thickness of Shell}}{\text{Area}}$$

In the case of ground resistance, uniform earth (or soil) resistivity throughout the volume is assumed, although this is seldom the case in nature. The equations for systems of electrodes are very complex and often expressed only as approximations. The most commonly used formula for single ground electrode systems, developed by Professor H. R. Dwight of the Massachusetts Institute of Technology, is the following:

$$R = \frac{\rho}{2\,\pi\,L}\ \frac{\{(\ln 4L) - 1\}}{r}$$

R   =   resistance in ohms of the ground rod to the earth (or soil)

L   =   grounding electrode length

r   =   grounding electrode radius

ρ =   average resistivity in ohms-cm

Effect of ground electrode size and depth on resistance

Size: Increasing the diameter of the rod does not materially reduce its resistance. Doubling the diameter reduces resistance by less than 10% (Figure 6-22).

Depth: As a ground rod is driven deeper into the earth, its resistance is substantially reduced. In general, doubling the rod length reduces the resistance by an additional 40% (Figure 6-23). The

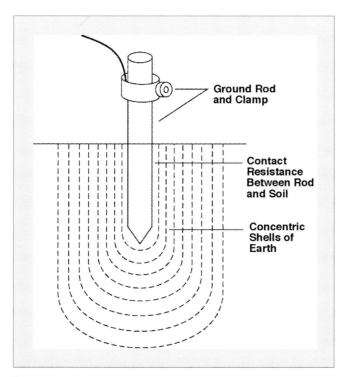

Ground Rod and Clamp

Contact Resistance Between Rod and Soil

Concentric Shells of Earth

Figure 6-21

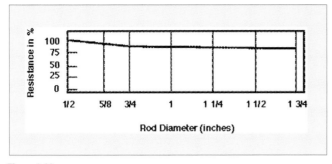

Figure 6-22

NEC® [1999, 250-52(c)(3)] requires a minimum of 8 ft (2.4 m) to be in contact with the soil. The most common is a 10 ft (3 m) cylindrical rod which meets the NEC® code. A minimum diameter of 5/8 inch (1.59 cm) is required for steel rods and 1/2 inch (1.27 cm) for copper or copper clad steel rods [NEC® 250-52(c)(2)]. Minimum practical diameters for driving limitations for 10 ft (3 m) rods are:

- 1/2 inch (1.27 cm) in average soil
- 5/8 inch (1.59 cm) in moist soil
- 3/4 inch (1.91 cm) in hard soil or more than 10 ft driving depths

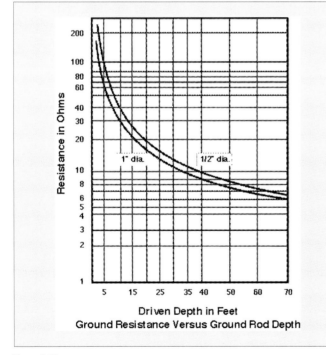

Figure 6-23

Ground resistance values

NEC® 250-56(1999): Resistance of man-made electrodes:

"A single electrode consisting of a rod, pipe, or plate that does not have a resistance to ground of 25 ohms or less shall be augmented by one additional of any of the types specified in Section 250-50 or 250-52. Where multiple rod, pipe or plate electrodes are installed to meet the requirements of this section, they shall be not less than 6 ft (1.83 m) apart."

The National Electrical Code® (NEC®) states that the resistance to ground shall not exceed 25 ohms. This is an upper limit and guideline, since much lower resistance is required in many instances.

"How low in resistance should a ground be?" An arbitrary answer to this in ohms is difficult. The lower the ground resistance, the safer, and for positive protection of personnel and equipment, it is worth the effort to aim for less than one ohm. It is generally impractical to reach such a low resistance along a distribution system or a transmission line or in small substations. In some regions, resistances of 5 ohms or less may be obtained without much trouble.

In other regions, it may be difficult to bring resistance of driven grounds below 100 ohms.

Accepted industry standards stipulate that transmission substations should be designed not to exceed one ohm resistance. In distribution substations, the maximum recommended resistance is for 5 ohms or even 1 ohm. In most cases, the buried grid system of any substation will provide the desired resistance.

In light industrial or in telecommunication central offices, 5Ω is often the accepted value. For lightning protection, the arrestors should be coupled with a maximum ground resistance of 1Ω.

These parameters can usually be met with the proper application of basic grounding theory. There will always exist circumstances which will make it difficult to obtain the ground resistance required by the NEC® or other safety standards. When these situations develop, several methods of lowering the ground resistance can be employed. These include parallel rod systems, deep driven rod systems utilizing sectional rods and chemical treatment of the soil. Additional methods, discussed in other published data, are buried plates, buried conductors (counterpoise), electrically connected building steel, and electrically connected concrete reinforced steel.

Electrically connecting to existing water and gas distribution systems was often considered to yield low ground resistance; however, recent design changes utilizing non-metallic pipes and insulating joints have made this method of obtaining

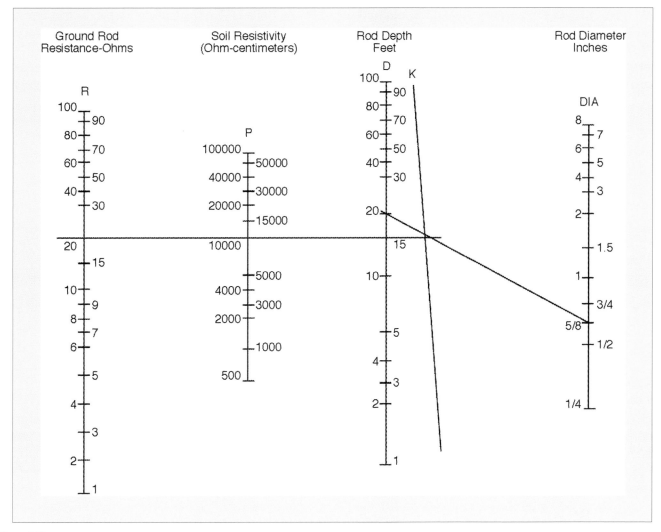

Figure 6-24   **Grounding Nomograph**

1. Select required resistance on R scale.
2. Select apparent resistivity on P scale.
3. Lay straightedge on R and P scale, and allow to intersect with K scale.
4. Mark K scale point.
5. Lay straightedge on K scale point & DIA scale, and allow to intersect with D scale.
6. Point on D scale will be rod depth required for resistance on R scale.

a low resistance ground questionable and in many instances unreliable.

The measurement of ground resistances may only be accomplished with specially designed test equipment. Most instruments use the fall-of-potential principle of alternating current (AC) circulating between an auxiliary electrode and the ground electrode under test; the reading will be given in ohms, and represents the resistance of the ground electrode to the surrounding earth. AEMC has also recently introduced clamp-on ground resistance testers.

Note: The National Electrical Code® and NEC® are registered trademarks of the National Fire Protection Association.

Ground resistance testing principle
(Fall of potential - 3-point measurement)

The potential difference between rods X and Y is measured by a voltmeter, and the current flow between rods X and Z is measured by an ammeter. (Note: X, Y and Z may be referred to as X, P and C in a 3-point tester or C1, P2 and C2 in a 4-point tester.) (See Figure 6-25.)

By Ohm's Law $E = RI$ or $R = E/I$, we may obtain the ground electrode resistance R. If $E = 20$ V and $I = 1$A, then

$$R = \frac{E}{I} = \frac{20}{1} = 20$$

It is not necessary to carry out all the measurements when using a ground tester. The ground tester will measure directly by generating its own current and displaying the resistance of the ground electrode.

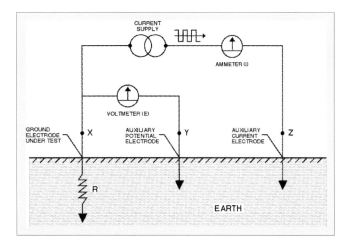

Figure 6-25

Position of the auxiliary electrodes on measurements

The goal in precisely measuring the resistance to ground is to place the auxiliary current electrode Z far enough from the ground electrode under test so that the auxiliary potential electrode Y will be outside of the effective resistance areas of both the ground electrode and the auxiliary current electrode. The best way to find out if the auxiliary potential rod Y is outside the effective resistance areas is to move it between X and Z and to take a reading at each location. If the auxiliary potential rod Y is in an effective resistance area (or in both if they overlap, as in Figure 6-26), by displacing it, the readings taken will vary noticeably in value. Under these conditions, no exact value for the resistance to ground may be determined.

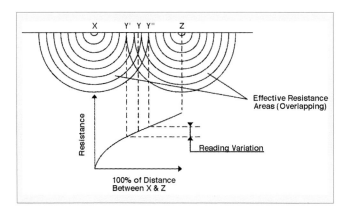

Figure 6-26

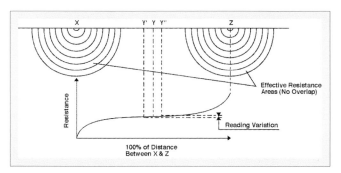

Figure 6-27

On the other hand, if the auxiliary potential rod Y is located outside of the effective resistance areas (Figure 6-27), as Y is moved back and forth the reading variation is minimal. The readings taken should be relatively close to each other, and are the best values for the resistance to ground of the ground X. The readings should be plotted to ensure that they lie in a "plateau" region as shown in Figure 6-27. The region is often referred to as the "62% area."

Measuring resistance of ground electrodes (62% method)

The 62% method has been adopted after graphical consideration and after actual test. It is the most accurate method but is limited by the fact that the ground tested is a single unit.

This method applies only when all three electrodes are in a straight line and the ground is a single electrode, pipe, or plate, etc., as in Figure 6-28.

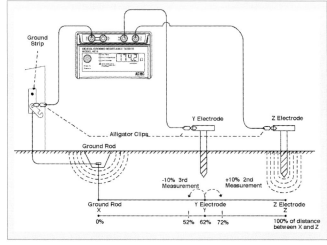

Figure 6-28

Consider Figure 6-29, which shows the effective resistance areas (concentric shells) of the ground electrode X and of the auxiliary current electrode Z. The resistance areas overlap. If readings were

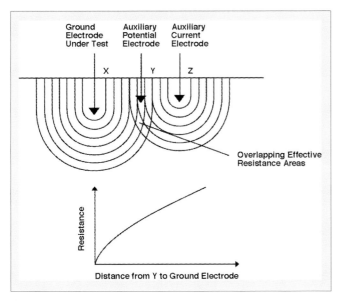

Figure 6-29

taken by moving the auxiliary potential electrode Y towards either X or Z, the reading differentials would be great and one could not obtain a reading within a reasonable band of tolerance. The sensitive areas overlap and act constantly to increase resistance as Y is moved away from X.

Now consider Figure 6-30, where the X and Z electrodes are sufficiently spaced so that the areas of effective resistance do not overlap. If we plot the resistance measured we find that the measurements level off when Y is placed at 62% of the distance from X to Z, and that the readings on either side of the initial Y setting are most likely to be within the established tolerance band. This tolerance band is defined by the user and expressed as a percent of the initial reading: ± 2%, ± 5%, ± 10%, etc.

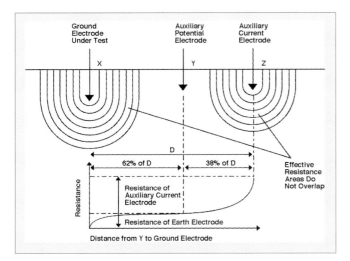

Figure 6-30

Auxiliary electrode spacing

No definite distance between X and Z can be given, since this distance is relative to the diameter of the electrode tested, its length, the homogeneity of the soil tested, and particularly, the effective resistance areas. However, an approximate distance may be determined from the following chart which is given for a homogeneous soil and an electrode of 1" in diameter. (For a diameter of 1/2", reduce the distance by 10%; for a diameter of 2" increase the distance by 10%.)

Approximate distance to auxiliary electrodes using the 62% method		
Depth Driven	Distance to Y	Distance to Z
6 ft	45 ft	72 ft
8 ft	50 ft	80 ft
10 ft	55 ft	88 ft
12 ft	60 ft	96 ft
18 ft	71 ft	115 ft
20 ft	74 ft	120 ft
30 ft	86 ft	140 ft

Multiple electrode system

A single driven ground electrode is an economical and simple means of making a good ground system. But sometimes a single rod will not provide sufficient low resistance, and several ground electrodes will be driven and connected in parallel by a cable. Very often when two, three or four ground electrodes are being used, they are driven in a straight line; when four or more are being used, a hollow square configuration is used and the ground electrodes are still connected in parallel and are equally spaced (Figure 6-31).

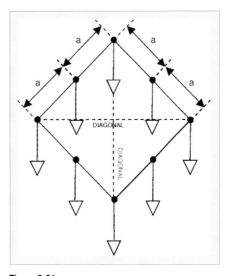

Figure 6-31

In multiple electrode systems, the 62% method electrode spacing may no longer be applied directly. The distance of the auxiliary electrodes is now based on the maximum grid distance (i.e., in a square, the diagonal; in a line, the total length. For example, a square having a side of 20 ft will have a diagonal of approximately 28 ft).

Multiple Electrode System		
Max Grid Distance	**Distance to Y**	**Distance to Z**
6 ft	78 ft	125 ft
8 ft	87 ft	140 ft
10 ft	100 ft	160 ft
12 ft	105 ft	170 ft
14 ft	118 ft	190 ft
16 ft	124 ft	200 ft
18 ft	130 ft	210 ft
20 ft	136 ft	220 ft
30 ft	161 ft	260 ft
40 ft	186 ft	300 ft
50 ft	211 ft	340 ft
60 ft	230 ft	370 ft
80 ft	273 ft	440 ft
100 ft	310 ft	500 ft
120 ft	341 ft	550 ft
140 ft	372 ft	600 ft
160 ft	390 ft	630 ft
180 ft	434 ft	700 ft
200 ft	453 ft	730 ft

[1]The text and graphics starting on page 88 to the Multiple Electrode System chart are taken from *Understanding Ground Resistance Testing Workbook E6*, AEMC Instruments.

Listed grounding electrode systems

In addition to ground rods and plates, grounding electrode systems consisting of 2.125 in. diameter copper pipe have been developed. These pipe elec-trodes are available in straight or L-shaped versions and in various lengths from 10 to 20 feet. Longer models are available on special order.

These grounding electrode systems are filled with a mixture of nonhazardous earth salts. Weep holes are provided at the top and various locations toward the bottom or end of the pipe. This allows the circulation of moisture through the pipe to enhance performance. In addition, a specially formulated backfill material is available. This material improves the earth-to-grounding electrode performance.

Grounding electrode study

The Southern Nevada Chapter of the IAEI has embarked on an ambitious study of the effectiveness of several types of grounding electrodes. The testing program began in 1992. It is planned that the five different sites with 18 electrodes will be monitored over a ten year period. (See the sidebar for a description of electrodes being studied.) Data on electrode resistance, soil moisture, area temperature and rainfall is being collected on a regular basis.

Soil conditions at the five different sites are reported to be as follows:

"Balboa" – predominately gravel rock loosely compacted and is normally dry.

"Lone Mountain" – normally dry with sandy/silty soil.

	Balboa	Lone Mountain	Pecos	Pawnee	Average
Electrode A	45.15	11.48	22.69	4.38	20.93
Electrode B	31.94	9.17	10.08	5.79	14.24
Electrode C	30.53	7.21	9.60	29.44	19.19
Electrode D	34.00	10.11	24.45	8.46	19.26
Electrode E	16.84	3.50	3.49	2.46	6.57
Electrode F	30.98	4.21	5.84	2.84	10.97
Electrode G	222.08	74.47	27.44	11.41	83.85
Electrode H	52.38	6.35	6.73	9.03	18.62
Electrode I	32.80	6.86	14.15	3.50	14.33
Electrode J	142.46	10.72	11.67	3.50	42.08
Electrode K	589.74	92.40	114.35	36.88	208.34
Electrode L	19.43	17.05	6.83	5.11	12.11
Electrode M	82.00	32.05	19.50	15.25	37.20
Electrode N	72.59	14.18	41.02	15.46	35.81
Electrode O	55.32	8.59	21.36	6.63	22.98
Electrode P	73.94	15.79	6.46	11.94	27.03
Electrode Q	184.98	26.15	20.63	19.62	62.84
Electrode R	15.41	---	---	---	15.41
Electrode S	---	---	13.70	3.60	8.65

Figure 6-32

"Pecos" – generally dry sand and silt down to a depth of about six feet and has clay and is wet down to about 12 feet.

"Pawnee" – a high water table with silt and clay.

Figure 6-32 gives results of actual on-site resistance measurements made of the various grounding electrodes at each location. The results are an average of the readings taken over the six-year period of 1993 through 1998. For example, electrode "E" (which happens to have the lowest average resistance over the period) has an average resistance of 16.84 ohms at the Balboa site and an average resistance of 2.46 ohms at the Pawnee site. The average for all sites for the six-year period is 6.57 ohms.

The averages are calculated from data furnished. The data is believed to be accurate and is subject to errors and omissions in collection and recording. In some cases, the data is incomplete as indicated for electrodes "R" and "S."

This is an active study and while valuable data is being collected on a regular basis, it may be too early to make conclusions. The sponsorship of the grounding electrode study has been assumed by the National Fire Protection Association Research Foundation. Under their sponsorship, the study has been expanded to four other locations including Illinois, Texas, Florida, and Virginia. As a result, additional and significant data from diverse locations with different soil conditions is being collected.

Also, the study now includes information on soil temperature and moisture content.

Lightning protection system

Any comprehensive discussion of lightning protection systems is beyond the scope of this text. Lightning protection systems should be installed in accordance with the *Standard for the Installation of Lightning Protection Systems*, NFPA-780.

The Code prohibits, in Section 250-60, the use of air terminal conductors and grounding electrodes such as driven pipes, rods, or other made electrodes in place of the grounding electrodes required by the Code for a wiring system and for equipment. Note that where two grounding electrodes are installed, they are required to be bonded together. See Section 260-106 for the requirement that lightning protection ground terminals (grounding electrodes) be bonded to the building or structure grounding electrode system.

The Fine Print Notes following Section 250-106

Description of Electrodes

A. No. 2 AWG stranded copper wire 50' in length centered in 12" of sand at 36" measured to the conductor.

B. No. 4 steel reinforcing bar located within and near the bottom of a concrete foundation consisting of 12" by 12" of concrete 2500 psi. Top of concrete located approx. 6" below grade.

C. No. 4 solid copper wire 25' in length centered in 6" by 6" of Erico ground enhancement material. Located at 20" to the bottom of the concrete.

D. No. 4 solid copper wire 25' in length centered in 6" by 6" of concrete 2500 psi; located at 20" to the bottom of the concrete. This electrode is designed to imitate a thickened edge of a post tensioned concrete construction.

E. 8' by 5/8" copper clad ground rod centered in a pre-drilled 9" dia. by 9' vertical hole, encased by Erico GEM.

F. 8' by 5/8" copper clad ground rod installed horizontally centered in a 6" Erico GEM located 20" to the bottom of the GEM.

G. 8' by 5/8" copper clad ground rod installed horizontally directly buried at 30".

H. 8' by 5/8" copper clad ground rod vertically driven.

I. 10' by 3/4" galvanized ground rod driven vertically.

J. 10' by 3/4" galvanized ground rod directly buried horizontally at a depth of 30".

K. Grounding plate directly buried at 30".

L. Lyncole XIT 10' vertical electrode in pre-drilled 9" hole.

M. Arrangement of steel and concrete designed to imitate a light pole base; six, 2 ft. No. 4 steel vertical rods tied with three No. 2 steel hoops separated 12" vertically in 2' dia. by 30" concrete 2500 psi.

N. No. 4 solid copper wire 20' roll shaped as a coil installed in 2' by 2' concrete 2500 psi.

O. No. 4 solid copper wire 20' roll shaped as a coil installed in 2' by 2' Erico GEM.

P. Wood pole 18" dia. by 8' with pole plate attached at bottom with No. 6 solid copper wire wrapped in a spiral for 6' at 6" spacing between wraps.

Q. 8' by 1/2" copper clad ground rod installed horizontally directly buried at 30".

R. Lightning Eliminators Chemrod vertical installed per manufacturer's recommendations.

S. Lyncole XIT horizontal rod installed per manufacturer's recommendations.

provide valuable information regarding lightning protection systems. The National Electrical Code® no longer has specific spacing requirements for separation of lightning protection system conductors from metal raceways and other metal enclosures of the building electrical system. The required spacing is typically 6 ft. through air and 3 ft. through dense construction materials such as concrete, brick or wood. Specific requirements are given in the *Lightning Protection Code*.

Conclusion

No potential exists between a conductor or equipment enclosure and earth if the system is properly and adequately grounded, except during a fault. By careful and intelligent design we must reduce the time of clearing a fault to a minimum.

While it is desirable to obtain a grounding electrode having a resistance as low as practical, it is more important that we provide a path of low impedance for the return of ground-fault current which will clear a ground fault when it occurs.

No hazard can exist in the distribution system if there is no insulation failure to create a ground fault. Further, the hazard can exist only for the period of time it takes to clear the fault. If a ground fault clears promptly, it is extremely unlikely that any loss of life would occur and property damage would be kept to a minimum.

For maximum safety, one rule that must be followed is; **ONLY ONE** grounding electrode system must be used in or on a building, with everything that should or is required to be grounded connected to that same grounding electrode system.

If more than one grounding electrode is required, they may be used providing **ALL** the grounding electrodes used are bonded together to form a grounding electrode system which in effect becomes **ONE** common grounding electrode.

Grounding electrodes should never be relied upon to provide the ground-fault return path for equipment. That is not their intended function in the electrical system. They cannot be relied upon to provide that path.

Chapter Six: The questions included here were developed using material included in this chapter. The answers can be found by reviewing the text. It is also important that students make use of the 1999 NEC®, where many answers can be found. See page 279 for answers.

1. A conducting element used to establish a ground for connecting electrical systems to the earth this conductor is defined as the ____.
   a. equipment grounding
   b. grounding electrode
   c. main bonding jumper
   d. earthing conductor

2. Intentionally connected to earth through a ground connection or connections of sufficiently low impedance and having sufficient current-carrying capacity that will prevent buildup of voltages that may result in undue hazard to connected equipment or to persons, defines ____.
   a. effectively grounded
   b. sufficiently grounded
   c. approved
   d. identified

3. One or more steel reinforcing bars or rods not less than $1/2$ inch in diameter, or 20 feet or more of bare copper conductor not smaller than No. 4 is defined as a ____.
   a. butt ground
   b. pole ground
   c. feeder electrode
   d. concrete-encased electrode

4. A concrete-encased electrode must be located within or near the bottom of a foundation or footing and is required to be encased by not less than ____ inches of concrete.
   a. 0
   b. 1
   c. 2
   d. 6

5. A copper conductor not smaller than No. 2, at least 20 feet long encircling a building or structure, and buried at not less than 2½ feet deep defines a ____.
   a. encased electrode
   b. ground ring
   c. Ufer ground
   d. common bonding grid

6. Where an underground metal water pipe is the only grounding electrode that is available, and is connected at the building or structure served, it must be supplemented by another ____.
   a. grounding electrode
   b. main bonding jumper
   c. equipment grounding conductor
   d. circuit bonding jumper

7. An electrode that is considered as being suitable for supplementing the metal underground water pipe, includes a concrete-encased electrode, ground ring, other local metal underground systems or structures, or ____ electrodes.
   a. rod
   b. pipe
   c. plate
   d. all of the above

8. If available on the premises, at each building or structure served, all grounding electrodes, including ____ electrodes, are required to be bonded together to form the grounding electrode system.
   a. identified
   b. approved
   c. made
   d. gas pipe

9. Where the electrodes described in Section 250-50 are not available at the service location, a grounding electrode must be made. It may be a ____.
   a. local metallic underground piping system, tanks, etc.
   b. pipe or conduit electrodes not less than 8 feet in length nor smaller than $5/8$ inch, and if of iron or steel, must be galvanized or metal-coated for corrosion protection.
   c. rod electrodes of steel or iron at least $5/8$ inch diameter.
   d. any of the above

10. Where two or more grounding electrodes are bonded together, they are considered to be ____ electrode.
   a. one
   b. an identified
   c. an approved
   d. a listed

11. The Code recognizes that conditions may exist that may cause an objectionable flow of current over the grounding electrode conductor, other than temporary currents that may be set up under accidental conditions. Permitted alterations include ____.
   a. if due to multiple grounds, one or more, but not all, of such grounds may be abandoned
   b. their location can be changed
   c. continuity of the grounding conductor may be suitably interrupted
   d. any of the above

12. Currents that introduce noise or data errors in electronic equipment are not considered to be ____.
   a. dangerous currents
   b. objectionable currents
   c. harmonic currents
   d. unsafe currents

13. Where a single rod, pipe, or plate electrode does not achieve a resistance to ground of ____ ohms or less, it is required to be supplemented by only one additional electrode.
   a. 50
   b. 40
   c. 30
   d. 25

14. The use of continuous metallic underground water and metal well casings, as well as the metal frames of buildings, should provide a ground resistance not exceeding ____ ohms.
   a. 3
   b. 6
   c. 12
   d. 25

15. In general, doubling the length of a rod type grounding electrode reduces the resistance by about ____ percent.
   a. 10
   b. 20
   c. 30
   d. 40

16. The use of lightning rod conductors and electrodes are ____ for grounding lightning rods in place of the grounding electrodes required by the Code for a wiring system and for equipment.
   a. permitted to be used
   b. prohibited from being used
   c. permitted by special permission
   d. identified when labeled to be used

Grounding Electrode Conductors

Objectives

After studying this chapter, the reader will be able to understand:

- General requirements and definitions for grounding electrode conductors.
- Functions of the grounding electrode conductor.
- Grounding electrode conductor connections.
- Material and protection for grounding electrode conductors.

Definition

Grounding electrode conductor: "The conductor used to connect the grounding electrode to the equipment grounding conductor, to the grounded conductor, or to both, of the circuit at the service equipment or at the source of a separately derived system."[N]

General requirements

The grounding electrode conductor is the conductor used to connect the electrical system, or the equipment grounding conductor, or both, to a grounding electrode, hence the specific term "Grounding electrode conductor." It must be properly sized based on the size of the service-entrance conductors but is not required to exceed No. 3/0 copper or 250 kcmil aluminum. Specific requirements are given in the Code regarding conductor material, how it is to be installed, protection from physical damage and how it must be connected.

Function of grounding electrode conductor

The grounding electrode conductor is the sole connection between the grounding electrode and the grounded system conductor (may be a neutral) and the equipment grounding conductor for a grounded system; or the sole connection between the grounding electrode and the service equipment enclosure and equipment grounding conductor for

an ungrounded system. See Section 250-24(c) of the NEC®.

A common grounding electrode conductor is required to ground both the system grounded conductor and the equipment grounding conductor. In other words, one grounding electrode conductor cannot be used to ground the system conductor and another grounding electrode conductor be used to ground the equipment grounding conductor even though both grounding conductors are connected to the same grounding electrode. A single grounding electrode conductor must be used for both the circuit and the equipment.

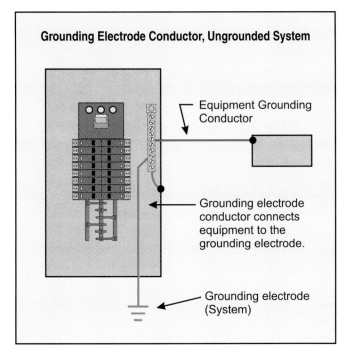

Figure 7-2

Maximum current on grounding electrode conductors

In all grounded systems of 600 volts or less, the maximum current that will flow in the grounding electrode conductor is limited by the impedance of the circuit. The circuit consists of the resistance of the service conductor, the grounding electrode at the service, the grounding electrode at the transformer bank plus the resistance of the earth path between the two grounding electrodes. In addition, there will be the resistance of the grounding electrode conductors themselves. In many installations, the length of the grounding electrode conductors is short enough that their resistance can be ignored. In addition to these resistances, there will be an inductive reactance that will vary as the spacing between the supply path and return path increases.

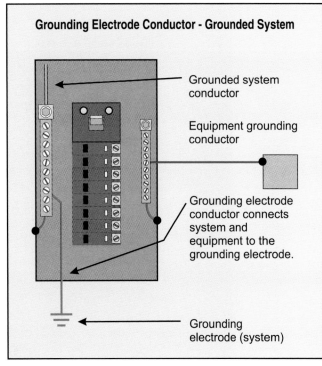

Figure 7-1

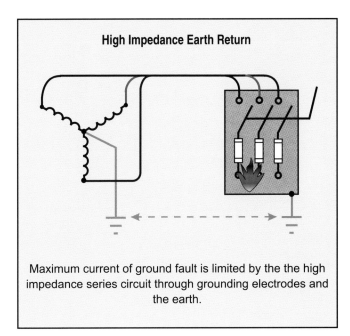

High Impedance Earth Return

Maximum current of ground fault is limited by the the high impedance series circuit through grounding electrodes and the earth.

Figure 7-3

As illustrated in Figure 7-3, if we assume that the sum of these resistances is equal to 22 ohms (a figure that may be lower than is found in actual practice), then, in a 120 volt-to-ground system, the maximum current through the grounding electrode conductor and grounding electrodes will be 5.5 amperes (120 volts ÷ 22 ohms = 5.5 amperes). As can be seen, this grounding connection is ineffective. Equipment grounded only in this manner is unsafe as a ground fault would not cause an upstream overcurrent protective device to clear the fault from the equipment. Obviously, the current through the high-impedance path provided by the grounding electrodes and the earth is not nearly enough to cause a 15-ampere circuit breaker or fuse to operate. Voltage levels that could be fatal could appear on equipment that is grounded in this manner.

This maximum current value is arrived at by considering resistance only. If the reactance of the circuit is included, the net result will be an impedance value much higher than the resistance value. The actual current that will flow in the grounding electrode conductor will thus be considerably less than the figures as calculated based on resistance only. Ground-fault currents through the grounding electrode conductor are thus not likely to attain a value anywhere near the continuous rating of the conductor even under ground-fault conditions.

As covered in Chapter 6 of this text in detail, the main purpose of the grounding electrode is to establish and maintain an earth reference for the system and noncurrent-carrying parts of the system and not to clear ground faults.

Sizing the grounding electrode conductor at a single service

The grounding electrode conductor must be sized in accordance with Table 250-66 of the NEC®. That conductor is required to be a minimum size of No. 8 and need not be larger than No. 3/0 copper. Where aluminum or copper-clad aluminum grounding electrode conductors are installed they are required to be not smaller than No. 6 nor larger than 250 kcmil. The size of this grounding electrode conductor is based upon the size of the service entrance conductor by reference to Table 250-66 and not on the rating of the circuit breaker or fuse in the service equipment. Table 250-66 is reprinted in Chapter 18 of this text.

Sizing Grounding Electrode Conductors

- **Refer to Table 250-66.**

- **Go down left column and find size of service-entrance conductor used.**

- **Go across to find size of copper or aluminium grounding electrode conductor.**

For example, where a No. 3/0 copper service-entrance conductor is installed, the minimum size grounding electrode conductor is No. 4 copper or No. 2 aluminum. If 750 kcmil aluminum service-entrance conductors are installed, a No. 1/0 copper or No. 3/0 aluminum grounding electrode conductor must be used.

Sizing grounding electrode conductor for service with parallel conductors

Where service-entrance conductors are installed in parallel as allowed by Section 310-4, the circular mil area of the conductors is added together and treated as a single conductor for purposes of sizing the grounding electrode conductor. For example, if four 250 kcmil copper conductors are installed, they are considered to be a single 1,000 kcmil conductor. In this case, by reference to Table 250-

66, we find the minimum grounding electrode conductor to be a No. 2/0 copper or No. 4/0 aluminum conductor.

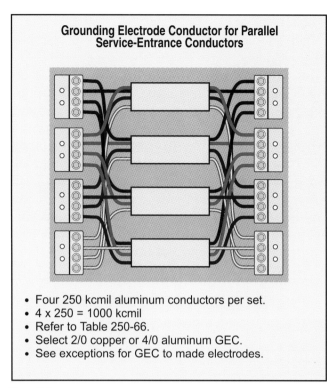

Grounding Electrode Conductor for Parallel Service-Entrance Conductors

- Four 250 kcmil aluminum conductors per set.
- 4 x 250 = 1000 kcmil
- Refer to Table 250-66.
- Select 2/0 copper or 4/0 aluminum GEC.
- See exceptions for GEC to made electrodes.

Figure 7-4

If No. 1/0 through No. 4/0 AWG conductors are installed in parallel, they must first be converted to circular mil area before applying Table 250-66. For example, if No. 2/0 aluminum service-entrance conductors are installed in parallel, refer to NEC® Table 8 of Chapter 9 for determining the conductor's circular mil area. (NEC® Table 8 of Chapter 9 is reprinted in Chapter 18 of this text.) There, we find it to have an area of:

133,100 circular mils;

133,100 cm x 2 = 266,200 cm.

By reference to Table 250-66 we find that a No. 2 copper or No. 1/0 aluminum grounding electrode conductor is required.

Sizing grounding electrode conductors for multiple enclosure services

Services are permitted to be installed in up to six enclosures installed at one location or at separate locations. The method of sizing grounding electrode conductors is a matter of choice. One basic rule must be followed: size the grounding electrode conductor for the circular mil area of service-entrance conductor(s) at the point of connection, or for the size of the main service-entrance

conductor(s). See Section 250-64(d).

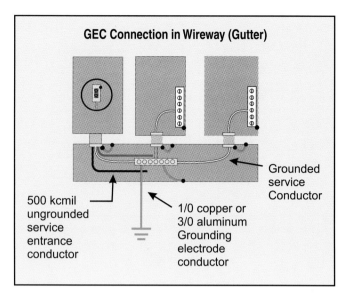

GEC Connection in Wireway (Gutter)

500 kcmil ungrounded service entrance conductor

1/0 copper or 3/0 aluminum Grounding electrode conductor

Grounded service Conductor

Figure 7-5

In Figure 7-5 with assumed 500 kcmil copper service-entrance conductors, the service-entrance grounded conductor is shown grounded inside the wireway. By reference to Table 250-66, we find that the grounding electrode conductor must be No. 1/0 copper or No. 3/0 aluminum.

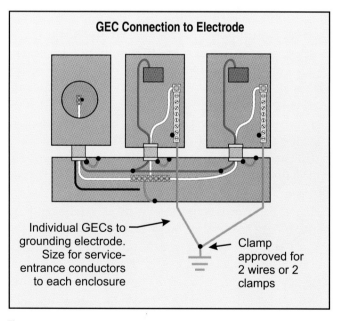

GEC Connection to Electrode

Individual GECs to grounding electrode. Size for service-entrance conductors to each enclosure

Clamp approved for 2 wires or 2 clamps

Figure 7-6

If the installer chooses to, an individual grounding electrode conductor may be installed from each service disconnecting means enclosure to the grounding electrode. In this case, the grounding electrode conductor is sized for the service-entrance conductor serving that enclosure. See Figure 7-6.

In addition, a single grounding electrode conductor is permitted to serve several enclosures. The **main** grounding electrode conductor is sized for the main service-entrance conductors. The NEC® does not use the term "main grounding electrode conductor" to identify the grounding electrode conductor that connects directly to the grounding electrode. We use it here to distinguish it from the grounding electrode tap conductor. Taps that are sized for the individual service-entrance conductors are connected from the service to the **main** grounding electrode.

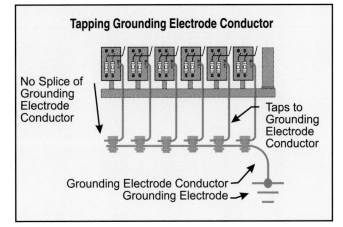

Tapping Grounding Electrode Conductor

No Splice of Grounding Electrode Conductor

Taps to Grounding Electrode Conductor

Grounding Electrode Conductor
Grounding Electrode

Figure 7-8

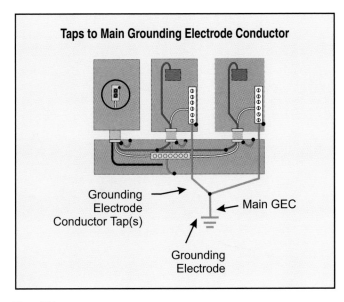

Taps to Main Grounding Electrode Conductor

Grounding Electrode Conductor Tap(s)

Main GEC

Grounding Electrode

Figure 7-7

In Figures 7-7 and 7-8, it is assumed that No. 2 copper conductors serve each service disconnecting means from the wireway. Shown are taps to a **main** grounding electrode conductor. The tap conductors must be connected to the main grounding electrode conductor in such a manner that it is not spliced. This means that the tap conductor must be connected with a split bolt or tap connector allowing the **main** grounding electrode conductor to pass through the connector unbroken. The concept of tapping the grounding electrode conductor applies whether the sets of service-entrance conductors are tapped from a wireway or are installed individually as an overhead or underground system.

For Figure 7-8, the size of the **main** grounding electrode conductor is determined as follows.

Assume that 500 kcmil copper service-entrance conductors supply the service and are connected in the wireway to the service-entrance conductors that serve each enclosure.

Refer to Table 250-66. The minimum size **main** grounding electrode conductor is No. 1/0 copper or No. 3/0 aluminum. This conductor is installed from the grounding electrode to the vicinity of the wireway.

The grounding electrode tap conductors from the individual enclosures are sized from Table 250-66 and are No. 8 copper or No. 6 aluminum. Note that there is no minimum or maximum length for these grounding electrode tap conductors.

If the installer chose to, the grounding electrode conductors from the service disconnects may be connected individually to the grounding electrode rather than being tapped to the **main** grounding electrode conductor.

Exceptions to size of grounding electrode conductor

What amounts to a three-part exception to the general rule for sizing the grounding electrode conductor is provided in Section 250-66(a) through (c).

Section 250-66(a) permits the grounding electrode conductor to be not larger than No. 6 copper or No. 4 aluminum wire where it is the sole connection to a rod, pipe or plate electrode. Section 250-66(b) provides that the grounding electrode conductor that is the sole connection to a concrete encased grounding electrode need not be larger than No. 4 copper wire. Note that aluminum wire is not permitted for this application. The use of this smaller No. 6 conductor is based on the fact that it can never be called upon, even under ideal conditions, to carry a current beyond its safe short-time rated capacity.

Section 250-66(c) provides that, where connected

to a ground ring, that portion of the grounding electrode conductor that is the sole connection to the ground ring need not be larger than the ground ring conductor. The minimum size and conductor material for a ground ring is No. 2 copper. That requirement assumes, and correctly so, that the resistance of such electrodes will limit the maximum current that may be expected to flow through them.

The term "sole connection" means that the grounding electrode conductor does not serve or connect to more than one grounding electrode. For example, a grounding electrode conductor does not connect to a concrete encased electrode then to building steel or an underground metal water pipe. In this example, a grounding electrode conductor larger than No. 4 may be required for the water pipe electrode based on the size of the service-entrance conductors.

Grounding electrode conductor connections

The Code requires generally that the point of connection of the grounding conductor to grounding electrodes shall be accessible and made in a manner that will assure a permanent and effective ground. An exception provides that a connection at a concrete-encased, driven, or buried grounding electrode is not required to be accessible. See Section 250-68(a).

Specific rules for connection of the grounding electrode conductor to grounding electrodes are found in Section 250-70. Grounding electrode conductor connections must be by exothermic welding, listed lugs, listed pressure connectors, listed clamps or other listed means. The only connection

means not required to be listed are those made by exothermic welding although listed connections are now available. Connections depending solely on solder shall not be used.

Other requirements state that not more than one conductor is permitted to be connected to the grounding electrode by a single clamp or fitting, unless the clamp or fitting is listed as being suitable for connecting multiple conductors.

Ground clamps must be listed for both the materials of the grounding electrode and the grounding electrode conductor. For example, ground clamps must be listed for aluminum conductors to be used for such connections. Clamps used on a pipe, rod or other electrode that is buried must be listed for direct earth burial. Typically, these clamps are identified by the manufacturer with "direct burial" or "dir bur" or similar.

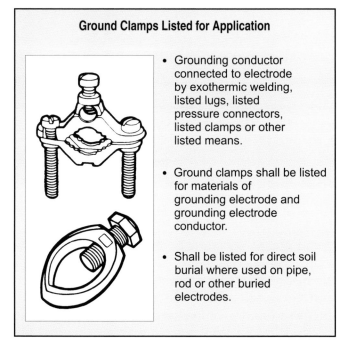

Ground Clamps Listed for Application

- Grounding conductor connected to electrode by exothermic welding, listed lugs, listed pressure connectors, listed clamps or other listed means.

- Ground clamps shall be listed for materials of grounding electrode and grounding electrode conductor.

- Shall be listed for direct soil burial where used on pipe, rod or other buried electrodes.

Figure 7-10

Sheet-metal-strap type ground clamps attached to a rigid metal base that are listed are permitted for indoor telecommunication purposes only. See Section 250-70(3). Also, Underwriters Laboratories Guide Card information (KDER) states, "Strap type ground clamps are not suitable for attachment of the grounding conductor of an interior wiring system to a grounding electrode." Use strap type ground clamps for only the type of conductor and grounding electrode, as well as the environment it is listed and labeled for, such as for communication circuits.

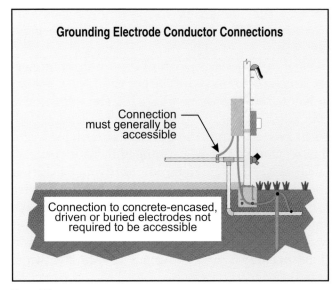

Grounding Electrode Conductor Connections

Connection must generally be accessible

Connection to concrete-encased, driven or buried electrodes not required to be accessible

Figure 7-9

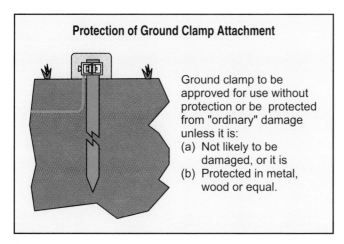

Protection of Ground Clamp Attachment

Ground clamp to be approved for use without protection or be protected from "ordinary" damage unless it is:
(a) Not likely to be damaged, or it is
(b) Protected in metal, wood or equal.

Figure 7-11

It is required that ground clamps be protected from physical damage unless the fittings are approved for use without protection or are installed in a location where they are not likely to be damaged. Protection of the ground clamp is permitted to consist of metal, wood or equivalent materials. See Section 250-10.

Clean surfaces

It is required in Section 250-12 that nonconductive coatings such as paint, lacquer and enamel be removed from threads and other contact surfaces of equipment to be grounded or be connected by means of fittings designed to make such removal unnecessary. (Underwriters Laboratories reported in a June 27, 1995 letter and confirmed on June 7, 1998 that no fittings listed by them incorporate this feature.) This will ensure a good electrical connection. See Chapter 8 of this text for additional discussion on the subject.

Here, again, it is important to consider that all grounding conductors and their connections form a part of a circuit which is required under certain conditions to carry fault current. In some cases the fault current is several times the full-load current rating of the conductor involved if the conductor was being used for continuous duty in an electrical circuit. Removal of paint under connections of raceways and cable fittings is critical to make a reliable connection that can carry fault current when necessary.

Grounding electrode conductor material

The grounding electrode conductor is required to be of copper, aluminum or copper-clad aluminum and may be solid, stranded, insulated, covered or bare. The material selected shall be resis-

tant to any corrosive condition it may be exposed to. As an option, it may be suitably protected against corrosion. See Section 250-62. Note that there is no color code for the grounding electrode conductor such as exists for grounded or equipment grounding conductors.

The required current-carrying capacity of the grounding electrode conductor is limited by the resistance of the grounding electrode(s) and the earth return path. In a grounded system, it is further limited by the fact that it completes a parallel circuit from the grounded conductor through both grounding electrodes. The other side of the parallel circuit is the grounded service conductor which is a circuit of much lower impedance.

Aluminum or copper-clad aluminum grounding electrode conductors

Specific limitations are placed on aluminum or copper-clad aluminum grounding electrode conductors. Note that these rules apply regardless of whether the conductors are insulated or bare. See Section 250-64(a).

Insulated or bare conductors are not permitted where they are in direct contact with masonry or the earth or where subject to corrosive conditions. This rule would not prohibit installing conduit on masonry and pulling the aluminum grounding electrode conductor in it.

Where used outside, aluminum or copper-clad aluminum grounding electrode conductors are not permitted to be installed within 18 in. of the earth.

Because of these restrictions on their installation, they cannot be used for connection to concrete encased electrodes, ground rods or pipes where the top of is within 18 in. of the earth or for connection to plate electrodes. Aluminum conductors are otherwise permitted to be used as grounding electrode conductors where the clamp or connector are listed for both the conductor material and the electrode.

Grounding electrode conductor installation

Section 250-64(b) requires that the grounding electrode conductor or its enclosure be securely fastened to the surface on which it is carried.

No. 4 or larger conductors require protection where exposed to "severe physical damage," a term that is not defined. This leaves it to a judgment call on the part of the inspector as to what constitutes "severe physical damage." No. 6 conductors must

---

Protection of Grounding Electrode Conductors

- GEC or enclosure to be securely fastened to surface.

- No. 4 or larger—protect from severe physical damage.

- No. 6 run along surface and be securely fastened or be protected.

- Smaller than No. 6—protect from damage.

- Aluminum—not allowed in contact with masonry, the earth, where subject to corrosive conditions or within 18 inches of the earth.

---

be run along the surface of the building construction and be securely fastened. Otherwise, they must be protected by installation in rigid or intermediate metal conduit, rigid nonmetallic conduit, electrical metallic tubing or cable armor. Grounding conductors smaller than No. 6 must be protected in one of the methods allowed for No. 6 conductors.

Splicing grounding electrode conductor

The general rule in Section 250-64(c) is that grounding electrode conductors are to be installed in one continuous length without a splice or joint.

The exception permits splices in busbars as they are at times used as a grounding electrode conductor. Busbars often come in standard lengths such as 10 ft. and have to be joined together to achieve the required length.

This section permits the grounding electrode conductor to be spliced by two methods; irreversible compression-type connectors listed for the purpose or by the exothermic welding process. It is vital that the manufacturer's instructions be carefully followed where either of these splicing methods is chosen.

Where compression-type connectors are used, the correct splicing sleeve for the conductor material and size must be selected. Then, the compression tool, and in some cases the proper die for the sleeve to be crimped, must be used. These compression-type connectors must be specifically listed by a qualified electrical testing laboratory for the purpose of splicing grounding electrode conductors.

Where exothermic welding of the grounding electrode is performed, the correctly sized form or mold for the conductor to be spliced must be used. Also, unless specifically permitted otherwise by the manufacturer, the conductors to be spliced by this method must be clean and dry. Inspect the resulting splice carefully to be certain that it has been made satisfactorily.

Protecting grounding electrode conductor from magnetic field

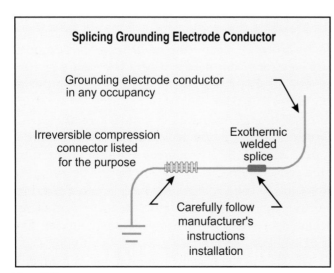

Splicing Grounding Electrode Conductor

Grounding electrode conductor in any occupancy

Irreversible compression connector listed for the purpose

Exothermic welded splice

Carefully follow manufacturer's instructions installation

Figure 7-12

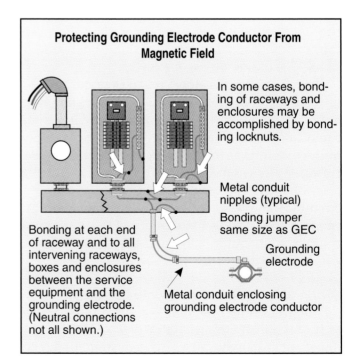

Protecting Grounding Electrode Conductor From Magnetic Field

In some cases, bonding of raceways and enclosures may be accomplished by bonding locknuts.

Metal conduit nipples (typical)

Bonding jumper same size as GEC

Grounding electrode

Metal conduit enclosing grounding electrode conductor

Bonding at each end of raceway and to all intervening raceways, boxes and enclosures between the service equipment and the grounding electrode. (Neutral connections not all shown.)

Figure 7-13

Where metal enclosures are provided for protection of the grounding electrode conductor, some special procedures must be followed. This is required by Section 250-64(e) and 250-92(a)(3).

1. Metal conduit must be electrically continuous from the point of attachment to cabinets or equipment to the grounding electrode.

2. The conduit must be securely fastened to the ground clamp or fitting.

3. Metal conduit that is not physically continuous from the enclosure to the ground clamp must be made continuous by bonding both ends of the conduit to the conductor.

4. Bonding of the metal raceway or armor that encloses a grounding electrode must be done at each end and to all intervening raceways, boxes and enclosures between the service equipment and the grounding electrode. [Section 250-92(a)(3)]

It is common practice to use a No. 8 grounding electrode conductor protected by metallic armored cable. The need for bonding the metallic armor of such cable is required in Section 250-64(e). Where that bonding procedure is not followed, the impedance of the grounding electrode conductor is approximately doubled with the result that its effectiveness is markedly reduced.

Impedance of conduit and conductor

Table Two of Chapter 18 of this text compares the continuous rating of copper grounding electrode conductors with the service conductors with which they must be used.

Table Three of Chapter 18 of this text compares the resistance and impedance of copper grounding electrode conductors where enclosed in a steel conduit for physical protection. The last two columns of that table show how the impedance of the conductor is approximately doubled where the conduit is not properly installed as required in Section 250-64(e). The Code requires that the conduit be bonded at both ends of the grounding electrode conductor to form a parallel circuit with the copper grounding electrode conductor. That important rule, if not observed, will result in doubling the impedance of the grounding conductor. Test data to confirm the above are available. Of course, where the impedance of the installation is increased, the effectiveness is reduced.

Where grounding electrode conductors are used

Table 7-14. Division of current in conduit.				
Conductor	Conduit Inches	Total Current Amperes	Current in Conductor, Amperes	Current in Conduit, Amperes
6	½	100	3	97
6	½	300	5	295
2	¾	90	7	83
2	¾	350	10	340
2/0	1	150	15	135
2/0	1	590	5	585
4/0	1¼	225	15	210
4/0	1¼	885	15	870

The above test data confirm that, for all practical purposes, the impedance of a conductor enclosed in a steel conduit, when the conduit is bonded to the conductor at both ends, is approximately equal to the impedance of the conduit.

on an alternating-current circuit and enclosed in conduit, it is necessary to compare the impedance values and not the resistance values.

The data in Table Three of Chapter 18 of this text shows that where a No. 8 copper conductor is installed in steel conduit and properly bonded at each end, the impedance of the circuit is about the same as where the No. 8 copper conductor was used alone.

For all other sizes of copper conductors installed in the proper sized conduit, the impedance of the circuit is greater where the conduit is used, as compared to using the copper conductor alone. The impedance values are from about 40 percent more for a No. 6 copper conductor in a ¾-inch conduit to about 500 percent more for a No. 3/0 copper conductor in a 1¼-inch conduit, as compared to not using a conduit for physical protection.

Over 100-foot grounding electrode conductor

The short-time rating of a copper conductor is related to the I^2t (current x current x time) rating of the conductor for a given temperature rise which will not damage adjacent insulated conductors or affect the continuity established by the bolted joints. For a period of five seconds the short-time rating may be taken as approximately 1 ampere for every 42.25 circular mils area. A No. 6 conductor has an area of 26,240 circular mils and is thus capable of carrying about 621 amperes for five seconds safely.

Based on the safe I^2t values for the circuit comprising the various grounding electrode conductors, it will be seen that for a five-second flow of current, the IR drop in the different sizes of grounding electrode conductors will be approximately 37 volts per 100 feet. Using that figure as a standard, it is recommended that where a grounding electrode conductor exceeds 100 feet in length, the conductor cross section be increased to keep the IR drop to not over 40 volts when carrying the maximum short-time current for the size conductor specified for five seconds (I^2t value). The National Electrical Code® does not, now, place a limit on the length of the grounding electrode conductor.

An example of selecting the proper size grounding electrode conductor for a run exceeding 100 feet is as follows.

Given: a No. 1/0 copper service-entrance conductor. The grounding electrode conductor specified in Table 250-66 is a No. 6 copper and the length of the grounding electrode conductor is 150 feet.

If a No. 6 conductor is used which has 26,240 circular mils, resulting in a short-time rating of 621 amperes, and a dc resistance of 0.0737 ohms for 150 feet (0.491 ohms/M ft.), the voltage drop would be 621 x 0.0737 or 46 volts.

We must thus select a larger grounding electrode conductor whose resistance times the short-time rating of the No. 6 conductor in amperes must not exceed 40 volts.

The next larger-sized grounding conductor, a No. 4, has a resistance of 0.0462 ohms for 150 feet (0.308 ohms/M ft.), so the voltage drop would be 621 x 0.0462 or 28.7 volts. That would make a No. 4 copper grounding conductor the proper size to use for a service using a No. 1 or No. 1/0 copper service-entrance conductor and having a grounding electrode conductor run of 150 feet.

Direct current systems

Where used on a direct-current circuit, we do not destroy the value of the grounding conductor if the conduit is properly installed, that is, bonded at both ends to the grounding conductor. The resultant resistance is lower where the conduit is used for physical protection. However, that assumes a steady direct current flow. Special considerations must be given if a direct-current circuit is to be properly protected with a grounding electrode conductor against transient currents such as are produced by lightning. It is necessary to treat the selection of the grounding electrode conductor as would be done for an alternating-current circuit.

Table Four of Chapter 18 of this text shows that where an aluminum conduit is used to enclose an aluminum grounding electrode conductor, the required conduit size has a lower resistance than the aluminum grounding electrode conductor in every case. In the case of aluminum wire size No. 6, the conduit is about one-tenth the resistance of the conductor. For the largest aluminum grounding electrode conductor size, 250 kcmil, the conduit is about half the resistance of the conductor. Thus, where an aluminum grounding electrode conductor is protected with an aluminum conduit and properly bonded at both ends, we will always have a much lower impedance than where the wire is not installed in an aluminum conduit. However, aluminum is subject to certain restrictions owing to chemical corrosion.

Since aluminum conduit is nonmagnetic and has lower resistance as well as impedance values compared to steel, the use of aluminum conduit for physical protection will provide lower impedance values.

Conclusion

All of the above means that we decrease the safety of an installation (which requires a grounding electrode conductor larger than No. 8) where we enclose the conductor in a steel conduit. Although we decrease physical damage to assure the integrity of the grounding electrode conductor, another hazard is introduced by decreasing the effectiveness of the grounding electrode conductor through increasing the voltage rise on it. No better case could thus be made for not permitting the use of a steel conduit on a grounding electrode conductor larger than a No. 8.

The obvious solution, where physical protection is necessary, is to use nonmetallic or aluminum conduit for enclosing the grounding electrode conductor. That is especially true in view of the improvement in the art of manufacturing nonmetallic conduit which now can be obtained in ample physical strength to meet the requirements of proper physical protection.

Chapter Seven: The questions included here were developed using material included in this chapter. The answers can be found by reviewing the text. It is also important that students make use of the 1999 NEC®, where many answers can be found. See page 279 for answers.

1. A conductor used to connect the grounding electrode to the equipment grounding conductor and/or to the grounded conductor of the circuit at the service equipment, or at the source of a separately derived system is defined as a ____.
  a. main bonding jumper
  b. grounding electrode conductor
  c. feeder bonding jumper
  d. grounded

2.   A grounding electrode conductor must be properly sized and is based on the size or rating of the:
  a. service breaker or fuse
  b. transformer
  c. service conductors
  d. grounding electrode

3.   For service conductors sized at over 1100 kcmil copper or 1750 kcmil aluminum or copper-clad aluminum, the grounding electrode conductor must be a ____ copper or ____ aluminum.
  a. No. 2/0 - No. 4/0
  b. No. 1/0 - No. 3/0
  c. No. 3/0 - 250 kcmil
  d. No. 3/0 - No. 4/0

3. In all grounded systems of ____ volts or less, the maximum current that will flow in the grounding electrode conductor is dependent on the sum of the resistance of the grounding electrode at the service, the grounding electrode at the transformer bank, plus the resistance of the earth path between the two grounding electrodes.
  a. 600
  b. 1,000
  c. 700
  d. 800

4.   Where of copper, the grounding electrode conductor is required to be sized at not less than No. ____.
  a. 10
  b. 14
  c. 8
  d. 12

5. Where a 3/0 copper service-entrance conductor is installed, the minimum size grounding electrode conductor is No. ____ copper or No. ____ aluminum, or copper-clad aluminum.
  a. 8 - 4
  b. 6 - 6
  c. 4 - 2
  d. 4 - 4

6. Where four 250 kcmil copper service entrance conductors are installed per ____, they are considered to be a single 1,000 kcmil conductor.
  a. service
  b. feeder
  c. cable
  d. phase

7.   The minimum size for a grounding electrode conductor for a 1,000 kcmil service entrance conductor is No. ____ copper or No. ____ aluminum, or copper-clad aluminum conductor.
  a. 1/0 - 2/0
  b. 3/0 - 4/0
  c. 1/0 - 3/0
  d. 2/0 - 4/0

8.   Services are permitted to be installed in up to ____ enclosures where they are installed at one location or at separate locations.
  a. one
  b. two
  c. six
  d. eight

9. A grounding electrode conductor is required to be sized at not less than No. ____ copper or No. ____ aluminum wire where it is the sole connection to a rod, pipe, or plate electrode.
  a. 8 - 8
  b. 6 - 4
  c. 6 - 6
  d. 4 - 8

10. A grounding electrode conductor that is the sole connection to a concrete-encased grounding electrode is not required to be larger than No. ____ copper wire.
  a. 6
  b. 4
  c. 8
  d. 10

11. Grounding electrode conductor connections are required to be by ____, listed clamps, or other listed means.
   a. exothermic welding
   b. listed lugs
   c. listed pressure connectors
   d. any of the above

12. Which one of the following statements is INCORRECT?
   a. Exothermic welding connections are not required to be listed.
   b. Soldered connections are permitted for connections to a grounding electrode.
   c. Unless listed, not more than one conductor can be connected to the grounding electrode by a single clamp or fitting.
   d. Clamps used on a pipe, rod, or other buried electrode must also be listed for direct earth burial.

13. Where exposed to severe physical damage, No. ____ or larger grounding electrode conductors require protection.
   a. 4
   b. 8
   c. 6
   d. 3

14. No. ____ grounding electrode conductors must be run along the surface of the building construction and be securely fastened, or they must be protected by installation in rigid or intermediate metal conduit, rigid nonmetallic conduit, electrical metallic tubing, or cable armor.
   a. 8
   b. 6
   c. 4
   d. 2

15. Aluminum or copper-clad aluminum conductors are not permitted to be installed outside within ____ inches of the earth.
   a. 12
   b. 18
   c. 14
   d. 16

Bonding Enclosures and Equipment

Objectives

After studying this chapter, the reader will be able to understand:

- Requirements for maintaining continuity of ground-fault path.
- Bonding of systems over 250 volts to ground.
- Bonding multiple raceway systems.
- Bonding of receptacles.
- Bonding of metal piping systems.
- Bonding of exposed structural steel.

Maintaining Continuity

Section 250-96(a) of the NEC® requires that bonding be done around connections of metal raceways, cable trays, cable sheaths, enclosures, frames, fittings and other metal noncurrent-carrying parts used as equipment grounding conductors where necessary. This may be necessary to assure that these systems have electrical continuity and the current-carrying capacity to safely conduct the fault current likely to be imposed on them. Note that this bonding of raceways, etc. must be performed regardless of whether or not an equipment grounding conductor is run within the raceway, etc. This will ensure that the raceway will not become energized by a line-to-enclosure fault without having the capacity and capability of clearing the fault by allowing sufficient current to flow to operate the overcurrent protective device on the line side of the fault.

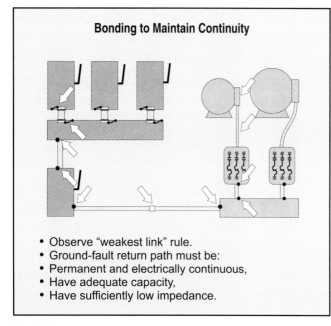

Bonding to Maintain Continuity

- Observe "weakest link" rule.
- Ground-fault return path must be:
- Permanent and electrically continuous,
- Have adequate capacity,
- Have sufficiently low impedance.

Figure 8-1

Keep in mind that the "weakest link rule" applies to the ground-fault return path. To provide adequate safety, the ground-fault return path must be (1) permanent and electrically continuous, (2) have the capacity to conduct safely any fault current likely to be imposed on it, and (3) have sufficiently low impedance to limit the voltage to ground and to facilitate the operation of the circuit protective devices. See Section 250-2(d). This ground-fault path must meet all three conditions from the farthest enclosure or equipment all the way back to the service equipment and ultimately to the

source. This path can be through many boxes, conduit or other raceways, pull boxes, wireways, auxiliary gutters, panelboards, motor control centers and switchboards. Every connection is important. It only takes one loose or broken fitting to break a link in the fault-current chain.

Section 250-96(a) also refers to conditions where a nonconducting coating might interrupt the required continuity of the ground-fault path, and it points out that such coatings must be removed unless the fitting(s) is (are) designed as to make such removal unnecessary. Underwriters Laboratories confirmed in writing that, as of May 7, 1998 no fittings having such a feature were listed by them.

In some cases, the locknut may pierce painted enclosures to establish a good electrical connection. This applies to the use of heavy-type, formed-steel locknuts. General instructions are that the locknuts be tightened by hand, then be further tightened ¼ turn by means of a screwdriver and hammer. At that point, examine the connection to be sure any paint under the locknut has been adequately broken. Remove the locknut and scrape the paint off or install a bonding bushing if there is any question about the adequacy of the connection.

Testing of conduit fittings

The importance of removing paint from enclosures where the conduit or raceway is intended to serve as the fault-current path is further emphasized in a report on "Conduit Fitting Ground-Fault Current Withstand Capability" issued by Underwriters Laboratories on June 1, 1992. Over 300 conduit fitting assemblies from ten different manufacturers were subjected to a current test to simulate performance under ground-fault conditions.

A sample assembly consisted of a conduit fitting secured to one end of a two-foot length of conduit and attached to a metal enclosure.

After securing the conduit fitting to the conduit properly, the conduit fitting was secured to the enclosure using the locknut provided by the manufacturer. The locknut was first hand-tightened and then further tightened ¼ turn with a hammer and standard screwdriver. The fittings were installed through holes in the enclosures that were punched rather than being installed in preexisting knockouts. A pipe clamp, wire connector, conductors and a power supply were assembled to complete the testing. Thermocouples were placed at strategic locations to record pertinent temperature data. Figure 8-2 is a drawing of the sample assembly.

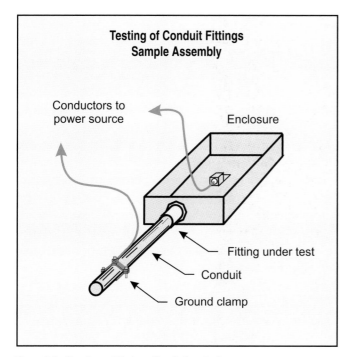

Testing of Conduit Fittings Sample Assembly

Conductors to power source

Enclosure

Fitting under test

Conduit

Ground clamp

Figure 8-2. *Courtesy of Underwriters Laboratories*

Fittings for conduit in the 3/8-inch through 3-inch trade sizes were tested. The appropriate current applied to the fittings in the test program is as shown in the following table.

TEST CURRENTS AND TIMES		
Conduit Trade Size, Inch	Test Time	Current, Amperes
3/8	4 Sec.	470
1/2	4 Sec.	1180
3/4, 1	6 Sec.	1530
1-1/4, 1-1/2	6 Sec.	2450
2	6 Sec.	3900
2-1/2	6 Sec.	4900
3, 3-1/2, 4	9 Sec.	5050
4-1/2	9 Sec.	6400
5, 6	9 Sec.	8030

This test should not be confused with a short-circuit withstand test and is not intended to test the maximum short-circuit current these fittings can withstand. Due to the time and current involved, a great deal of heat is generated in the test assembly.

Seven of the more than 300 assemblies tested sustained damage. A visual examination of sample assemblies that failed showed that melt-

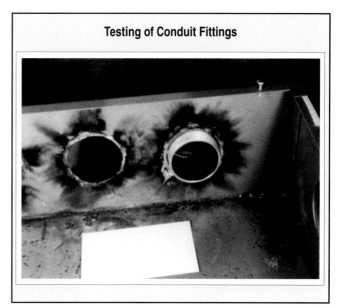

Testing of Conduit Fittings

Figure 8-3. *Courtesy of Underwriters Laboratories*

ing of the die cast zinc locknuts occurred as a result of the fault current. Melting of the die-cast zinc body occurred on five sample assemblies. The painted enclosures on which the fittings were tested were also examined. The examination indicated that melting of the die-cast zinc was probably due to the inability of the locknut to penetrate through the enclosure paint and provide good electrical contact between the fitting and metal of the enclosure.

A visual examination of all the conduit fittings with die-cast zinc locknuts showed that there were three different constructions of the locknuts. The three constructions differed in that the surface of the locknut contacting the enclosure was either flat, nibbed, or serrated. The sample assemblies with die-cast zinc locknuts that did not complete the Current Test with acceptable results had locknuts with flat or ribbed surfaces. All fittings having die-cast zinc locknuts with serrations completed the test with acceptable results. It appeared as though locknuts with serrations consistently penetrated through the enclosure paint and provided better electrical contact between the fitting and the metal of the enclosure than did the locknuts with flat or ribbed surfaces.

The fittings investigated in this work were formed of die-cast zinc, steel and malleable iron. The melting point of zinc is 420°C while the melting point of steel and malleable iron is much higher, typically greater than 1400°C. Heat generated from the fault current in some sample assemblies was obviously greater than the melting point of the die-cast zinc fittings and locknuts, but not greater than the melting point of steel or malleable iron since no melting of the steel enclosure occurred. This was further evidenced by tests of two sample assemblies where the die cast zinc body of the fittings melted, but the steel locknuts did not.

All of the conduit fittings that were constructed of steel bodies and steel locknuts completed the test with acceptable results. A visual examination of the steel locknuts indicated that the nibs on these locknuts, which in most cases were sharp and well defined from the metal forming process, provided for better penetration through the enclosure paint than the nibs on the die-cast zinc locknuts.

For most of the sample assemblies which completed the Current Test with acceptable results, the maximum temperatures on the fitting bodies and locknuts were about the same as or less than the temperature of the conduit. In the case of the flexible metal conduit, the temperatures on the fittings were much less than on the conduit. This would seem to indicate that if the fitting can provide good electrical contact to the enclosure metal the fitting will provide for adequate equipment grounding.

Conclusions reached by Underwriters Laboratories as a result of the testing are as follows:

"1. Over 300 conduit fitting assemblies from ten different manufacturers were subjected to the Current Test to simulate performance under ground-fault conditions. As a result of the tests, only seven assemblies representing four different conduit fittings and three different manufacturers did not withstand the fault current without breaking or melting of the conduit fitting assembly. All seven of these sample assemblies were compression type connectors with die-cast zinc bodies, and all but one of these assemblies utilized a die-cast zinc locknut.

"2. An examination of the seven sample assemblies that did not complete the Current Test with acceptable results showed that the failures were probably due to high resistance from the inability of the fitting locknut to penetrate through the enclosure paint and provide good electrical continuity between the fitting and the metal enclosure. Heat generated by the high-resistance arcing was sufficient to melt the zinc, but not steel or iron.

"3. Some of the sample assemblies that did not exhibit breaking or melting did show signs of arcing and welding between the locknut and the enclosure and/or the fitting and the conduit. These sample assemblies usually had higher temperatures during the Current Test, however, the temperatures were not sufficient to cause melting of the zinc or steel parts nor loss of continuity between the conduit, fitting, and enclosure.

"4. Most of the sample assemblies that were subjected to the Current Test attained maximum temperatures on the fitting bodies and locknuts that were about the same as or less than the temperature of the conduit. For the tests with flexible metal conduit, the temperatures of the fittings were much less than the temperatures of the flexible conduit.

"5. As a result of the tests, it was observed that if the fitting provides good electrical contact to both the enclosure and the conduit, the fitting will provide a suitable equipment ground path for fault current."

Bonding for over 250 volts

For systems having a voltage exceeding 250 volts to ground, the electrical continuity of metal raceways or metal-sheathed cables that are not service-entrance cable must also be assured. See

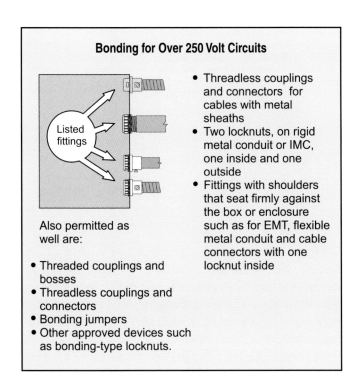

Figure 8-4

Section 250-97. Acceptable methods include any of the methods approved for bonding at service equipment found in Sections 250-94(2) through (4). Note that standard locknuts and bushings without additional bonding means are not generally permitted for bonding equipment over 250 volts to ground.

The bonding methods permitted include those for services including:

(2) For rigid and intermediate metal conduit, connections made up wrenchtight with threaded couplings or threaded bosses on enclosures.

(3) Threadless couplings and connectors made up tight for rigid metal and intermediate metal conduit and electrical metallic tubing.

(4) Other approved devices like bonding type locknuts and bushings.

In addition, bonding jumpers are permitted around concentric or eccentric knockouts that are punched or formed so as to impair the electrical connection to ground.

An exception to the requirement for bonding for these circuits provides that where oversized, concentric or eccentric knockouts are not encountered, or where concentric or eccentric knockouts have been tested and the box or enclosure is listed for the use, the following methods of ensuring continuity for these connections are permitted:

(a) threadless couplings and connectors for cables with metal sheaths.

(b) for rigid and intermediate metal conduit, two locknuts, one inside and the other outside the boxes and enclosures.

(c) fittings that seat firmly against the box or enclosure or cabinet such as for electrical metallic tubing, flexible metal conduit and cable connectors, with one locknut inside the enclosure, or

(d) listed fittings.

Outlet boxes are available that have specially designed and tested knockouts that perform satisfactorily for over 250 volt-to-ground applications. These boxes also are required to be listed for this purpose. It is required that listed fittings that are used to comply with the exception to the rule for bonding comply with the requirements of suitability found in Section 110-3.

Oversized knockouts

The installer needs to be cautious in the use of equipment that has concentric or eccentric knockouts as their ability to carry fault current must be of concern. It is very common to find nibs of adjacent rings damaged during removal of the desired knockout. This leaves less material available for carrying fault current. The safest practice is to install bonding bushings around concentric and eccentric knockouts where there is any question about their integrity.

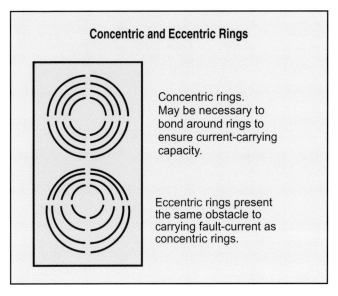

Figure 8-5

In other areas, where oversized, concentric or eccentric knockouts are not present, threadless fittings which are made up tight with conduit or armored cable or the use of two locknuts, one inside and one outside of boxes and cabinets, are acceptable for bonding.

Concentric and eccentric knockouts in equipment such as panelboards, auxiliary gutters and wireways are not tested or certified by an electrical products testing laboratory for their current-carrying ability. As such, they should not be relied upon for such a purpose and should have bonding bushings and bonding jumpers installed around them.

Where loosely-jointed metal raceways are used and especially where there may be expansion joints or telescoping sections of raceways, the Code requires that they be made electrically continuous by the use of equipment bonding jumpers or other means. See Section 250-98.

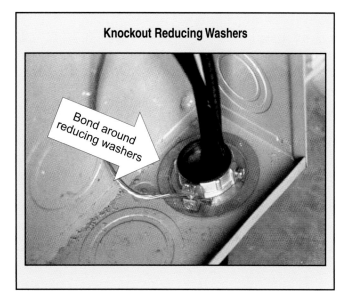

Knockout Reducing Washers

Bond around reducing washers

Figure 8-6

Reducing washers

Reducing washers are commonly used in electrical installations where it is desirable or necessary to install conduit or fittings of a size that is smaller than the knockout available in the enclosure. These reducing washers are not suitable for bonding and must always be bonded around to provide an adequate fault-return path.

Fault currents

Those equipment bonding jumpers form a part of the equipment grounding conductor and are, therefore, called upon to carry the same fault current that the equipment grounding conductor would carry and must be the same size.

The size of the bonding jumper will depend on its location and is always based on the size of the nearest overcurrent device in the circuit ahead of the equipment. See Section 250-102(d). Column 1 of Table 250-122 gives the size or setting of the overcurrent device in the circuit ahead of the equipment. Columns 2 and 3 give the minimum size of the equipment grounding conductor whether copper or aluminum.

Attaching jumpers

Where bonding jumpers are used between grounding electrodes or around water meters and the like, the Code requires that good electrical contact be maintained and that the arrangement of conductors be such that the disconnection or removal of equipment will not interfere with or interrupt the grounding continuity of the jumper. See Section 250-68(b).

Bonding jumpers shall be attached to circuits and equipment by means of exothermic welding, listed pressure connectors, listed clamps or other suitable and listed means. A connection that depends solely on solder is not acceptable. Sheet metal screws are not permitted to connect grounding conductors to enclosures. See Section 250-8.

Bonding multiple raceway systems

Where more than one raceway enters or leaves a switchboard, pull or junction box or other equipment, it is permissible to use a single conductor to bond these raceways to the equipment. The equipment grounding (bonding) conductor is sized for the largest overcurrent device ahead of conductors contained in the raceways. See Section 250-102(d). These are feeder or branch circuit conductors and not service-entrance conductors.

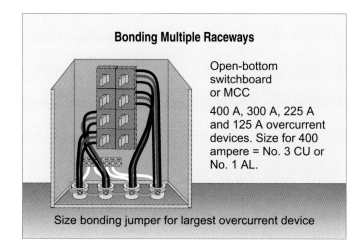

Bonding Multiple Raceways

Open-bottom switchboard or MCC

400 A, 300 A, 225 A and 125 A overcurrent devices. Size for 400 ampere = No. 3 CU or No. 1 AL.

Size bonding jumper for largest overcurrent device

Figure 8-7

For example, as shown in Figure 8-7, four metallic raceways leave the bottom of an open switchboard or motor control center. The overcurrent protective devices ahead of the raceways are 400, 300, 225 and 125 amperes respectfully. According to Table 250-122, the minimum size equipment bonding jumper for the raceway having conductors protected at 400 amperes is No. 3 copper or No. 1 aluminum. If this conductor were looped through a grounding bushing on each raceway, compliance with the Code would be obtained. Of course, the grounding bushings would need to be listed for both the size of conduit and conductor.

In addition, as shown in Figure 8-8, it is acceptable to install an equipment bonding jumper individually from each raceway to the equipment grounding terminals of the equipment. Each bonding conductor is sized for the overcurrent device

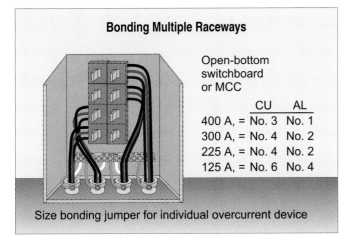

Figure 8-8

ahead of the conductors in the raceway per Table 250-122. In some cases, this method may require a larger conductor to bond some conduits than where individual bonding jumpers are installed.

Nonmetallic enclosures

Section 370-3 permits metal raceways or metal-armored cables to be used with nonmetallic enclosures only where:

1. internal bonding means are provided between all raceways or metal-armored cables.
2. internal bonding means with provision for attaching a grounding jumper inside the box are provided.

The bonding jumpers must, of course, be sized in accordance with Table 250-122. It should be noted that the size of the bonding jumpers given in Section 250-122 is the minimum size. Larger bonding jumpers may be required to comply with the available fault-current requirements in Section 110-10.

Bonding receptacles

An equipment bonding jumper is required to connect the grounding terminal of a grounding type receptacle to a grounded box. See NEC® Section 250-146. Where more than one equipment grounding conductor enters a box, they must be spliced or joined inside the box with suitable devices. Four exceptions to the general rule are provided in subsections (a) through (d).

(a) Where the box is mounted on or at the surface and direct metal-to-metal contact is made between the receptacle and the box. Note that the rule will now permit receptacles without an equipment bonding jumper where the box is mounted "at" the

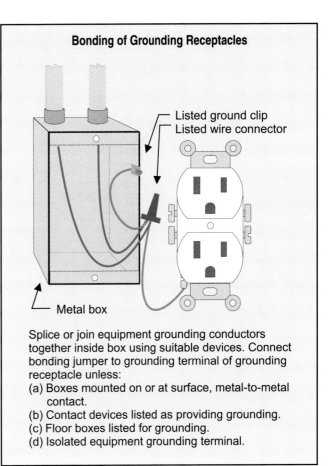

Splice or join equipment grounding conductors together inside box using suitable devices. Connect bonding jumper to grounding terminal of grounding receptacle unless:
(a) Boxes mounted on or at surface, metal-to-metal contact.
(b) Contact devices listed as providing grounding.
(c) Floor boxes listed for grounding.
(d) Isolated equipment grounding terminal.

Figure 8-9

surface in addition to being installed "on" the surface. Cover mounted receptacles such as in a raised cover on 4-in. square boxes are not approved unless the box and cover combination are listed as providing adequate ground continuity.

(b) Contact devices or yokes designed and listed as providing bonding with the mounting screws to establish the grounding circuit

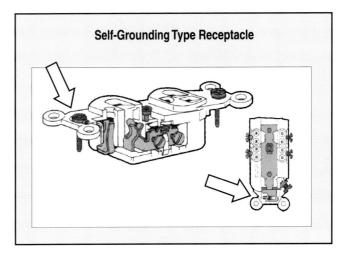

Figure 8-10. *Courtesy of Eagle Electric Co.*

between the device yoke and flush-type boxes. This is the device commonly referred to as a "self-grounding receptacle." The device is designed and listed as maintaining good electrical contact between the yoke and box by means of a spring-type device that maintains good continuity between the device and the mounting screws.

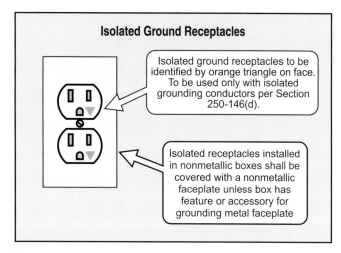

Isolated Ground Receptacles

Isolated ground receptacles to be identified by orange triangle on face. To be used only with isolated grounding conductors per Section 250-146(d).

Isolated receptacles installed in nonmetallic boxes shall be covered with a nonmetallic faceplate unless box has feature or accessory for grounding metal faceplate

Figure 8-11

(c) Floor boxes that are listed as providing satisfactory grounding continuity.

(d) A receptacle having an isolated equipment grounding terminal for reduction of electromagnetic interference. In this case, the equipment grounding terminal must be grounded by an insulated equipment grounding conductor that is run with the circuit conductors. This equipment grounding conductor is permitted to pass through one or more panelboards to terminate at an equipment grounding terminal of the separately derived system or service. However, this insulated equipment grounding conductor must be connected to the equipment grounding terminal bar at the building disconnecting means or the source of a separately derived system within the same building or structure.

To provide effective grounding, the insulated equipment grounding conductor must never pass without being terminated at the separately derived system that is the source of power for the equipment being grounded.

Also note that special rules are provided in Section 410-56(c) for grounding metal receptacle covers where the isolated ground receptacle is installed in a nonmetallic outlet box.

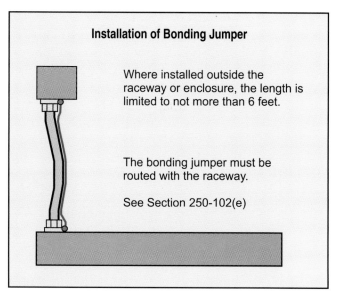

Installation of Bonding Jumper

Where installed outside the raceway or enclosure, the length is limited to not more than 6 feet.

The bonding jumper must be routed with the raceway.

See Section 250-102(e)

Figure 8-12

Installation of equipment bonding jumper

The equipment bonding jumper is permitted to be installed either inside or outside of a raceway or enclosure. See Section 250-102(e).

Where the jumper is installed on the outside, the length is limited to not more than 6 feet. In addition, the bonding jumper must be routed with the raceway. This is vital to keep the impedance of the equipment bonding jumper as low as possible. See Figure 8-12.

Bonding of piping systems

Section 250-104 requires that interior metal

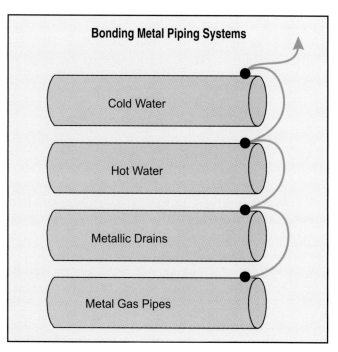

Bonding Metal Piping Systems

Cold Water

Hot Water

Metallic Drains

Metal Gas Pipes

Figure 8-13

water piping and other metal piping systems be bonded. This requirement for bonding is not to be confused with the requirement in Section 250-50 that metal underground water piping is to be used as a grounding electrode. Some requirements change depending upon whether the piping is interior metal water piping or other metal piping systems.

Included among the items within a building that should be adequately grounded to the one common grounding electrode system are the interior water piping system (hot and cold) and gas and sewer piping, any metallic air ducts within the building, as well as such devices as TV towers, gutters provided with a deicing system, etc.

The bonding conductor is generally required to be sized in accordance with Table 250-66. Also it must be installed in accordance with the general rules for installing grounding conductors in Sections 250-64(a), (b) and (e). The point(s) of attachment of the bonding conductor to the water piping system is required to be accessible.

enclosure, the grounded conductor at the service, the grounding electrode conductor where large enough, or to one or more grounding electrodes.

For example, if Table 250-66 requires a No. 2 bonding jumper to the metal water piping, it cannot be connected to a No. 6 or No. 4 grounding electrode conductor.

Where a metallic underground water piping system exists and is connected to a metallic interior water-piping system and there is not an insulated coupling, the interior water piping system is automatically and adequately grounded (bonded) when the metallic underground water piping system is used as the grounding electrode. However, with the expanding use of nonmetallic piping and insulated couplings it becomes more important to be sure that the interior piping not only is electrically continuous, but that also it is adequately grounded by bonding it to the same grounding electrode used for the premises. That is a mandatory and essential requirement of the Code.

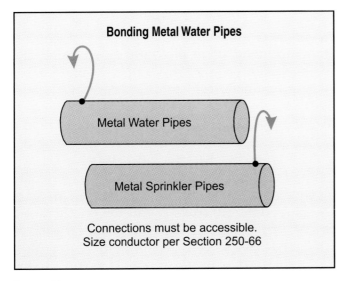

Bonding Metal Water Pipes

Metal Water Pipes

Metal Sprinkler Pipes

Connections must be accessible.
Size conductor per Section 250-66

Figure 8-14

Metal water piping

This requirement for bonding applies to all interior metal water piping systems. See Section 250-104(a). The bonding jumper in the building where the service is located is generally required to be sized in accordance with Table 250-66 and, thus, is based on the size of the service-entrance conductor and not on the rating of the service overcurrent device. In addition, the points of attachment of the bonding jumper to the metal water piping system are required to be accessible. The piping system is permitted to be bonded to the service equipment

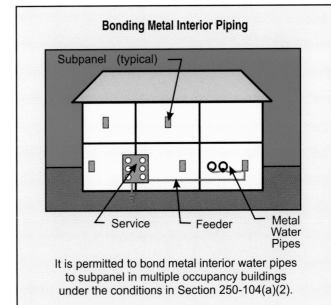

Bonding Metal Interior Piping

Subpanel (typical)

Service   Feeder   Metal Water Pipes

It is permitted to bond metal interior water pipes to subpanel in multiple occupancy buildings under the conditions in Section 250-104(a)(2).

Figure 8-15

Multiple occupancy building

Section 250-104(a)(2) allows the metal interior water piping system to be bonded to the panelboard or switchboard enclosure (other than service equipment) under specific conditions. The conditions are as follows:

(1) The building is multiple occupancy, and
(2) The metallic water piping is isolated from all other occupancies by nonmetallic water piping.

In this case, the bonding jumper to the water piping is sized in accordance with Table 250-122. Thus, the ampere rating of the overcurrent device supplying the feeder to the unit or occupancy determines the minimum size of the bonding conductor. The bonding jumper for the interior metal water piping runs from the equipment grounding terminal bar in the panelboard serving the unit to the piping. In this case, the bonding jumper does not connect to the neutral terminal bar in the panelboard.

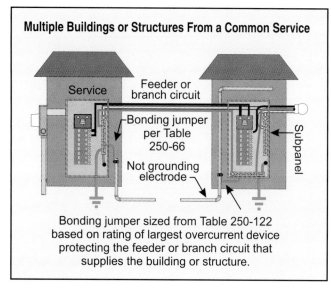

Figure 8-16

Multiple buildings or structures supplied from a common service

Where a building or structure is supplied from a common service by a feeder or branch circuit, bonding of interior metal water piping systems must be ensured by one of the following methods:

(1) Bonding to the building or structure disconnecting means where it is located at the building or structure.

(2) Bonding to the equipment grounding conductor that is run with the supply conductors to the building or structure. This connection would usually be made inside the building or structure disconnecting means enclosure on the equipment grounding terminal bar. Note that the equipment grounding conductor is permitted to consist of the wiring method that supplies the building or structure if recognized in Section 250-118.

(3) Bonding to the one or more grounding electrodes (grounding electrode system) used.

The bonding jumper to the interior water piping is required to be sized according to Table 250-122 based on the rating of the overcurrent device on the line side of the feeder or branch circuit.

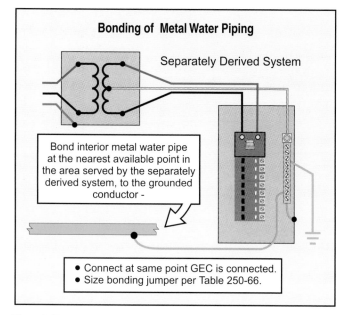

Figure 8-17

Bonding metal water piping to a separately derived system

The grounded conductor of a separately derived system is required to be bonded to the nearest available point on the interior metal water piping system in the area served by the separately derived system. See Section 250-104(a)(4).

The bonding connection must be made to the same point on the separately derived system where the grounding electrode conductor is connected. See the rules on separately derived systems in Chapter 12 of this text.

This requirement assures that metal piping, such as process or manufacturing equipment piping, in the area served by the separately derived system is bonded. This ensures an adequate and low-impedance path for fault current to return to its source which is the separately derived system, not the service to the building or structure.

Bonding metal gas piping

Each aboveground portion of a metal gas piping system upstream (on the supply side of) the equipment shutoff valve must be made electrically continuous and is required to be bonded to the grounding electrode system. See Section 250-104(b) of the NEC®. This requirement in the

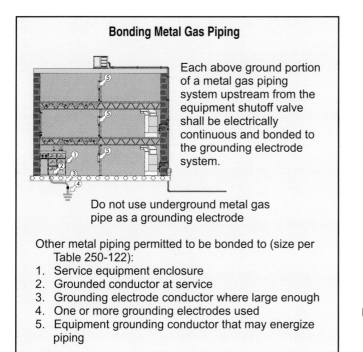

Bonding Metal Gas Piping

Each above ground portion of a metal gas piping system upstream from the equipment shutoff valve shall be electrically continuous and bonded to the grounding electrode system.

Do not use underground metal gas pipe as a grounding electrode

Other metal piping permitted to be bonded to (size per Table 250-122):
1. Service equipment enclosure
2. Grounded conductor at service
3. Grounding electrode conductor where large enough
4. One or more grounding electrodes used
5. Equipment grounding conductor that may energize piping

Figure 8-18

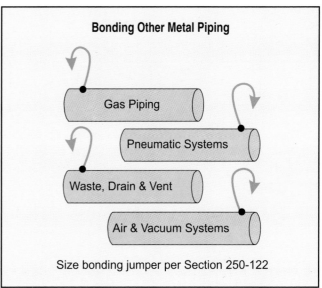

Bonding Other Metal Piping

Gas Piping

Pneumatic Systems

Waste, Drain & Vent

Air & Vacuum Systems

Size bonding jumper per Section 250-122

Figure 8-19

NEC® is intended to support a similar requirement in Section 3.14(a) of NFPA 54-1996, the *National Fuel Gas Code.*

Note that this requirement applies to both indoor and outdoor metal gas piping systems.

The Code rule is silent on requirements for sizing the bonding conductor. It is recommended that where metal gas piping is installed on the exterior of buildings or structures, the bonding conductor be sized according to Table 250-66 to allow for exposure to lightning storms. In addition, compliance with NFPA 780, the *Standard for the Installation of Lightning Protection Systems,* is recommended.

Where the metal gas piping is installed entirely within a building or structure, the sizing rules of Section 250-104(c) can be considered. This subsection permits the bonding conductor to be based on the circuit that is likely to energize the piping. In this case, the bonding jumper is determined from Table 250-122 based on the rating of the overcurrent device ahead of the circuit.

Bonding other metal piping

Other interior metal piping which may become energized must be bonded. The Code does not give guidance on how to determine the conditions under which metal piping is likely to become energized. Since metal piping systems are conductive,

from a safety standpoint, all metal piping systems should be bonded.

Common systems that must be bonded include interior metal:

Pneumatic systems
Waste, drain and vent lines
Oxygen, air and vacuum systems.

The bonding conductor is sized from Table 250-122 using the rating of the overcurrent device in the circuit ahead of the equipment. The equipment grounding conductor for the circuit that may energize the equipment can be used as the bonding conductor. The point of connection of these bonding conductors to the metal piping systems is not required to be accessible as the connections to metal water piping systems are.

The bonding conductor is permitted to be connected to the:
(1) Service equipment enclosure.
(2) Grounded conductor at the service.
(3) Grounding electrode conductor, where of adequate size.
(4) One or more of the grounding electrodes used.

Bonding structural steel

Exposed interior structural steel that is interconnected to form a steel building frame and is not intentionally grounded and may become energized is required to be bonded. Bonding must be to the service equipment enclosure, the grounded conductor at the service, the grounding electrode

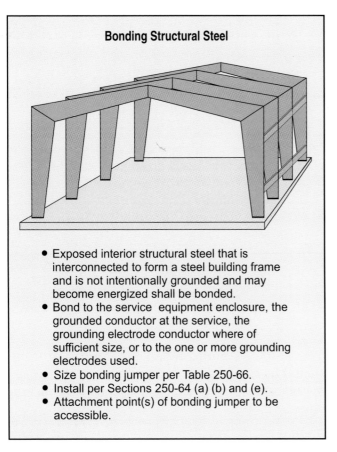

Bonding Structural Steel

- Exposed interior structural steel that is interconnected to form a steel building frame and is not intentionally grounded and may become energized shall be bonded.
- Bond to the service equipment enclosure, the grounded conductor at the service, the grounding electrode conductor where of sufficient size, or to the one or more grounding electrodes used.
- Size bonding jumper per Table 250-66.
- Install per Sections 250-64 (a) (b) and (e).
- Attachment point(s) of bonding jumper to be accessible.

Figure 8-20

conductor where it is large enough, or to the one or more grounding electrodes used.

The bonding jumper must be sized in accordance with Table 250-66 and installed in accordance with the rules in Section 250-64(a), (b) and (e). The points of attachment of the bonding jumper to the structural steel shall be accessible.

Chapter Eight: The questions included here were developed using material included in this chapter. The answers can be found by reviewing the text. It is also important that students make use of the 1999 NEC®, where many answers can be found. See page 279 for answers.

1. For systems having a voltage exceeding ____ volts to ground, the electrical continuity of metal raceways or metal-sheathed cables, that are not service-entrance cable, must also be assured.
   a. 125
   b. 150
   c. 100
   d. 250

2. Bonding jumpers are required to be attached to circuits and equipment by means of ____, or other listed means.
   a. exothermic welding
   b. listed pressure connectors
   c. listed clamps
   d. any of the above

3. Metal raceways or metal-jacketed cables are permitted to be used with nonmetallic enclosures only where ____.
   a. internal bonding means are provided between all raceways
   b. internal bonding means with provision for attaching a grounding jumper inside the box are provided
   c. internal bonding means are provided between all cables
   d. any of the above

4. Without exception, where more than one equipment grounding conductor enters a box, they must be spliced or joined with____ devices.
   a. labeled
   b. listed
   c. suitable
   d. bonding

5. An equipment grounding conductor is permitted to pass through one or more panelboards within the same ____ to terminate at an equipment grounding terminal of the separately derived system or service.
   a. equipment
   b. building
   c. enclosure
   d. cabinet

6. Where the equipment bonding jumper is installed on the outside of a raceway, the length is limited to not more than ____ feet. In addition, the bonding jumper must be routed with the raceway.
   a. 7
   b. 8
   c. 6
   d. 10

7. The metal water piping system is permitted to be bonded to the service equipment enclosure, the grounded conductor at the service, the grounding electrode conductor where large enough, or to one or more grounding ____.
   a. clamps
   b. devices
   c. fittings
   d. electrodes

8. Common systems that must be bonded include interior metal ____.
   a. gas piping and pneumatic systems
   b. waste, drain and vent lines
   c. oxygen, air and vacuum systems
   d. all of the above

9. The equipment grounding conductor for the circuit that may energize the equipment can be used as the ____ conductor.
   a. equipment bonding
   b. grounded
   c. bonding
   d. identified

10. Expansion joints or telescoping sections of metal raceways must be made electrically continuous by the use of a ____.
   a. steel strap
   b. equipment bonding jumper
   c. welding cable
   d. equipment grounding conductor

11. The locknut/bushing and double-locknut types of installations are not acceptable for bonding in ____.
   a. hazardous (classified) locations
   b. commercial locations
   c. industrial location
   d. computer rooms

12. If four metal raceways have their conductors protected by overcurrent protective devices sized at 400, 300, 225 and 125 amperes and leave the bottom of an open switchboard or motor control center, the minimum size of a single equipment bonding jumper to be used to bond them together is ____.
   a. No. 4 copper or No. 4 aluminum
   b. No. 6 copper-clad aluminum
   c. No. 3 copper or No. 1 aluminum
   d. No. 4 copper or No. 3 aluminum

13. The points of attachment of the bonding jumper to the interior metal water piping system are required to be ____.
   a. acceptable
   b. marked
   c. accessible
   d. soldered

14. Concentric and eccentric knockouts in enclosures like wireways and panelboards:
   a. are always suitable for grounding and bonding.
   b. are tested by a qualified electrical testing laboratory for their current-carrying ability.
   c. are not tested by a qualified electrical testing laboratory for their current-carrying ability.
   d. are capable of carrying large amounts of fault current.

15. Bonding connections to interior metal water piping systems:
   a. are permitted to be made with solder.
   b. must be accessible.
   c. are permitted to be connected to the neutral in a subpanel.
   d. are not required.

16. Structural steel must be bonded where:
   a. it is exposed on the interior of the building.
   b. it is interconnected to form a steel building frame.
   c. it is not intentionally grounded.
   d. all of the above.

Equipment Grounding Conductors

Objectives

After studying this chapter, the reader will be able to understand:

- General requirements for equipment grounding conductors on grounded and ungrounded systems.
- Rules applied to multiple raceways or cables.
- Rules for flexible cords.
- Use of building steel that is properly grounded by an equipment grounding conductor.
- Grounding of equipment by the grounded circuit conductor.

Definition

"**Grounding conductor, equipment.** The conductor used to connect the noncurrent-carrying metal parts of equipment, raceways, and other enclosures to the system grounded conductor, the grounding electrode conductor, or both, at the service equipment or at the source of a separately derived system."[N]

Equipment Grounding Conductor Path

Must:

• be permanent and electrically continuous,
• have ample capacity to conduct safely any currents likely to be imposed on it,
• be of the lowest practical impedance.

General requirements

Equipment grounding conductors are intended to prevent an objectionable potential above ground on conductor and equipment enclosures and to provide a low-impedance path for fault-currents by connecting the equipment grounding conductor to the grounded conductor if it is a grounded system. The equipment grounding conductor or path must also:

(1) Be permanent and electrically continuous;
(2) Have ample capacity to conduct safely any currents likely to be imposed on it; and
(3) Be of the lowest practical impedance. See Section 250-2(d).

The equipment grounding conductor or path must thus connect with the grounding electrode in a low-impedance path, and if the system is grounded, it must also be connected through a low-impedance path to the grounded service conductor (often a neutral).

Grounded system

The equipment grounding conductor or path must extend from the furthermost point on the circuit to the service equipment where it is connected to the grounded conductor. This connection is made through the main bonding jumper. Often, this equipment grounding conductor or path is the conductor enclosure (conduit, cable jacket, etc.).

For enclosed panelboards typically installed at dwelling services, the grounded and equipment

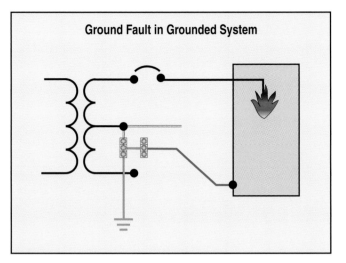

Ground Fault in Grounded System

Figure 9-1

grounding conductors connect to the same terminal bar. See Section 384-20. Where this type of equipment is installed, the main bonding jumper connects the enclosure to the grounded service conductor as well as a separate equipment grounding conductor terminal bar where one is installed.

Should insulation failure occur anywhere on a phase conductor and a ground fault develop between the energized conductor and the conductor enclosure, a ground-fault circuit will be established.

The ground-fault circuit will thus be from the source, through the supply conductors, through the overcurrent devices, to the point of fault on the phase conductor, usually through an arc, through the equipment grounding conductor, through the main bonding jumper, to the grounded conductor or neutral and back to the source.

If this circuit is complete, of adequate capacity and low impedance, the equipment and persons who may contact it are protected. A break in this equipment grounding circuit or other grounding system failure will expose persons to possibly lethal shocks if a ground fault from a source having sufficient potential (usually more than 50 volts) energizes the enclosure and the person provides the path for current to flow through.

Ungrounded systems

In an ungrounded system, the equipment grounding conductor or path must be permanent and continuous to ground to keep all equipment and conductor enclosures at or near ground potential. The equipment grounding conductor must be the same size as called for in a grounded system if we are to have maximum safety. The sizes of equip –

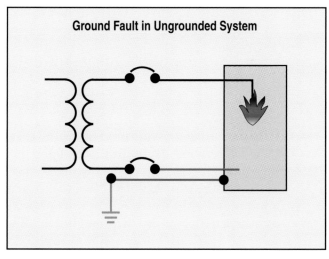

Figure 9-2

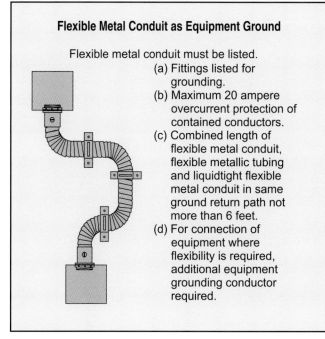

Figure 9-3

ment grounding conductors given in Table 250-122 apply equally to an ungrounded system and a grounded system.

In general, it may be said that any conductor or equipment enclosure, be it conduit, electrical metallic tubing, raceway or busway enclosure, provides a satisfactory equipment grounding conductor for an ungrounded system if all joints are made electrically continuous. It may be necessary to use bonding jumpers at certain points. Such bonding jumpers must also be sized per Table 250-122.

Equipment grounding conductor material

The Code specifies in Section 250-118 the conductors that are permitted to be used for equipment grounding conductors. They are as follows:

(1) A conductor of copper or other corrosion-resistant material such as aluminum. The conductor is generally permitted to be solid, stranded, insulated, covered or bare in the form of a wire or busbar of any type (some sections of the Code may specifically require an equipment grounding conductor to be insulated or solid);

(2) Rigid metal conduit.

(3) Intermediate metal conduit.

(4) Electrical metallic tubing.

(5) Flexible metal conduit where both the conduit and fittings are listed for grounding.

(6) Listed flexible metal conduit that is not listed for grounding, meeting all the following conditions:

    (a) The conduit is terminated in fittings listed for grounding

    (b) The circuit conductors contained in the conduit are protected by overcurrent

devices rated at 20 amperes or less.

    (c) The combined length of flexible metal conduit and flexible metallic tubing and liquidtight flexible metal conduit in the same ground return path does not exceed 6 feet.

    (d) The conduit is not installed for flexibility. Where installed for flexibility, NEC® Section 350-14 requires that an equipment grounding conductor be installed.

Flexible metal conduit is commonly available as a listed product but has not been listed for grounding by a nationally recognized electrical testing laboratory. That is the reason for the phrase, "Listed flexible metal conduit that is not listed for grounding." The conduit has been recognized for several years for use as an equipment grounding conductor under the limitations indicated above. Note, however, the flexible metal conduit must be terminated in fittings that are listed for grounding.

(7) Listed liquidtight flexible metal conduit meeting all the following conditions:

    (a) The conduit is terminated in fittings that are listed for grounding.

    (b) For trade sizes 3/8 in. through ½ in., the circuit conductors are protected by overcurrent devices rated at 20 amperes or less.

    (c) For trade sizes ¾ in. through 1¼ in. the circuit conductors contained in the con-

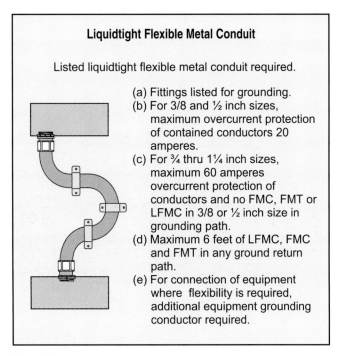

Liquidtight Flexible Metal Conduit

Listed liquidtight flexible metal conduit required.

(a) Fittings listed for grounding.
(b) For 3/8 and ½ inch sizes, maximum overcurrent protection of contained conductors 20 amperes.
(c) For ¾ thru 1¼ inch sizes, maximum 60 amperes overcurrent protection of conductors and no FMC, FMT or LFMC in 3/8 or ½ inch size in grounding path.
(d) Maximum 6 feet of LFMC, FMC and FMT in any ground return path.
(e) For connection of equipment where flexibility is required, additional equipment grounding conductor required.

Figure 9-4

duit are protected by overcurrent devices rated not more than 60 amperes and there is no flexible metal conduit, flexible metallic tubing, or liquidtight flexible metal conduit in trade sizes 3/8 in. or ½ in. in the grounding path.

(d) The combined length of flexible metal conduit and flexible metallic tubing and liquidtight flexible metal conduit in the same ground return path does not exceed 6 feet.

(e) The conduit is not installed for flexibility. Where installed for flexibility, NEC® Section 351-9 requires that an equipment grounding conductor be installed.

(8) Flexible metallic tubing where the tubing is terminated in fittings listed for grounding and meeting all the following conditions:

(a) The circuit conductors contained in the tubing are protected by overcurrent devices rated at 20 amperes or less.

(b) The combined length of flexible metal conduit and flexible metallic tubing and liquidtight flexible metal conduit in the same ground return path does not exceed 6 feet.

(9) Armor of Type AC cable as provided in Section 333-21.

(10) The copper sheath of mineral-insulated, metal sheathed cable.

(11) The metallic sheath or the combined metallic sheath and grounding conductors of Type MC cable.

(12) Cable trays as permitted in Sections 318-3(c) and 318-7.

(13) Cablebus framework as permitted in Section 365-2(a).

(14) Other electrically continuous metal raceways listed for grounding. Included are auxiliary gutters (not specifically a raceway as defined in Article 100 but the same equipment as wireways), wireways with associated fittings; busway enclosures, and in some cases and additional ground bus, surface metal raceways and pull and junction boxes that installed in the ground fault path.

It should be noted here that these requirements or provisions in Section 250-118 are general in nature. Many sections of the Code contain specific requirements that must be complied with. A few examples follow. Section 501-16 does not recognize double locknut-type conduit connections for Class I hazardous (classified) locations. Section 517-13(a) requires a supplementary insulated equipment grounding conductor in patient care areas of health care facilities. Section 550-24(a) generally requires the equipment grounding conductor for the feeder to a mobile home to be insulated. Several sections of Article 680 require an insulated equipment grounding conductor. It is always best to carefully examine the specific requirements for the equipment grounding conductor for the type of installation being made.

Conductor enclosures

Conduit runs of rigid or intermediate metal, which are properly threaded and in which the couplings are set up tightly, preferably using a joint sealer that will not reduce continuity, may be expected to perform satisfactorily as an equipment grounding conductor for runs of limited length. A joint sealer aids in assuring an effective equipment grounding path because it acts as a lubricant and permits the joint to be screwed up tighter. Under poor conditions, the conduit impedance with couplings should not show an increase of over 50 percent when compared with a straight run of conduit. The use of this higher impedance figure would provide a factor of safety. In the case cited, there is no economic justification for using an additional equipment grounding conductor.

The Code further requires that where conduit is used as a grounding conductor, all joints and fittings shall be made up tight using suitable tools. See Section 250-120(a). This calls attention to the fact that conduit, where used as an equipment grounding means, is a current-carrying conductor under fault conditions and must be made electrically continuous by having joints made up tight.

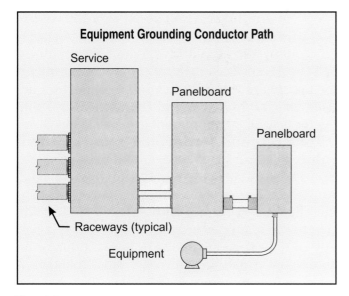

Equipment Grounding Conductor Path

Service

Panelboard

Panelboard

Raceways (typical)

Equipment

Figure 9-5

Usually, large and often parallel conduits are installed from the transformer to the service equipment. Then, smaller and smaller conduits are installed for feeders and branch circuits. For instance, at one point the equipment grounding path may be three 4-inch conduits in parallel; at another point, two 4-inch conduits in parallel while down the line it may be only one 1¼-inch conduit, all being connected together to form a permanent and continuous path. As the circuit changes from large overcurrent protective means to smaller ones, the conductivity of the equipment grounding path becomes smaller and smaller. The conduit or tubing at the end of the circuit may be no larger than ½-inch electrical metallic tubing or 3/8-inch flexible metal conduit.

The NEC® does not dictate any particular size of conduit or tubing to serve as the equipment grounding conductor for an upstream overcurrent device, other than as mentioned in the previous section of this text. It is felt that the metallic raceway that is sized properly for the conductor fill will provide an adequate equipment ground fault return path.

Cables as equipment grounding conductors

Several cables used as wiring methods are suitable for use as an equipment grounding conductor or contain an equipment grounding conductor. These include:

Figure 9-6. Type AC cable. *Photo courtesy of AFC Cable Systems.*

Type AC cable. Armored cable (Article 333 of the NEC®) is manufactured with conductors in sizes from No. 14 through No. 1 copper and from No. 12 through No. 1 aluminum. Type AC cable is required to have an armor of flexible metal tape. The insulated conductors are required to be in accordance with Section 333-20. Cables of the AC type are required to have an internal bonding strip of copper or aluminum in intimate contact with the

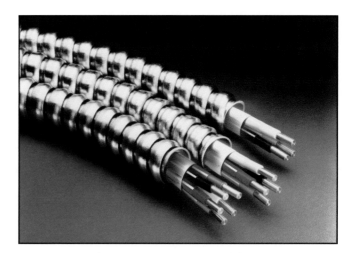

Figure 9-7. Type MC cable. *Photo courtesy of AFC Cable Systems.*

armor for its entire length. It is suitable as an equipment grounding conductor in accordance with Section 250-118(9).

In addition to the internal bonding strip, another means of identification of Type AC cable is the conductors are required to have an overall moisture-resistant and fire-retardant fibrous covering. Additional constructions include an insulated equipment grounding conductor and is acceptable in patient care areas as provided in Section 517-13 and for isolated equipment ground receptacles.

Type MC cable. Type MC cable is produced in three configurations: spiral interlocking metal tape, corrugated metal tube and a smooth metal tube.

(a) The spiral-interlocking-metal-tape Type MC cable must always have an equipment grounding conductor which may be insulated or bare. The jacket itself is not suitable as an equipment grounding conductor. The principal equipment grounding conductor may be divided (sectioned) into more than one conductor, often to facilitate spacing in the cable construction. Additional equipment grounding conductors have green insulation and either a yellow stripe or other identification.

(b) The sheath of the smooth or corrugated tube Type MC cable or a combination of the sheath and a supplemental bare or unstriped green insulated conductor is suitable for the required equipment grounding conductor. The principal equipment grounding conductor may be divided (sectioned) into more than one conductor, often to facilitate spacing in the cable construction. Additional equipment grounding conductors have green insulation and either a yellow stripe or other identification.

Nonmetallic-sheathed cable. This cable is permitted to be produced in three styles, Type NM, Type NMC and Type NMS. The power conductors are permitted to be in sizes No. 14 through No. 2 copper and No 12 through No. 2 aluminum. Type NMS is permitted to contain signaling conductors in compliance with Section 780-5. These cables typically contain an equipment grounding conductor sized in compliance with Table 250-122.

Service-entrance cable. Type SE cable is produced in a variety of configurations. The type most commonly used for internal wiring is Type SE style U and Type SE style R. Specific rules for Type SE cables are contained in Section 338-3.

(a) Type SE service-entrance cables are permitted in interior wiring systems where all of the circuit conductors of the cable are of the rubber-covered or thermoplastic type.

(b) Type SE service-entrance cables without individual insulation on the grounded circuit conductor are not to be used as a branch circuit or as a feeder within a building, except a cable that has a final nonmetallic outer covering and is supplied by alternating current at not over 150 volts to ground is permitted as a feeder to supply only other buildings on the same premises.

Type SE service-entrance cables are permitted for use where the fully insulated conductors are used for circuit wiring and the uninsulated conductor is used for equipment grounding purposes.

Underground feeder and branch circuit cable. Type UF cable is permitted to be produced in sizes No. 14 copper or No. 12 aluminum through No. 4/0. Mulitconductor cables are permitted to be installed in accordance with Article 336. In addition to the insulated conductors, the cable is permitted to have an insulated or bare conductor for equipment grounding purposes only. As such, it must be in compliance with Table 250-122.

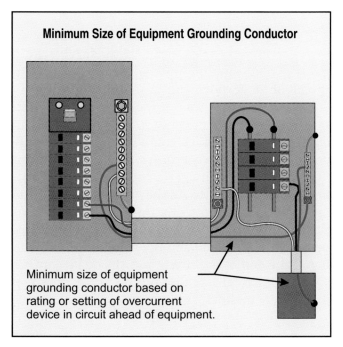

Minimum Size of Equipment Grounding Conductor

Minimum size of equipment grounding conductor based on rating or setting of overcurrent device in circuit ahead of equipment.

Figure 9-8

Size of equipment grounding conductor

The entire equipment grounding conductor or path of any raceway system will be as shown in Figure 9-8. Starting at the service we have a large overcurrent protective device which is in series with other and usually smaller feeder or branch overcurrent protection devices. The ungrounded (hot) conductor usually decreases in size as it progresses through smaller and smaller overcurrent devices.

Thus, a ground fault may be cleared through a 15 ampere overcurrent device so that the equipment ground path below the 15 ampere overcurrent device only needs to be a No. 18 copper wire (subject to Code limitations) as given in Section 250-122(e). That conductor has a short-time rating of about 38 amperes and it should thus be adequate to operate the 15-ampere overcurrent device if it operates in a reasonable time.

Section 250-122(a) provides the general rules for sizing the equipment grounding conductor. It refers to Table 250-122 for determining the minimum size of conductor that is permitted to be used as an equipment grounding conductor. The size is based on the ampere rating of the overcurrent protective device ahead of the conductor. (Table 250-122 is reprinted as Table 6 in Chapter 18 of this text.)

For example, if the overcurrent protection ahead of the circuit or feeder is 225 amperes, the minimum size equipment grounding conductor is found as follows:

In Table 250-122, follow the first column, which gives the rating of the overcurrent device, down to find the rating which equals or exceeds 225 amperes. Since 225 amperes is not found, go to the next larger size which is 300 amperes. Follow that line across to find the minimum size copper wire to be No. 4 and for aluminum, a No. 2 wire.

Follow a similar process to determine the minimum size conductor for any installation. In addition, the note below Table 250-122 requires that, "Where necessary to comply with Section 250-2(d), the equipment grounding conductor shall be sized larger than this table."[N] A comprehensive analysis of the withstand rating of these equipment grounding conductors can be found in Chapter 11 of this text.

Specific requirements are provided for: adjusting equipment grounding conductors for voltage drop in Section 250-122(b), for multiple circuits in Section 250-122(c), for motor circuits in Section 250-122(d), for flexible cord and fixture wire in Section 250-122(e), and for conductors in parallel in Section 250-122(f).

Adjusting size of equipment grounding conductor for voltage drop

Section 250-122(b) requires that, "Where conductors are adjusted in size to compensate for voltage drop, equipment grounding conductors, where installed, shall be adjusted proportionately according to circular mil area."[N] This means that where a feeder or branch-circuit conductor is increased in size to compensate for voltage drop, the equipment grounding conductor, where run, must be increased at not less than the same ratio the feeder or circuit conductors are increased.

For example, a 200 ampere feeder is to be installed. It is determined that the voltage drop would be excessive. A 250 kcmil conductor is selected for the feeder rather than installing the 3/0 copper conductor as is permitted by Table 310-16. Table 250-122 requires a No. 6 equipment grounding conductor for the 200 ampere overcurrent device.

Determine the minimum size equipment grounding conductor required for the feeder by the following formula: (Use Table 8 of Chapter 9 to determine the area in circular mils where the conductor size is given by a non-circular mil designation.)

$$\frac{\text{Selected Feeder Conductor Area}}{\text{Required Feeder Conductor Area}} = \text{Ratio}$$

Table 250-122 Equipment Grounding Conductor X Ratio = Required EGC

$$\frac{250000 \text{ kcmil}}{167800 \text{ kcmil}} = 1.49$$

.027 (Circular mil area of No. 6) x 1.49 = .04023 circular mils

Next larger size = No. 3 copper required equipment grounding conductor.

Equipment grounding conductors for multiple circuits

The Code permits a single equipment grounding conductor to serve several circuits that are in the same raceway or cable. To use this concept, the equipment grounding conductor must be sized for the rating of the largest overcurrent device of the group.

For example, a conduit contains branch circuit conductors that have conductors having overcurrent protection rated, 20-amperes, 30-amperes, 50-amperes and 60-amperes. A single No. 10 equipment grounding conductor is permitted to serve all the branch circuits in the raceway. The minimum size is determined from Table 250-122 based on the rating of the 60-ampere overcurrent device.

Equipment grounding conductors for motor circuits

The general rule for sizing the equipment grounding conductor for motor circuits is contained in Section 250-122 – determine the minimum size conductor from Table 250-122 based on the rating of the overcurrent protective device. In some cases, this will result in an equipment grounding conductor that is the same size as the branch circuit conductors. This is illustrated as follows: a 30 hp, 460-volt motor is being installed. From Table 430-150, the full-load amperes of the motor is 40 amperes. The minimum size branch circuit conductors is 40 amperes x 1.25 = 50 amperes, from Table 310-16 = No. 8 copper conductors (75°C insulation and terminations). Maximum rating of overcurrent device of a circuit breaker type is 250 percent of the motor full load amperes = 250 x 40 = 100 amperes (Table 430-152) unless the one of the exceptions to Section 430-52(c) applies. From Table 250-122, the minimum size of equipment grounding conductor based on a 100-ampere overcurrent device is No. 8 copper which is the same size as the branch circuit conductors.

Note that Section 250-122(a) provides that the size of the equipment grounding conductor is not required to be larger than the branch circuit conductors.

Where the overcurrent device for the motor consists of an instantaneous trip circuit breaker (rather than a more standard inverse-time circuit breaker) or a motor short-circuit protector, the equipment grounding conductor size is permitted to be based on the rating of the motor overload protective device but not less than Table 250-122. Note that the instantaneous trip circuit breaker is permitted to be used only if it is a part of a listed combination motor controller having coordinated motor overload protection. Using the above example, the instantaneous-trip circuit breaker that serves as the branch-circuit, short-circuit and ground-fault protective device is permitted to be up to 1300 percent of the motor full-load current (up to 1700 percent for Design E and Design B energy efficient motors). The minimum size branch circuit conductors is determined as 40 amperes x 1.25 = 50 amperes. The minimum conductor from Table 310-16 is a No. 8 copper conductors with

75°C insulation and terminations. The maximum rating of an overcurrent device of an instantaneous circuit breaker type is 1300 percent of the motor full load amperes = 40 x 13 = 520 amperes unless one of the exceptions following Section 430-52(c)(3) applies. However, assume the motor FLA on the nameplate is 38.2 amperes and the running overload protection for the motor is set at 115 percent. See Section 430-32(a). The equipment grounding conductor is permitted to be based on 43.93 amperes (38.2 x 1.15). From Table 250-122, the minimum size equipment grounding conductor based on a 50-ampere overcurrent device is No. 10 copper.

Equipment grounding conductors for flexible cord and fixture wire

The use of an equipment grounding conductor in a supply cord is permitted providing the cord is connected as specified in Section 400-7. The method of grounding noncurrent-carrying metal parts of portable equipment may be by means of the equipment grounding conductors in the flexible cord feeding such equipment. The proper type attachment plug must be used to terminate the conductors and the attachment plug must also have provision to make contact with a grounding terminal in the receptacle.

For the grounding of portable or pendant equipment, the conductors of which are protected by fuses or circuit breakers rated or set at not exceeding 20 amperes, the Code (Section 240-4) permits the use of a No. 18 copper wire as an equipment grounding circuit conductor, provided the No. 18 grounding conductor is a part of a listed flexible cord assembly.

See Section 250-122(a) where the equipment grounding conductor is not required to be larger than the circuit conductors.

Equipment grounding conductors in parallel

Special rules apply where more than one raceway or cable is installed with parallel conductors and an equipment grounding conductor is installed in the raceway. (Parallel conductors consist of two or more conductors that comply with Section 310-4 and are connected together at each end to form a single conductor.) In this case, Section 250-122(f) requires that an equipment grounding conductor be installed in each raceway or cable. Generally, each equipment grounding conductor must be sized in compliance with the ampere rating of the overcurrent device protecting the conductors in the raceway or cable.

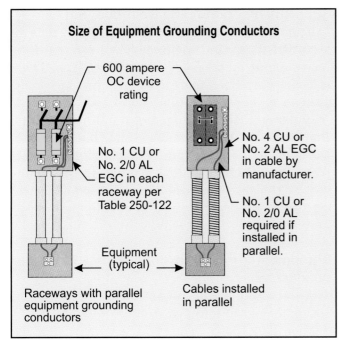

Size of Equipment Grounding Conductors

600 ampere OC device rating

No. 1 CU or No. 2/0 AL EGC in each raceway per Table 250-122

No. 4 CU or No. 2 AL EGC in cable by manufacturer.

No. 1 CU or No. 2/0 AL required if installed in parallel.

Equipment (typical)

Raceways with parallel equipment grounding conductors

Cables installed in parallel

Figure 9-9

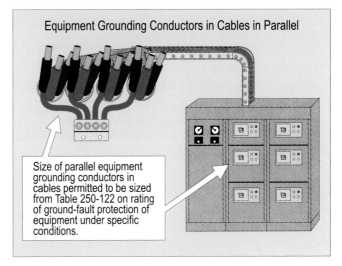

Equipment Grounding Conductors in Cables in Parallel

Size of parallel equipment grounding conductors in cables permitted to be sized from Table 250-122 on rating of ground-fault protection of equipment under specific conditions.

Figure 9-10

Section 310-4 permits equipment grounding conductors, smaller than 1/0, to be sized in compliance with Table 250-122. However, all other requirements for installing conductors in parallel must be met. These rules require that each set: (1) be the same length; (2) be of the same conductor material (all copper or all aluminum); (3) be the same size in circular mil area; (4) have the same insulation type; (5) be terminated in the same manner; and (6) the raceways or cables must have the same physical properties.

One reason for this requirement for installing equipment grounding conductors in parallel is shown in Figure 9-9. In the event of a line-to-ground fault in the equipment supplied by the circuit, the fault current will divide equally between the equipment grounding conductors. However, should a line-to-ground fault occur in the raceway or cable, current will be fed to the fault from both directions. The equipment grounding conductor will thus be called upon to carry the entire amount of fault current until the overcurrent protective device ahead of the fault opens.

Equipment grounding conductors in cables in parallel

In some cases, where cables are installed in parallel, special constructions will be required to get the larger equipment grounding conductor in the cable. Listed cables are generally produced

with the equipment grounding conductors sized in compliance with a construction standard that complies with Table 250-122. For example, a copper cable construction suitable for a 300 ampere overcurrent device will have a No. 4 copper equipment grounding conductor within the cable by the manufacturer. If two of these cables are installed in parallel and connected to a 600 ampere overcurrent protective device, a No. 1 copper equipment grounding conductor would be required in each cable to comply with Table 250-122. These "special" cables can be ordered from the manufacturer although conditions such as minimum length requirements may apply.

Rules are now in place that permit an equipment ground fault protection system to provide protection for conductors in cables where the equipment grounding conductors are too small for the overcurrent device on the line side of the conductors. Conditions that must be met to use this concept include:

1. Conditions of maintenance and supervision ensure that only qualified persons will service the installation.
2. The ground-fault protection equipment is set to trip at not more than the ampacity of a single ungrounded conductor of one of the cables in parallel.
3. The ground-fault protection is listed for the purpose.

This is an effort to mitigate the impact of requiring, that where multiconductor cables were installed in parallel, they were required to have an equipment grounding conductor that is based on the rating of the overcurrent device. This requirement resulted in an equipment grounding conductor larger than normally is available in listed cables.

For example, a cable rated for 200 amperes will typically have a No. 6 copper or No. 4 aluminum equipment grounding conductor in it. If three of these cables are installed in parallel and have overcurrent protection rated at 600 amperes, Table 250-122 requires a No. 1 copper or No. 2/0 aluminum equipment grounding conductor in each cable.

This new provision allows the standard size equipment-grounding conductor in multiconductor cables that are connected in parallel to be protected by ground fault protection of equipment (not GFCI protection for personnel). The ground-fault protection is required to be set at the ampacity of a single ungrounded circuit conductor not on the ampacity of the equipment-grounding conductor.

For example, a 1600 ampere feeder is installed with four cables each having an ampacity of 400 amperes. The equipment ground fault protection for these cables would have to be set at not more than 400 amperes. These cables would typically have a No. 3 copper or No. 1 aluminum equipment-grounding conductor according to Table 250-122. The overcurrent protection for the feeder would be 1600 amperes.

In a circuit without equipment ground fault protection, the equipment grounding conductor would typically be called upon to carry several times the rating of the overcurrent device on the line side of the circuit until the overcurrent device clears the fault. As a result, this new provision is more conservative than circuits installed without equipment ground fault protection.

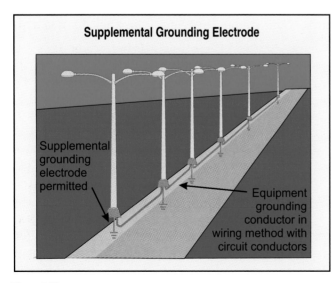

Supplemental Grounding Electrode

Supplemental grounding electrode permitted

Equipment grounding conductor in wiring method with circuit conductors

Figure 9-11

Supplemental grounding electrode

Engineers often specify that ground rods be installed to ground metal lighting standards or poles and at metal poles for electric signs. These rods are permitted to be used but must be considered to be supplemental ground rods. They can "supplement" the equipment grounding conductor that is run with the branch circuit but cannot be the only means of grounding this or similar equipment. See Section 250-54. To use these ground rods as the only means of grounding would constitute an earth return which is unsafe and prohibited by Code.

Equipment grounding conductor with circuit conductors

A very important requirement for installing equipment grounding conductors is contained in Section 250-134(b). The requirement is that the equipment grounding conductor must generally be installed in the same raceway, cable or cord, or otherwise be run with the circuit conductors. This requirement is repeated in Section 300-3(b) where, in addition to the requirement for raceways, equipment grounding conductors are required to be contained in the same trench with other circuit conductors. This requirement is critical for the installation of alternating current systems.

It has been proven that separating the equipment grounding conductor from the circuit conductors greatly increases the impedance of the circuit. This excessive separation can render an adequately-sized equipment grounding conductor ineffective in carrying enough current to operate the circuit protective device and clear the faulted equipment. In this case, providing a properly-sized equipment grounding conductor, but installing it improperly, results in an ineffective and possibly unsafe installation.

See Chapter 11 of this text for additional information on this subject.

Nonmetallic raceway

Where the wiring system is nonmetallic, it is necessary to run an equipment grounding conductor along with the circuit conductors. Do not separate them at any point in the circuit by any metallic material regardless of whether the metallic material is magnetic or not. It is true that if the material is nonmagnetic, the increase in impedance of the circuit will not be as great as if the material was magnetic. In any case such separation must be avoided.

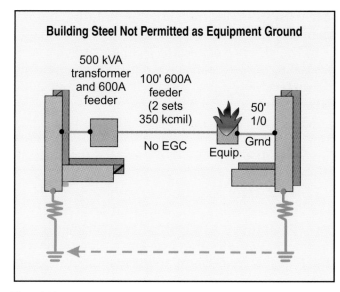

Figure 9-12

Use of building steel for grounding

Section 250-136(a) permits a metal rack or structure to ground electric equipment that is secured to it and in electrical contact, provided the support means is grounded by an equipment grounding conductor as specified by Section 250-134(b). However, the building steel is not permitted to serve as an equipment grounding conductor to ground equipment. That is due to the uncertain path that ground fault current must take in an effort to clear a fault.

This section emphasizes the requirement in Sections 250-134(b) and 300-3(b) that the equipment grounding conductor must be in the same raceway, cable or cord, or otherwise be run with the circuit conductors. Again, this is so the grounding circuit impedance will be as low as possible to allow adequate ground-fault current to flow so the circuit protective device will clear the fault.

In the same manner, and on the same basis, metal car frames, supported by metal hoisting cables attached to or running over sheaves or drums of elevator machines, are considered grounded when the machine is grounded as required by the Code.

Grounding for direct-current circuits

All of the above applies to the grounding of alternating-current systems where reactance of the circuit plays a large part in the impedance of the ground-return path. In the case of direct-current circuits the concern is with ohmic resistance only. Owing to that fact, the current-carrying capacity of the grounding conductor for a di-rect-current supply system must be equal to that of the largest conductor of the system. However, if the grounded circuit conductor is a neutral derived from a balancer winding or a balancer set which has overcurrent protection as required in the Code under Section 445-4(d), then the grounding conductor size shall be not less than the size of the neutral. The requirement for overcurrent protective devices here states that the two-wire direct-current generators used in conjunction with balancer sets shall be equipped with overcurrent protective devices which will disconnect the 3-wire system in the case of excessive unbalancing of voltages or currents.

Grounding conductors for direct-current circuits

For direct-current circuits, the size of the grounding conductor is specified in Section 250-93. The size may be larger than would be required for the same size alternating-current circuit. That is because resistance is the only factor in determining current flow in a direct-current circuit. Therefore, fault-currents are larger in a dc circuit. In an alternating-current circuit, impedance becomes the important factor for not only resistance, but reactance also must be taken into account.

Long term reliability of metal raceways

In the above discussions, it is assumed that a conductor enclosure (conduit or other raceway) has been properly installed with good tight joints that will provide a permanent and continuous electrical circuit when it is first installed. However, time and corrosion will take its toll and will tend to destroy the continuity of the conduit.

The safety of an electrical system will thus depend on how long we can expect the equipment grounding conductor of the conduit to remain permanent and continuous. The answer will vary depending on the type of metal raceway, the environment it is installed in and the quality of the installation.

For design purposes, two categories can be created:

(1) Where little corrosion will exist and where it can be reasonably expected that the equipment grounding conductor, in the form of a metal raceway, will remain permanent and continuous for a period of fifty years or more.

(2) Where corrosion in varying degrees will exist and where the permanency of the equipment grounding conductor provided by the metal raceway can be questioned.

Most commercial and residential buildings are in the first category. That being the case, conductor enclosures, which are approved for the purpose can be used as part of the equipment grounding conductor (with the use of bonding jumpers where required).

Industrial and petrochemical plants are in the second category where a metallic equipment grounding conductor, sized per Table 250-122, must be run in parallel with and within the conductor enclosure so as to ensure continuity if the conduit circuit is broken owing to corrosion sooner or later.

Some electrical design engineers and local electrical inspection agencies require that an equipment grounding conductor be installed in each metal conduit and tubing to help ensure the reliability of the equipment grounding conductor path.

Metal Conduit Corrosion Protection Required

In Concrete:	Required	Optional
Rigid Steel		X
Intermediate Steel		X
Aluminum Rigid	X	
Steel EMT	Below grade may need	On or above grade
Alum. EMT	X	
In Soil:		
Rigid Steel		X
Intermediate Steel		X
Aluminum Rigid	X	
Steel EMT	Generally required	
Alum. EMT	X	

Metal conduit underground

Care must be taken when installing metallic conduit and electrical metallic tubing in the earth, in concrete on or below grade, or where exposed to moisture. [Section 346-1(c)] The Underwriters Laboratories' 1998 guide card information for rigid metal conduit (DYIX) and intermediate metal conduit (DYBY) contains the following information regarding corrosion protection:

"Galvanized rigid (and intermediate) steel conduit installed in concrete does not require supplementary corrosion protection. Galvanized rigid (and intermediate) steel conduit installed in con-

tact with soil does not generally require supplementary corrosion protection.

In the absence of specific local experience, soils producing severe corrosive effects are generally characterized by low resistivity (less than 2000 ohm-centimeters).

Wherever ferrous metal conduit runs directly from concrete encasement to soil burial, severe corrosive effects are likely to occur on the metal in contact with the soil.

Conduit that is provided with a metallic or nonmetallic coating, or a combination of both, has been evaluated for resistance to atmospheric corrosion. Nonmetallic outer coatings that are part of the required resistance to corrosion have been additionally evaluated for resistance to the effects of sunlight.

Rigid metal conduit with or without a nonmetallic coating has not been evaluated for severely corrosive conditions."[U]

In addition, experience has shown that steel conduit fails rapidly where exposed to corrosive environments found at some seacoast marinas, boatyards and plants as well as at some chemical plants. In addition, experience has shown that metal conduit systems are particularly vulnerable to failure from corrosion where they pass from concrete that is on or below grade to exposure to an atmosphere containing corrosive elements, particularly in combination with atmospheres containing oxygen.

For electrical metallic tubing, (See Section 348-1) the following instructions are given in the UL 1998 guide card (FJMX):

"Galvanized steel electrical metallic tubing installed in concrete, on grade or above, generally requires no supplementary corrosion protection. Galvanized steel electrical metallic tubing in concrete slab below grade level may require supplementary corrosion protection.

In general, galvanized steel EMT in contact with soil requires supplementary corrosion protection. Where galvanized steel electrical metallic tubing without supplementary corrosion protection extends directly from concrete encasement to soil burial, severe corrosive effects are likely to occur on the metal in contact with the soil.

Aluminum electrical metallic tubing used in concrete or in contact with soil requires supplementary corrosion protection.

Supplementary nonmetallic coatings presently

used have not been investigated for resistance to corrosion."[U]

As a result, the authority having jurisdiction must make a decision regarding the suitability of these raceways for these applications. This, of course, affects the reliability of the raceway serving as an equipment grounding conductor. Several reports have been made where electrical metallic tubing installed to provide an equipment grounding means has failed due to corrosion. To maintain the integrity of the equipment grounding means, some inspection agencies require that a copper equipment grounding conductor be installed in parallel with the electrical metallic tubing.

Also, the authority having jurisdiction must make a decision regarding the suitability of supplementary nonmetallic coatings intended for resistance to corrosion.

Grounding of equipment by the grounded circuit conductor

The Code does not generally permit the grounded circuit conductor (often a neutral) to be grounded

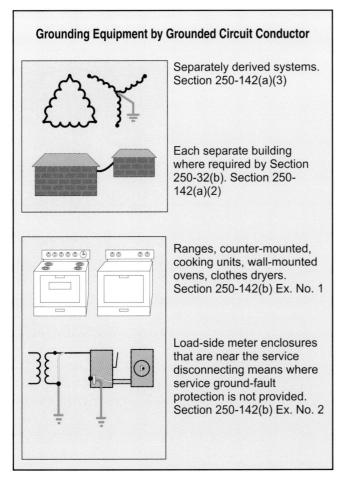

Grounding Equipment by Grounded Circuit Conductor

Separately derived systems. Section 250-142(a)(3)

Each separate building where required by Section 250-32(b). Section 250-142(a)(2)

Ranges, counter-mounted, cooking units, wall-mounted ovens, clothes dryers. Section 250-142(b) Ex. No. 1

Load-side meter enclosures that are near the service disconnecting means where service ground-fault protection is not provided. Section 250-142(b) Ex. No. 2

Figure 9-13

again on the load side of the service disconnecting means. See Sections 250-24(a)(5) and 250-142(b). Four exceptions to this rule exist:

(1) Grounding separately derived systems.
(2) Grounding the grounded circuit conductor at a remote building or structure.
(3) For existing branch circuits only, grounding the frame of an electric range or electric clothes dryer.
(4) Grounding meter enclosures that are near the service disconnecting means.

Where the electrical system produced by a separately derived system meets the conditions of Sections 250-20(a) or (b), the system is required to be grounded according to Section 250-30. Systems that fall within the parameters that require it to be grounded must have a grounding electrode conductor connected to the grounded conductor of the separately derived system. See Chapter Twelve of this text for additional information on this subject.

Under the specific conditions given in Section 250-32, a grounded circuit conductor is permitted to be grounded again at a separate building or structure. Where so installed, the grounded conductor serves as both a grounded conductor and an equipment grounding conductor between the buildings or structures. See Chapter Thirteen of this text for additional information on this subject.

Section 250-142(b) covers rules on the use of the grounded circuit conductor for grounding equipment on the load side of the service equipment. As stated previously, such practice is generally prohibited. Three exceptions to the general rule are provided.

Exception No. 1: For existing circuits only, the frames of ranges, counter-mounted cooking units, wall-mounted ovens, and clothes dryers, as well as outlet or junction boxes that are a part of the circuit for these appliances are permitted to be grounded by the grounded circuit conductor under the conditions provided in Section 250-140. See Chapter Ten of this text for additional information on this subject.

Exception No. 2: Meter enclosures that are on the load side of the service equipment are permitted to be grounded by connection to the grounded circuit conductor provided three conditions are met:

(1) Ground-fault protection of equipment is not provided. This condition is important

as grounding the grounded circuit conductor downstream from the service will desensitize the equipment ground-fault protection system.

(2) All meter enclosures are "near" the service equipment. (No distance is given to explain what is meant by the term "near.")

(3) The size of the grounded circuit conductor is not smaller than the size specified in Table 250-122 for equipment grounding conductors.

Exception No. 3 permits direct-current systems to be grounded on the load side of the disconnecting means or overcurrent device as provided in Section 250-164. Rules are different depending on whether the direct-current supply is from an off-premises or on-premises source.

Underground metal-sheathed cable system

The Code recognizes that if an underground service originates from a continuous underground metal sheath cable system and the sheath or armor is metallically connected to the underground system, or if an underground service conduit is used which contains a metal-sheathed cable which is itself bonded to the underground system, then the service run need not be grounded at the building and may be insulated from the interior conduit or piping. (Section 250-84)

The Code here recognizes that since the service armor is already bonded to a continuous underground metal sheath cable system, a path for fault-currents in the service cable is provided so there is no necessity to bond back to the service equipment in order to provide a return path to the neutral.

Chapter Nine: The questions included here were developed using material included in this chapter. The answers can be found by reviewing the text. It is also important that students make use of the 1999 NEC®, where many answers can be found. See page 279 for answers.

1. The conductor used to connect the noncurrent-carrying metal parts of equipment, raceways, and other enclosures to the system grounded conductor and/or the grounding electrode conductor at the service equipment, or at the source of a separately derived system is defined as ____.
   a. equipment grounding conductor
   b. main bonding jumper
   c. grounding systems conductor
   d. circuit bonding jumper

2. An equipment grounding conductor is intended to prevent an objectionable voltage above ground on conductor and equipment enclosures and to provide a low-impedance path for fault-currents. This path must also ____.
   a. be permanent and electrically continuous
   b. have ample capacity to conduct safely any currents likely to be imposed on it
   c. be of the lowest practical impedance
   d. all of the above

3. The equipment grounding conductor or path must extend from the ____ point on the circuit to the service equipment where it is connected to the grounded conductor.
   a. closest
   b. service
   c. furthermost
   d. bonding

4. Where the overcurrent protection ahead of a copper branch circuit or feeder is sized at 225 amperes, the minimum size of the equipment grounding conductor is to be a No. ____.
   a. 4 copper
   b. 6 copper
   c. 8 copper
   d. 10 copper

5. Which of the following is NOT recognized as a conductor or raceway for use as an equipment grounding conductor?
   a. a conductor of copper or other corrosion-resistant material such as aluminum
   b. rigid or intermediate metal conduit
   c. electrical metallic tubing
   d. approved surface metal raceways

6. Listed flexible metal conduit and listed flexible metallic tubing are permitted to be used for equipment grounding purposes. Which of the following statements is NOT true?
   a. The total combined ground return in the same path cannot exceed 6 feet
   b. They must be terminated with fittings listed for grounding
   c: They cannot be used on a circuit exceeding 15 amperes
   d. They cannot be used on a circuit exceeding 20 amperes

7. Listed liquidtight flexible metal conduit in the ¾-inch through 1¼-inch trade size is to be used as an equipment grounding conductor. Which of the following statements is NOT true? ____
   a. The total length of the combined ground return in the same path cannot exceed 6 feet.
   b. Listed fittings must be used for grounding at connections.
   c. The circuit is permitted to be protected by a 100 ampere or less overcurrent device.
   d. The circuit is permitted to be protected by a 60 ampere or less overcurrent device.

8. Where rigid metal conduit is used as an equipment grounding conductor, all joints and fittings are required to be ____.
   a. approved
   b. tested
   c. made up wrenchtight
   d. sealed

9. For the grounding of portable or pendant equipment and where protected by fuses or circuit breakers rated or set at not over ____ amperes, the Code permits the use of a No. 18 copper wire as an equipment grounding conductor provided it is a part of a listed flexible cord assembly.
   a. 25
   b. 20
   c. 30
   d. 35

10. Under what conditions can building steel serve as an equipment grounding conductor to ground equipment? ____
   a. Where it is effectively grounded.
   b. Never.
   c. Where approved.
   d. By special permission.

11.   Where equipment grounding conductors are installed in parallel in separate nonmetallic raceways, which of the following statements is true?

    a. A full-size equipment grounding conductor is required in only one of the conduits.

    b. A smaller equipment grounding conductor than required by Table 250-122 is permitted if the total area is not less than given in the table.

    c. A full size equipment grounding conductor is required in each of the conduits.

    d. Various size copper and aluminum conductors can be used together so long as they are not smaller than given in Table 250-122.

12.   Equipment grounding conductors in parallel listed cables are permitted to be smaller than given in Table 250-122 if

    a. they are protected by an equipment ground fault protection device

    b. the total area of the conductors is not less than the area required divided by the number for conductors

    c. they do not leave the building or structure they originate in

    d. not more than four cables are installed in parallel.

Enclosure and Equipment Grounding

Objectives

After studying this chapter, the reader will be able to understand:

- General requirements and definitions for enclosure and equipment grounding.
- Grounding of fixed and specific equipment.
- Grounding of nonelectrical equipment.
- Methods of effective grounding and grounding of equipment.
- Grounding of metal enclosures and panelboards.
- Installation of grounding-type receptacles.
- Grounding of cord- and plug-connected equipment.
- Equipment spacing from lightning rods.

Definitions

Bonding Jumper: "A reliable conductor to ensure the required electrical conductivity between metal parts required to be electrically connected."[N]

Bonding Jumper, Equipment: "The connection between two or more portions of the equipment grounding conductor."[N]

General requirements

Both enclosures for service conductor and other conductor enclosures, where of metal, are required to be grounded. (See Sections 250-80 and 250-86) That does not mean connecting equipment to a grounding electrode is acceptable or permitted in the NEC®. The installation must comply with the requirements of Section 250-2(d) where the concept of effective grounding is carefully outlined. It is important to realize that wherever the Code states, "shall be grounded" it means effectively grounded as spelled out in Section 250-2(d). Note that nothing in the Code permits equipment that is supplied by a grounded system to be grounded to only a grounding electrode. "Effective grounding" always includes providing a low-impedance path consisting of an equipment grounding conductor that has adequate capacity to conduct the maximum fault current it is likely to carry. It also must be permanent and electrically continuous.

While the grounding methods are different, electrical equipment associated with both grounded and ungrounded systems must be effectively grounded. Where grounding is not effectively accomplished, the situation, while bad in an ungrounded system, becomes worse in a grounded system.

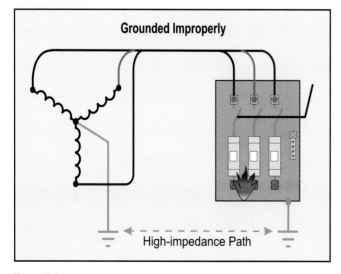

Figure 10-1

Equipment grounding conductor

It is important to recall that an equipment grounding conductor must be used for grounding equipment. This equipment grounding conductor is permitted to consist of any of the conductors or wiring methods identified in Section 250-118.

Installing a grounding electrode to ground equipment without having an equipment grounding conductor connected to the grounded service conductor is unsafe and not permitted by the Code. See Sections 250-2(d) and 250-54.

By referring to Figure 10-1, it can be seen that the grounding connections as shown comply literally with the wording of the Code in that the enclosures are "grounded" which means "connected to earth …" But the wiring does not comply with Section 250-2(d) as an effective fault current return path has not been provided. In addition, the grounding shown in this figure violates Sections 250-2(d) and 250-54 as an earth return grounding circuit is indicated.

The wiring in Figure 10-2 complies literally with Sections 250-80 and 250-86 and also complies with Section 250-2(d) as an equipment grounding conductor is installed.

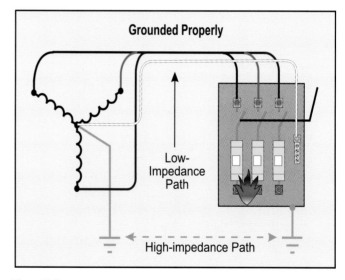

Figure 10-2

It is obvious that only a high-impedance fault current return path is indicated in Figure 10-1, while in Figure 10-2 there is a path having sufficiently low impedance to limit the voltage to ground and to facilitate the operation of the circuit protective devices in the circuit. Note that in Figure 10-2 there are two paths for current to return to the source. The primary and low impedance path is over the grounded system conductor (often a neutral) while a second, high-impedance path in parallel with the grounded service conductor is through the grounding electrodes and the earth.

Service raceways and enclosures

We have dealt in an earlier chapter of this text with the requirement for bonding service raceways and equipment. By bonding the service equipment to the grounded service and grounding electrode, we have complied with the requirements of Section 250-80.

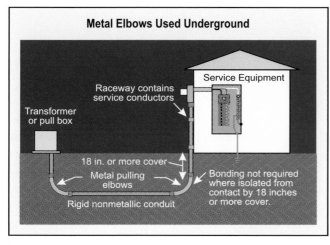

Figure 10-3

An exception has been added to Section 250-80 which exempts a metal elbow from the requirement that it be grounded where it is installed in an underground run of rigid nonmetallic conduit and is isolated from possible contact by a minimum cover of 18 inches to any part of the elbow. These metal elbows are often referred to as "pulling elbows" and are commonly installed in duct banks or other underground runs of rigid nonmetallic conduit and are more durable than PVC elbows during the cable-pulling process.

A similar exception regarding metal elbows used in underground runs of rigid nonmetallic conduit (Exception No. 3) for other than service raceways has been added to Section 250-86.

Exceptions from grounding requirements

For other than service conductor enclosures, four exceptions are provided from the requirement that metallic conductor enclosures be grounded. (Section 250-86).

Exception No. 1 covers the metal enclosures and raceways for conductors that are added to existing installations of open wire, knob-and-tube wiring,

and nonmetallic-sheathed cable. Conditions that must be met are as follows:

(1) An equipment grounding conductor is not provided by the wiring method.

(2) The metal enclosure or raceway must be less than 25 feet long.

(3) The metal enclosure or raceway must be free from probable contact with ground or a grounded object.

(4) The metal enclosure or raceway is guarded against contact by persons.

Exception No. 2 exempts short sections of metal enclosures from the requirement to be grounded where used to protect cable assemblies from physical damage. No explanation is given in the Code for the meaning of "short sections" of metal enclosures. Since the standard length is 10 feet, "short sections" probably are less than 10 feet long.

Exception No. 3 permits a metal elbow to not be grounded where it is installed in an underground run of rigid nonmetallic conduit and is isolated from possible contact by a minimum cover of 18 inches to any part of the elbow. This provision is identical to that for services as provided in Section 250-80, Exception. See Figure 10-3.

Grounding of fixed equipment

It is mandatory that noncurrent-carrying metal parts of fixed equipment, which are likely to become energized, be grounded under the six conditions cited in Section 250-110. Here, again, the term "shall be grounded" must not be interpreted literally to mean "connect to a grounding electrode" but must be interpreted in the light of all of Article 250 where providing an effective fault current path is outlined in Section 250-2(d). The Code does not define what is meant by "likely to become energized." Generally, if the equipment has exposed noncurrent-carrying metal parts and is supplied by electric current it should be grounded where any of the six conditions exist at the equipment.

The six conditions under which exposed noncurrent-carrying metal parts of equipment must be grounded are:

(1) Where within 8 feet vertically or 5 feet horizontally of ground or grounded metal objects and subject to contact by persons.

(2) Where located in a wet or damp location and not isolated.

(3) Where in electrical contact with metal.

(4) Where in hazardous (classified) locations as covered by Articles 500 through 517.

(5) Where supplied by a metal-clad, metal-sheathed, metal raceway, or other wiring method which provides an equipment ground. Short sections of metal enclosures covered in Section 250-86 are exempted from this requirement.

(6) Where equipment operates with any terminal at more than 150 volts to ground.

The three exceptions from the requirement that this equipment must be grounded are as follows:

Exception No. 1. Metal frames of electrically heated devices, exempted by special permission (the written approval of the Authority Having Jurisdiction), in which case the frames are required to be permanently and effectively insulated from ground;

Exception No. 2. Distribution apparatus such as transformers and capacitor cases that are mounted on wooden poles at a height of more than 8 feet from the ground or grade level;

Exception No. 3. Listed equipment that is protected by a system of double insulation. Such equipment must be distinctively marked.

Grounding specific equipment

The Code requires that exposed, noncurrent-carrying metal parts of certain specific equipment, regardless of voltage, shall be grounded. Those items are spelled out in Section 250-112 and include:

(a) Switchboard frames and structures supporting switching equipment other than frames of 2-wire dc switchboards where effectively insulated from ground.

(b) Generator and motor frames in an electrically-operated pipe organ unless the generator is effectively insulated from ground and from the motor driving it.

(c) Frames of stationary motors where: (Section 430-142)
  (1) Supplied by metal-enclosed wiring;
  (2) In a wet location and not isolated or guarded;
  (3) In a hazardous (classified) location as covered in Articles 500 through 517; and
  (4) The motor operates with any terminal at over 150 volts to ground.

(d) Enclosures for motor controllers. Exempted are enclosures attached to ungrounded portable equipment.

(e) Electrical equipment for elevators and cranes.

(f) Electrical equipment in commercial garages, theaters and motion picture studios. Exempted are pendant lampholders supplied by circuits not over 150 volts to ground.

(g) Electric signs, outline lighting and associated equipment as provided in Article 600. See Chapter Sixteen of this text for additional information on the subject.

(h) Motion picture projection equipment.

(i) Equipment supplied by Class 1 power-limited circuits and by Class 1, Class 2 and Class 3 remote-control and signaling circuits and by fire alarm circuits where these systems are required to be grounded by Part B of Article 250.

(j) Lighting fixtures as required by Part E of Article 410 as follows:
  (1) Exposed metal parts of lighting fixtures are either required to be grounded or be insulated from ground and other conductive surfaces or be inaccessible to unqualified personnel. Lamp tie wires, mounting screws, clips, and decorative bands on glass spaced at least 1½ inches from lamp terminals are not required to be grounded. See Section 410-18(a).
  (2) Fixtures that are supplied by a wiring method that does not provide an equipment ground must be made of insulating material and have no exposed conductive parts. See Section 410-18(b).
  (3) Fixtures with exposed metal parts must be provided with a means for connecting an equipment grounding conductor. See Section 410-20.

(k) Equipment mounted on skids and the (metal) skids are required to be grounded with an equipment bonding jumper that is sized in accordance with Section 250-122.

(l) Motor-operated water pumps including the submersible type. Also, see Sections 547-4(f) and 547-9 for specific requirements for grounding in agricultural buildings.

(m) Metal well casings must be bonded to the pump circuit equipment grounding conductor where a submersible type pump is installed.

Grounding of cord- and plug-connected equipment

Section 250-114 covers grounding of equipment connected by cord and plug. It is mandatory under certain conditions that noncurrent-carrying metal parts of cord- and plug-connected equipment, which are liable to become energized, be grounded. Listed tools, listed appliances, and listed equipment that is protected by a system of double insulation are not required to be grounded. This double-insulated equipment is required to be distinctively marked.

Specifically, cord- and plug- connected equipment must be grounded where located in:

(1) hazardous (classified) locations,

(2) if the equipment operates at more than 150 volts to ground. Exempted from this requirement are;

(1) motors, where guarded, and

(2) metal frames of electrically-heated appliances exempted by special permission in which case the frames shall be permanently and effectively insulated from ground, and

The Code cites in Section 250-114(3) specific equipment of the cord- and plug-connected type which must be grounded in residential occupancies. In addition, Section 250-114(4) lists specific cord- and plug-connected equipment in other than residential occupancies that must be grounded. In this section, recognition is given to the use of a listed system of double insulation which would provide the safety required without the use of grounding.

For tools and portable handlamps in other than residential occupancies, an exception from the requirement for grounding is provided for cord- and plug-connected equipment that is supplied through an isolating transformer with an ungrounded secondary of not over 50 volts.

Nonelectric equipment

The grounding of nonelectric equipment is covered by the Code in Section 250-116. The equipment mentioned is considered as being likely to become energized and is thus required to be grounded as a safety measure. Included are:

(1) Frames and tracks of electrically operated cranes and hoists;

(2) Frames of nonelectrically driven elevator cars to which electric conductors are attached.

(3) Hand-operated metal shifting ropes or cables of electric elevators,

The Fine Print Note following this section recommends that where extensive metal in or on buildings may become energized and is subject to personal contact, adequate bonding and grounding will provide additional safety.

Methods of grounding

To provide the reliable and effective ground-fault return path required by the Code, it is important to connect the equipment grounding conductor recognized in Section 250-118 to the equipment in such a manner that the requirements for "effective grounding" are met.

For equipment that is fastened in place or connected by permanent wiring methods, the equipment grounding conductor must be connected to the enclosure in a proper manner.

Grounding conductors and bonding jumpers must be connected to equipment that is required to be grounded by exothermic welding, listed pressure connectors, listed clamps or other listed means. See Section 250-8.

The only connecting means not required to be listed is the exothermic welding method. It is most important that equipment to be welded be clean and dry and that manufacturer's instructions be followed to ensure a satisfactory connection. These welds must be examined and tested after completion to be certain that a reliable connection has been made. It is common to test the welds in the field by x-ray where required by the job specifications or by striking the weld with a hammer after it has cooled for less demanding installations.

In the case of a metallic raceway being used as the equipment grounding conductor, the raceway must be connected to the enclosure by using fittings designed for the purpose. All connections must be made up tight using proper tools. This includes locknuts, bushings, conduit and electrical metallic tubing couplings and connectors. See Section 250-120(a).

Solder connections

Connections that depend solely upon solder cannot be used for grounding connections. See Sections 250-8 and 250-148.

The reason for this prohibition is that when equipment grounding conductors carry fault-current, they can get very hot. This elevated temperature may exceed the melting point of the solder and weaken or destroy the connection. This may create a hazard by leaving equipment at a dangerous potential above ground, creating a shock hazard to those who may contact the equipment.

It is permissible to secure the connections mechanically and then apply solder to the joint to make the electrical connection.

Special provisions for grounding certain appliances

Special requirements are set forth in Section 250-140 for the grounding of frames of electric ranges, electric clothes dryers and similar appliances. These provisions apply to only existing branch-circuit installations. New installations must comply with the requirements for an insulated neutral as well as an equipment grounding conductor given in Sections 250-134 and 250-138.

Appliances or equipment to which Section 250-140 applies include:

(a)   electric ranges;

(b)   wall-mounted ovens;

(c)   counter-mounted cooking units;

(d)   clothes dryers, and

(e)   outlet or junction boxes that are part of the circuit for these appliances.

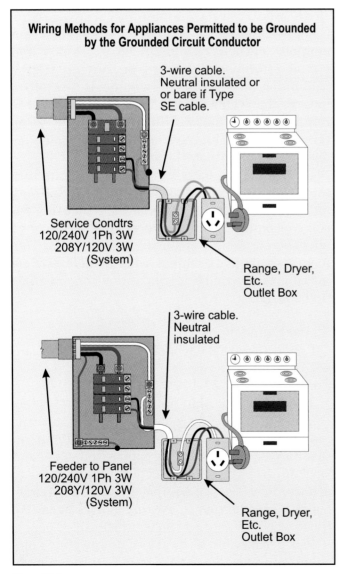

Wiring Methods for Appliances Permitted to be Grounded by the Grounded Circuit Conductor

3-wire cable.
Neutral insulated or
or bare if Type
SE cable.

Service Condtrs
120/240V 1Ph 3W
208Y/120V 3W
(System)

Range, Dryer,
Etc.
Outlet Box

3-wire cable.
Neutral
insulated

Feeder to Panel
120/240V 1Ph 3W
208Y/120V 3W
(System)

Range, Dryer,
Etc.
Outlet Box

Figure 10-4

These appliances and equipment, such as the junction boxes in the circuit, are permitted to be grounded in two ways, either by use of an equipment grounding conductor or, except for mobile homes and recreational vehicles, by the use of the grounded circuit (neutral) conductor, provided all the conditions of this section are complied with. If a 3-wire with ground circuit is installed, simply use the equipment grounding conductor for grounding metal equipment and install a grounding type receptacle.

The conditions that must be met for grounding the above equipment to the grounded circuit conductor in existing installations are as follows:

(1)   The appliances are supplied by a 120/240-volt, single-phase, 3-wire circuit or by a 3-wire circuit derived from a 208Y/120-volt, 3-phase, 4-wire, wye-connected system.

(2)   The grounded (neutral) conductor is No. 10 copper or No. 8 aluminum or larger.

(3)   The grounded (neutral) conductor is insulated. The grounded (neutral) conductor is permitted to be uninsulated if it is part of a type SE service-entrance cable and it originates at the service equipment, and

(4)   Grounding contacts of receptacles that are a part of the appliance are bonded to the equipment.

It is important to note that all of these special conditions must be met before the appliances, and outlet and junction boxes that are a part of the circuit to the appliances, are permitted to be grounded by use of the circuit neutral. Note that the supply cable must have an insulated neutral conductor where supplied from a sub-panel.

Use caution when applying the provisions of Section 250-140(3) which permits the use of a Type SE cable having a bare neutral for wiring these appliances when the circuit originates at service equipment. Section 338-3(b) no longer permits Type SE cable without an insulated neutral conductor to be used for the purposes of Section 250-140.

Wiring for new ranges and dryers

New installations for ranges, dryers and similar appliances require a three-wire with ground circuit having an insulated neutral conductor and an equipment grounding conductor. The equipment grounding conductor is permitted to be any of those included in Section 250-118 including conduit an cables. Receptacles, where installed, must be of the 3-pole, 4-wire grounding type. Supply cords, where used, must be of the 3-wire with ground type.

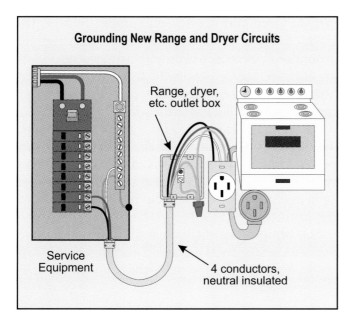

Figure 10-5

Neither the frame of the appliances nor the outlets or junction boxes that are a part of the supply to the appliances are permitted to be grounded to the circuit neutral. The frame of the appliances and junction boxes must be grounded by means of an equipment grounding conductor run with the branch circuit.

Care must be taken when these appliances are moved from a location employing one grounding scheme to a location having a different one. In the case where a 3-wire cable is used, the bonding jumper in the appliance junction box must be connected between the frame of the appliance and the neutral. Where a four-wire supply is used, the bonding jumper must be removed or disconnected, the neutral conductor be isolated from the appliance frame and the equipment grounding conductor connected to the frame.

Outlet, device, pull and junction boxes

Metal outlet, device, pull and junction boxes are required to be grounded in accordance with Article 250. See Section 370-4. It is important that the grounding of these enclosures be accomplished by the wiring method that supplies the enclosure. See Section 250-134. Under no circumstances should a metal enclosure be connected to a local grounding electrode in lieu of grounding them by means of the suitable wiring method or equipment grounding conductor unless specifically permitted by Code rules.

The metal raceway containing circuit conductors is permitted to be used for grounding these enclosures if installed properly with all connections made

up wrenchtight. In addition, where fittings are used for connecting cables or raceways being used for grounding, the fittings used for connecting the wiring methods must be suitable for grounding. Some wiring methods have restrictions on their use as an equipment grounding conductor. These include flexible metal conduit, liquidtight flexible metal conduit and flexible metallic tubing.

Equipment grounding conductors that are supplied for grounding these enclosures must be sized in accordance with Table 250-122 based upon the rating of the overcurrent device protecting the circuit conductors in the raceway.

Section 370-40(d) requires that a means be provided by the manufacturer in each metal box for the connection of the equipment grounding conductor. This provision for grounding the box is often a tapped hole for a grounding screw.

Grounding of panelboards

Section 215-6 requires that "where a feeder supplies branch circuits in which equipment grounding conductors are required, the feeder shall include or provide a grounding means, in accordance with the provisions of Section 250-134 to which the equipment grounding conductors of the branch circuit shall be connected."[N]

Specific requirements for grounding panelboards are contained in Section 384-20. All panelboard cabinets and panelboard frames, if of metal, are required to be in physical contact with each other and must be grounded. Section 384-3(c) requires that where used as service equipment, the panelboard must be provided with a main bonding jumper located inside the cabinet for the purpose

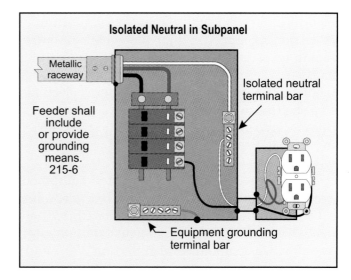

Figure 10-6

of bonding the enclosure to the grounded service conductor.

Where the panelboard is used with nonmetallic raceway or cable or with separate grounding conductors, a terminal bar for terminating the equipment grounding conductor must be secured inside the cabinet. See Section 384-20. This terminal bar must be bonded to the cabinet if it is metal. If the enclosure is made of nonmetallic material, the terminal bar must be connected to the equipment grounding conductor that supplies the panelboard. Usually, the manufacturer provides matching and tapped holes along with appropriate screws or bolts for attaching the bar to the enclosure.

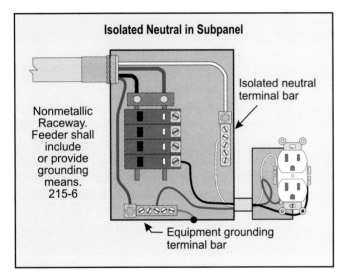

Isolated Neutral in Subpanel

Nonmetallic Raceway. Feeder shall include or provide grounding means. 215-6

Isolated neutral terminal bar

Equipment grounding terminal bar

Figure 10-7

Grounding and equipment grounding conductors are not permitted to be connected to the terminal bar for grounded (may be a neutral) conductors unless the terminal bar is identified for the purpose. This identification will usually be on the manufacturer's label that is located within the cabinet. Another requirement is that the panelboard is used at a location where the grounded conductor terminal (neutral) bar is connected to a grounding electrode, as permitted or required by Article 250. These locations include at services, at the building disconnecting means (Section 225-32) and for separately derived systems (Section 250-30).

An exception permits an insulated equipment grounding conductor for an isolated grounding scheme to pass through the panelboard without being connected to the equipment grounding terminal bar in compliance with Section 250-146(d).

Generally, the grounded (may be a neutral) conductor is isolated from the enclosure down-

stream from the service equipment. Often, the term "floating neutral" is used to describe the grounded conductor's relationship to the enclosure as it is isolated electrically from the enclosure. Section 250-24(b)(5) generally prohibits a grounding connection to a grounded conductor (may be a neutral) on the load side of the service disconnecting means. Exceptions to the general rule are provided for separately derived systems, at separate buildings, as well as for certain appliances such as electric ranges and clothes dryers on existing branch circuits.

See Chapter 12 of this text for additional information on separately derived systems and Chapter 13 for grounding at more than one building on the premises.

Grounding-type receptacles

Where grounding-type receptacles are installed, the grounding terminal of the receptacle must be connected to an equipment grounding conductor of the circuit supplying the receptacle. See Section 210-7(c). Where more than one equipment grounding conductor enters a box, they must be connected together using a suitable and listed connector. Also, it is permitted to connect each of the equipment grounding conductors to the metal box individually using a listed clip or screw. A jumper is used to the receptacle so grounding continuity is not disturbed if the device is removed.

A grounding jumper from the receptacle to the box is not required where a device with listed grounding means is installed in a metal box that is properly grounded. These devices are often referred to as

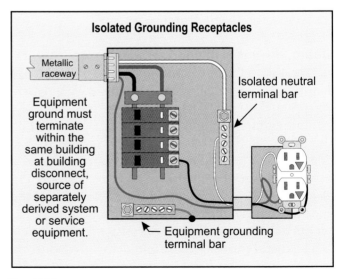

Isolated Grounding Receptacles

Metallic raceway

Equipment ground must terminate within the same building at building disconnect, source of separately derived system or service equipment.

Isolated neutral terminal bar

Equipment grounding terminal bar

Figure 10-8

"self-grounding receptacles" and are specially designed so one or more of the mounting screws are maintained in contact with the device's metal yoke.

See the requirements in Chapter 8 of this text for additional information on this subject.

Isolated grounding receptacles

Receptacles that have the equipment grounding terminal isolated from the mounting strap, and thus from the box, are commonly installed at computer terminals and cash registers. This is permitted by Section 250-146(d) for the purpose of reducing electrical noise (electromagnetic interference). The grounding terminal of the receptacle must be grounded by means of an insulated equipment grounding conductor that is run with the circuit conductors.

This insulated equipment grounding conductor is permitted to pass through one or more panelboards without connection to the terminal bar within the panelboard on its way back to, usually, the service disconnecting means. See Sections 250-146(d) and 384-20 Exception. However, the insulated equipment grounding conductor must terminate within the same building at the building or structure disconnecting means or source of a separately derived system. See Chapter 12 of this text for additional information on the subject of grounding separately derived systems.

Isolated equipment grounding

Section 250-96(b) permits an equipment enclosure supplied by a branch circuit to be grounded by means of an insulated equipment grounding conductor contained within the raceway with the branch circuit that supplies the equipment. A listed isolating raceway fitting at the point of connection to the equipment must be installed. This provision typically applies to listed data processing (information technology) equipment.

Underwriters Laboratories performed tests of a similar grounding scheme to determine whether isolating the metal conduit from the equipment had an adverse effect on the grounding circuit impedance. They found no appreciable increase in impedance with the metal conduit isolated from the equipment and being grounded by means of the insulated equipment grounding conductor.

Short sections of raceway

Where isolated sections of metal raceway or cable armor are required to be grounded, the Code requires in Section 250-132 that grounding of such sections be performed in accordance with the requirements of fixed equipment found in Section 250-134. While the Code does not identify what is meant by a "short section," perhaps this is a length less than the standard length of 10 ft. These short sections of raceways are often installed as physical protection of cables.

As mentioned previously, "short sections" of metal enclosures or raceways used to provide support or protection of cable assemblies from physical damage are not required to be grounded due to Exception No. 2 to Section 250-86.

Where these "short sections" of raceways or cable armor are required to be grounded, Section 250-134 generally requires that it be grounded by one of the equipment grounding conductors recognized by Section 250-118.

Spacing from lightning rods

Section 250-106 no longer requires that metal raceways, enclosures, frames and other noncurrent-carrying metal parts of electric equipment be kept at least 6 feet away from lightning rod conductors. It is recognized that the subject of lightning protection systems is covered extensively in the *Standard for the Installation of Lightning Protection Systems*, NFPA 780.

Fine Print Note No. 2 of Section 250-106 points out that the *Standard for the Installation of Lightning Protection Systems* may require bonding of adjacent portions of the electrical wiring system to

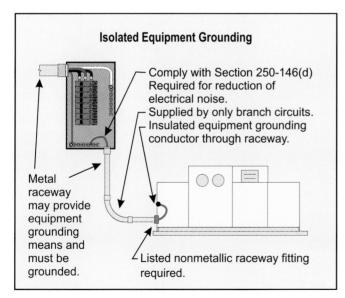

Isolated Equipment Grounding

— Comply with Section 250-146(d) Required for reduction of electrical noise.
— Supplied by only branch circuits.
— Insulated equipment grounding conductor through raceway.

Metal raceway may provide equipment grounding means and must be grounded.

— Listed nonmetallic raceway fitting required.

Figure 10-9

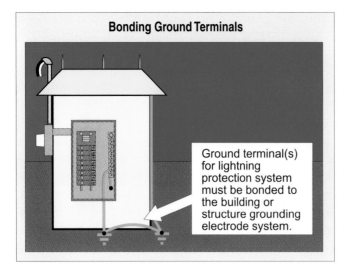

Bonding Ground Terminals

Ground terminal(s) for lightning protection system must be bonded to the building or structure grounding electrode system.

Figure 10-10

the lightning protection system. The spacing differs depending on whether the separation is through air or through differing construction materials.

Section 250-106 now simply requires that the ground terminals for the lightning protection system be bonded to the building or structure grounding electrode system.

Chapter Ten: The questions included here were developed using material included in this chapter. The answers can be found by reviewing the text. It is also important that students make use of the 1999 NEC®, where many answers can be found. See page 279 for answers.

1. A reliable conductor to assure the required electrical conductivity between metal parts required to be electrically connected is defined as a ____.
   a. grounding electrode conductor
   b. grounded conductor
   c. bonding jumper
   d. identified conductor

2. The connection between two or more portions of the equipment grounding conductor is defined as ____.
   a. equipment bonding jumper
   b. grounded
   c. neutral
   d. main bonding jumper

3. Exposed noncurrent-carrying metal parts of equipment must be grounded where within____ feet vertically or ____ feet horizontally of ground or grounded metal objects, and subject to contact by persons.
   a. 9 - 6
   b. 8 - 5
   c. 8 - 8
   d. 9 - 9

4. Exposed noncurrent-carrying metal parts of equipment are required to be grounded where located in wet or damp locations and are not ____.
   a. guarded
   b. identified
   c. shielded
   d. isolated

5. Where other than short sections of metal enclosures are installed as per Section 250-33, exposed noncurrent-carrying metal parts of equipment are required to be grounded where supplied by any of the following wiring methods EXCEPT ____.
   a. metal-clad, metal-sheathed cables
   b. knob-and-tube wiring
   c. other wiring methods that provide an equipment ground
   d. approved metal wireways

6. Exposed noncurrent-carrying metal parts of equipment not required to be grounded include enclosures for switches or circuit breakers used for other than ____ equipment where accessible to qualified persons only.
   a. welding
   b. service
   c. meters
   d. emergency

7. Exposed noncurrent-carrying metal parts of equipment not required to be grounded include metal frames of electrically heated devices unless exempted by ____ in which case the frames must be permanently and effectively insulated from ground.
   a. the product instructions
   b. the local government
   c. special permission
   d. the code

8. Exposed noncurrent-carrying metal parts of equipment not required to be grounded include distribution apparatus, such as transformers and capacitor cases, that are mounted on wooden poles at a height of more than ____ feet from the ground or grade level.
   a. 6
   b. 8
   c. 5
   d. 7

9. Exposed noncurrent-carrying metal parts of equipment not required to be grounded include ____ equipment that is distinctively marked.
   a. hospital
   b. triple insulated
   c. single insulated
   d. double insulated

10. Where added to existing installations of open wire, knob-and-tube wiring, and nonmetallic-sheathed cable without an equipment grounding conductor, a metal enclosure for conductors run in lengths not to exceed 25 feet, free from probable contact with ground, grounded metal, metal lath, or other conductive material, and if guarded against contact by persons are ____.
   a. required to be grounded
   b. not required to be grounded
   c. required to be protected by a GFCI
   d. not permitted unless approved

11. Motor-operated water pumps including the submersible type are required to be grounded when they operate at ____.
   a. any voltage
   b. 120 volts
   c. 240 volts
   d. 208 volts

12. All of the connecting means included below are required to be listed by a qualified electrical testing lab EXCEPT ____.
   a. pressure connectors
   b. lugs
   c. clamps
   d. exothermic welding

13. Connections that depend solely upon solder for grounding connections ____.
   a. are only permitted when approved
   b. cannot be used
   c. are permitted on the load side of the service
   d. are permitted on the line side of the service

14. Metal raceways that enclose service conductors are required to be grounded unless it is:
   a. installed on a pole
   b. more than 8 feet above the ground
   c. an elbow in PVC run covered by 18 in. of earth
   d. buried at a depth given in Table 300-5

15. All the following statements about grounding the frame of ranges and dryers are true EXCEPT:
   a. A three-wire with ground circuit is required for existing installations.
   b. The frame of the appliances is permitted to be grounded to the neutral in new installations.
   c. A two-wire with ground circuit is required for new installations.
   d. New installations must be grounded to comply with Sections 250-57 and 250-59.

16. All the following statements about grounding of panelboards supplied by a feeder in the same building as the service are true EXCEPT:
   a. The feeder must supply or provide the grounding means.
   b. The neutral conductor is permitted to ground the enclosure.
   c. The neutral conductor must connect to a neutral bar that is isolated from the enclosure.
   d. An equipment grounding conductor must be connected to an equipment grounding terminal bar.

Clearing Ground Faults and Short Circuits

Objectives

After studying this chapter, the reader will be able to understand:

- Ground faults and circuit impedance.
- Fundamentals of equipment grounding, circuit design, and test procedures.
- Common elements in clearing ground faults and short circuits.
- Sizing of equipment grounding conductors.
- Purposes served by neutral on grounded systems.
- Conductor withstand ratings.

Ground faults and circuit impedance

Section 310-2(a) of the NEC® requires, generally, that conductors be insulated. The exception allows covered or bare conductors to be used where specifically permitted in the Code. Ungrounded or hot conductors must be insulated for the applied voltage. Typical voltage rating of conductors is 300, 600, 2,000, 5,000 and 15,000. Conductors are available with much higher rated insulations, although these are primarily used for distribution of electrical energy rather than for premises wiring.

Table 310-13 gives conductor application and insulations. It gives the trade name, type letter such as MI, maximum operating temperature and application provisions. It also gives the conductor insulation, size in AWG (American Wire Gage), thickness in mils and outer covering, if any.

Bare or covered conductors are permitted to be used in certain cases. Grounded service-lateral conductors are permitted to be uninsulated under the conditions given in Section 230-30. Uninsulated service-entrance conductors are permitted to be used as provided in Section 230-41. Of course, insulated, covered or bare equipment grounding conductors are permitted to be used by Sections 250-118(1) and 250-119. Some Code sections require insulated conductors for specific applications such as for certain electrical equipment associated with swimming pools and in patient care areas of certain health care facilities.

If no hazard can exist on a distribution system unless there is an insulation failure of the ungrounded or hot conductor, it follows that every precaution should be taken to furnish the best possible insulation consistent with an economical installation. Good installation practices should be followed carefully to ensure that cable insulation is not damaged during the installation process. Cable insulation can be easily damaged by pulling operations if raceways are not cleaned prior to the installation of conductors, if excessive bends are in the run or where long or heavy pulls require the use of heavy-duty pulling equipment.

Further, if and when insulation does break down, the hazard can exist only as long as the circuit remains energized. Every effort should be made to deenergize, as quickly as is practical, any circuit on which a fault to ground has developed.

When first considered, the use of a copper or aluminum equipment grounding conductor may seem to offer an ideal method of getting a low-impedance path. It will in most cases, but that is not always true. Each individual circuit must be studied in relation to the size and length of the circuit, the size of its overcurrent device and the size of the conductor enclosure before any conclusion can be arrived at.

Fundamentals of equipment grounding circuit design

Briefly, it can be stated that to get low impedance of the grounding system in an ac system, the circuit conductors and the equipment grounding conductor must be kept intimately together at all times. See Sections 250-134(b) and 300-3(b). The equipment grounding conductor may, of course, be a copper or aluminum conductor or the metal enclosure of the conductors such as conduit or wireways, where acceptable by Code for such use.

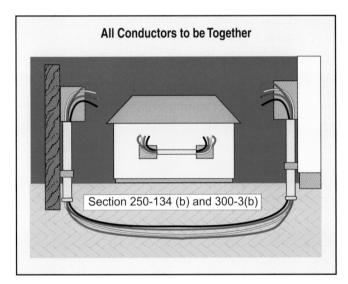

All Conductors to be Together

Section 250-134 (b) and 300-3(b)

Figure 11-1

In an ac electrical distribution system, whether grounded or ungrounded, inductive reactance exerts a powerful influence by directing the return current to flow in a path closely paralleling the outgoing power conductor. In addition to resistance, an inductive reactance value is associated with every conductor in an ac system. The inductive reactance (expressed in ohms) increases as the spacing between the conductors increases. This indicates that the inductive reactance of a current path closely paralleling the phase conductors offers a lower impedance to the ground-fault current than any other current path regardless of a lower resistance of the other current paths. Inductive reactance will be the predominate factor in determining current division in parallel ground return paths in heavy or larger circuit constructions. This has been proven in actual tests where it is shown that there is a strong tendency for ground return currents to seek a path physically close to the power conductor over which the outgoing current flows.[1]

Usually, the conduit or metallic raceway that encloses the conductors provides an excellent fault return path. The presence of magnetic material in the power conductor enclosure (conduit or raceway) introduces additional inductive effects tending to confine the return ground currents within the magnetic enclosure. Installation of an external equipment grounding conductor is quite ineffective. Connections to nearby structural building steel members, in an effort to serve as an equipment grounding means, are equally ineffective as well.

The intentional or accidental omission of using the metallic conductor enclosure as an equipment grounding conductor may lead to large induced voltages in nearby metallic structures, which may appear as a dangerous shock hazard or unwanted circulating currents. Only by the installation of an internal equipment grounding, in parallel with the raceway, can the current carried by the raceway be reduced. Joints in conduit and raceways must be connected wrenchtight, using proper tools, for the raceway to function effectively as an equipment grounding conductor.

Fault-current test procedure

As illustrated in Figure 11-2, a special installation of 2½-inch rigid steel conduit and 4/0 copper cable was made for this investigation. It was installed in a building previously used for short-circuit testing because the building had heavy steel column construction, and all columns were tied to an extensive grounding mat composed of 250 kcmil copper cables. The conduit was supported on insulators throughout the 100-foot length. The conduit was about 5 feet from a line of building columns. The external 4/0 cable was spaced about 1 foot from the conduit on the side opposite the building columns.

The setup was intended to simulate an electrical feeder circuit as may be found in many commercial and industrial plants. It allowed a study of a wide variety of equipment grounding arrangements. In every case the supply current was through the "A" cable run through the conduit. The "B" conductor served as an internal equipment grounding conductor. The "C" conductor was connected to the run of steel metal conduit, the "G" conductor was

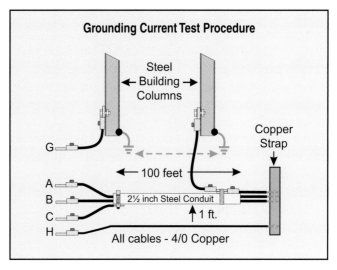

Figure 11-2

connected to the steel building columns and the "H" conductor was the external equipment grounding conductor spaced about 1 foot away from and parallel to the conduit. With this arrangement it was possible to simulate various fault conditions at the right end and to verify various fault current return paths.

One series of tests was made at low current of 200 and 350 amperes using an ac welding transformer as a source of 60 cycle power. At these low-current magnitudes, the current could be maintained for extended periods. Voltage measurements were made with high quality indicating meters. Current measurements were made with a clip-on ammeter.

A second series of tests was made at high current, around 10,000 amperes using a 450 kVA, 3-phase, 60 cycle transformer with a 600 volt secondary as a source of power. Switching was done at the 13,800 primary voltage. An induction relay was used to control the duration of current flow to about ¼ second. An oscillograph was used for all measurements of current and voltage.

Tests performed, low-current

A1 - 350 amperes out on internal conductor (A), return on internal conductor (B).

A2 - 350 amperes out on internal conductor (A), return on conduit (C).

A3 - 200 amperes out on internal conductor (A), return on conduit (C).

A4 - 350 amperes out on internal conductor (A), return on conduit (C) and external conductor (H).

A5 - 200 amperes out on internal conductor (A), return on conduit (C) and external conductor (H).

A6 - 350 amperes out on internal conductor (A), return on conduit (C) and building steel (G).

A7 - 200 amperes out on internal conductor (A), return on conduit (C) and building steel (G).

A8 - 350 amperes out on internal conductor (A), return on conduit (C) and internal equipment grounding conductor (B).

A9 - 200 amperes out on internal conductor (A), return on conduit (C) and internal equipment grounding conductor (B).

A10 - 350 amperes out on internal conductor (A), return on building steel (G) and external equipment grounding conductor (H).

A11 - 200 amperes out on internal conductor (A), return on building steel (G) and external equipment grounding conductor (H).

Tests performed, high current

B2 - 11,200 amperes out on internal conductor (A), return on conduit (C).

B3 - 11,070 amperes out on internal conductor (A), return on conduit (C).

B4 - 11,070 amperes out on internal conductor (A), return on conduit (C) and external equipment grounding conductor (H).

B5 - 11,080 amperes out on internal conductor (A), return on conduit (C) and external equipment grounding conductor (H).

B6 - 10,830 amperes out on internal conductor (A), return on conduit (C) and building steel (G).

B7 - 10,910 amperes out on internal conductor (A), return on conduit (C) and building steel (G).

B8 - 11,620 amperes out on internal conductor (A), return on conduit (C) and internal equipment grounding conductor (B).

B9 - 11,380 amperes out on internal conductor (A), return on conduit (C) and internal equipment grounding conductor (B).

B10 - 8,710 amperes out on internal conductor (A), return on building steel (G) and external equipment grounding conductor (H).

Fault-current test results

The results of the tests performed are shown in Table 11-1. The test number is for reference purposes. The next two columns show the cable connections used to determine current flow. Next, the current values are shown, first the total input current through cable A, and next the return current through the conduit and its percentage of the total. The other columns with an "I" heading, along with a subscript indicating the circuit, show the amount of current returning over other possible paths.

"No further analysis is needed to show conclusively that only by the use of an internal grounding

Current Magnitudes

Test No.	Current Flow Out On	Return On	IA Total	Ic Amperes	Ic Per Cent of Total	I_B	I_H	I_G	E_{AC}	E_{CG}	E_{AB}	E_{AH}	E_{AG}	E_{GB}
Low-Current Tests														
A1	A	B	350	0		350	0	0	2.47		4.85			
A2	A	C	350	350	100	0	0	0	15.9	0.45	2.5			
A3	A	C	200	200	100	0	0	0	9.05	0.15	1.51			
A4	A	CH	350	340	97	0	12	0	16.0	0.05	2.55			
A5	A	CH	200	190	95	0	8	0	9.13	nil	1.55			
A6	A	CG	350	340	97	0	0	12	14.6		2.54	14.4		
A7	A	CG	200	180	90	0	0	8	9.5		1.50	9.4		
A8	A	CB	350	62	18	290	0	0	4.55	nil	4.55			
A9	A	CB	200	40	20	150	0	0	2.68	nil	2.68			
A10	A	GH	350	0	0	0	160	160	14.0	12.5	2.5	26.4	26.4	24.6
A11	A	GH	200	0	0	0	98	98	9.2	8.1	1.5	17.1	17.1	15.1
High-Current Tests														
B2	A	C	11,200	11,200	100	0	0	0	168	36*				
B3	A	C	11,070	11,070	100	0	0	0	173	38*				
B4	A	CH	11,070	11,200	101	0	1,140	0	173	18*				
B5	A	CH	11,080	11,090	100	0	1,220	0	173	17*				
B6	A	CG	10,830	10,770	99	0	0	1,080	168		71			
B7	A	CG	10,910	10,780	99	0	0	1,145	173	9*				
B8	A	CB	11,620	5,810	50	5,660	0	0			27*		155	
B9	A	CB	11,380	6,070	53	5,620	0	0	146	25*				
B10	A	CH	8,710	0	0	0	4,300	4,500	146					268

*Distorted wave shape. Tabulated values are crest $/\sqrt{2}$ |

Table 11-1. Measured Electrical Quantities.   [2]

conductor can any sizable fraction of the return current be diverted from the raceway. In spite of the extremely low resistance of the building structural frame, it was ineffective in reducing the magnitude of the return current in the conduit. See tests A6, A7, B6 and B7.

"Some interesting secondary effects were observed in the course of the tests. The first high-current test produced a shower of sparks from about half of the couplings in the conduit run. From one came a blowtorch stream of sparks which burned out many of the threads. Several small fires set in nearby combustible material would have been serious if not promptly extinguished.

"The conduit run had been installed by a crew regularly engaged in such work and they gave assurance that the joints had been pulled up to normal tightness and perhaps a little more. A short 4/0 copper (wire) jumper was bridged around this joint but, even so, some sparks continued to be expelled from this coupling on subsequent tests. Other couplings threw no more sparks during subsequent tests. Apparently, small tack welds had occurred on the first test.

"In one high-current test, the conduit termination was altered to simulate a connection to a steel cabinet or junction box (see Figure 11-3). The bushing was applied finger-tight. In one test, with about 11,000 amperes flowing for about ¼ second,

a fan-shaped shower of sparks occurred parallel to the plate. In the process, a weld resulted and the parts were separated only with considerable difficulty, with the use of wrenches and a hammer. This suggested that a repeat shot (of current) would have produced no disturbance (shower of sparks).

"During high-current test B10 (conduit circuit open), a shower of sparks was observed at an intermediate building column. Careful inspection disclosed that the origin was at a spot at which a water pipe passed through an opening cut in the web of the steel beam involved. Here is evidence of the objectionable effects of forcing the short-circuit current to seek return paths remote from the outgoing conductor. The large spacing between outgoing and returning current creates a powerful magnetic field which extends far out in space around the current-carrying conductors."[2]

Fault-current study circuit analysis

"The reactance of the circuit including the 'B' conductor will be the lowest. Next will be the innermost tube of the conduit, followed by others in successive order until the outer tube is reached. The inductance of these tubular elements of the steel conduit assumes unusual importance because of the high magnetic permeability. Next in spacing (impedance) is the external grounding conductor (H cable) and last, the structural members of the building frame and their interconnecting grounding cables buried below floor level."[2]

"In test B10, both the exterior grounding conductor (H cable) and the building frame (G terminal) were connected to provide parallel paths for the return current, but the conduit circuit was left open (C terminal not connected). The test results clearly evidence the powerful forces tending to maintain current flow in the conduit circuit. Note that across the open connections at the 'C' terminal, a voltage of 146 volts (or, more significantly, over 50 percent of the impressed driving voltage), is required to force the current to return via the 'H' cable and the building frame in parallel. Such a voltage could be a serious shock hazard.

"Furthermore, unless the conduit was well insulated throughout its entire length (which is usually impossible or impractical in typical commercial or industrial installations), there would be a significant number of sparks at stray points to constitute a serious fire hazard. It was during this test that a shower of sparks occurred between

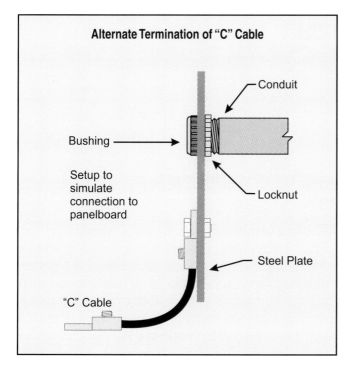

Alternate Termination of "C" Cable

Conduit

Bushing

Setup to simulate connection to panelboard

Locknut

Steel Plate

"C" Cable

Figure 11-3

magnetic members in the building system that was caused simply by the strong magnetic field extending far out from the power conductors."[2]

Conclusions on fault-current path study

"The significance of this investigation points unmistakably to the conclusions presented earlier in this paper. Effective use of the conduit or raceway in the equipment-grounding system is paramount. Additional work is needed to develop joints which will not throw fire during faults. To improve effectiveness requires greater conductivity in the conductor enclosure or the use of an internal (equipment) grounding conductor. Grounding cables (grounding electrode conductors) connecting building structure with grounding electrodes (connection to earth) are needed to convey lightning currents or similar currents seeking a path to earth, but these conductors will play a negligible part in the performance of the equipment grounding system. Of course, the importance of proper equipment grounding becomes greater with the larger size feeder circuits and the availability of higher short-circuit currents."[2]

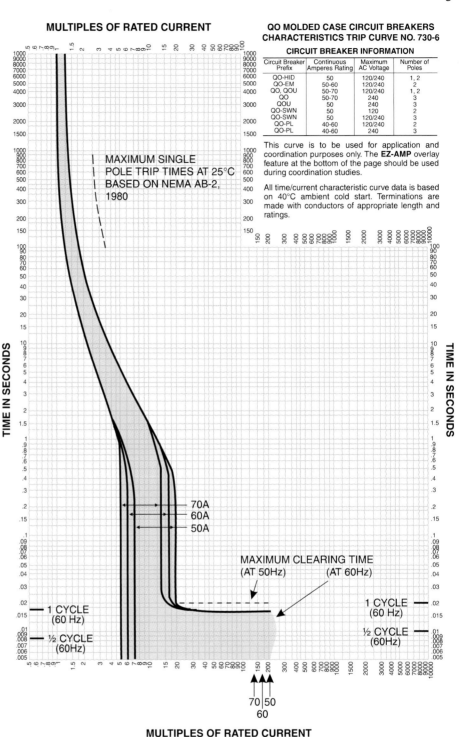

Figure 11-4. *Courtesy of Square D*

Common elements of fault-current path

Clearing faults involves one or two parts, depending on whether the service is grounded or ungrounded. In a grounded distribution system there are two parts; (a) the grounded system conductor and (b) the equipment and conductor enclosure grounding system. In an ungrounded system, the grounding system covers only (b) the equipment and conductor enclosure grounding system.

Effective grounding

To have effective grounding for both a grounded and an ungrounded system, the grounding system must: (1) provide a permanent and continuous path; (2) have ample carrying capacity to conduct safely any fault currents likely to be imposed on it; and (3) have impedance sufficiently low to limit the voltage to ground and to facilitate the operation of the circuit protective devices in the circuit. See Section 250-2(d).

"Impedance sufficiently low" means, for all practical purposes, that the grounding path and the circuit conductors for an ac system must always be within the same metallic enclosure such as a conduit, wireway or cable. The metallic enclosure (such as a conduit or cable) under certain conditions may be used to provide the equipment grounding path for the circuit conductors within it. In addition, where a nonmetallic wiring method is used for an ac system, all the circuit conductors, including the equipment grounding conductor, must be in the same raceway.

Overcurrent Device - Amps	Max Z of Ground-Fault Circuit in Ohms. Includes Fault Impedance.
600	0.4
100	0.24
200	0.12
400	0.06
600	0.04
800	0.03
1200	0.02
1600	0.015
2000	0.012
2500	0.0096
3000	0.008
4000	0.006

The Code gives no maximum impedance of the ground-fault circuit but states in Section 250-2(d) that it shall be sufficiently low to facilitate the operation of the circuit-protective devices. This path should have an impedance no greater than will allow the circuit breaker or fuse to reach its instantaneous pickup operating range. This will cause the overcurrent device to operate quickly to remove the fault from the system. Any fault current less than the instantaneous operating value will extend or delay, by some value, the opening time of the overcurrent device.

Every manufacturer of circuit breakers and fuses publishes operating or trip-curve charts for their products. See Figure 11-4 for one example. These charts should be carefully reviewed to be certain the ground-fault path has an impedance low enough to allow the overcurrent device to operate quickly to reduce thermal damage to the circuit and equipment. As can be seen by reviewing these charts, circuit breakers have a different trip curve depending on whether they are single-, double- or three-pole configurations. In addition, the same

family of circuit breakers will have different trip curves for different ampere-rated circuit breakers. Look for the multiple of the circuit breaker rating at which it reaches its instantaneous pickup rating which is no time delay in the operation of the breaker.

A similar review of the fuse operating characteristics chart should be performed to select a ground-fault circuit that will result in instantaneous operation of the fuse.

Some engineering designs may incorporate the concept of the fault current being not less than five times the rating of the overcurrent device. For example, if the overcurrent device is 400 amperes, not less than 2,000 amperes must flow in the circuit. For some overcurrent devices, current of five times the device rating may not reach the instantaneous trip range of the device. It can generally be said that the longer the excessive current flows in the circuit, the greater the thermal and mechanical stress to the conductor insulation.

Clearing short circuits

The method employed for clearing a short circuit is the same regardless of whether the system is grounded or ungrounded. Essentially, it involves placing an overcurrent device in series with each ungrounded system conductor.

In the event of a "short circuit," which is a fault from conductor to conductor, the words "as quickly as is practical" means a very short period of time down to as low as a fraction of a cycle depending on the amount of short-circuit current that will flow and the characteristics of the overcurrent device. The high-interrupting capacity current-limiting fuse does that automatically within virtually all ranges of short-circuit current values. In addition current-limiting circuit breakers and circuit breaker current limiters are available from a wide variety of manufacturers. Where properly applied, these current-limiting devices significantly reduce the thermal and mechanical damage to electrical equipment.

Current-limiting fuses and their correct application are covered in Underwriters Laboratories Electrical Construction Materials Directory Guide Card (JCQR) information. This section reads in part: "The term 'current limiting' indicates that a fuse, when tested on a circuit capable of delivering a specific short-circuit current (RMS amperes symmetrical) at rated voltage, will start to melt within

90 electrical degrees and will clear the circuit within 180 electrical degrees (½ cycle).

Because the time required for a fuse to melt is dependent on the available current of the circuit, a fuse which may be current limiting when subjected to a specific short-circuit current (RMS amperes symmetrical) may not be current limiting on a circuit of lower maximum available current."[U]

A current-limiting circuit breaker is defined in the Underwriters Laboratories Electrical Construction Materials Directory Guide Card (DIVQ) information as, "A current-limiting circuit breaker is one that does not employ a fusible element and that when operating within its current-limiting range, limits the let-through I^2t (current squared time) to a value less than the I^2t of a ½ cycle wave of the symmetrical prospective current. Current-limiting circuit breakers are marked 'current-limiting' and are marked either to indicate the let-through characteristics or to indicate where such information may be obtained." [U]

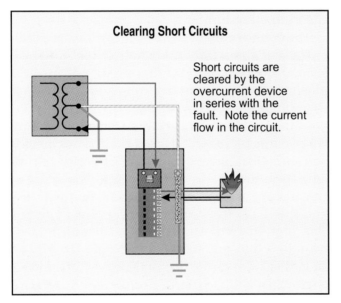

Clearing Short Circuits

Short circuits are cleared by the overcurrent device in series with the fault. Note the current flow in the circuit.

Figure 11-5

Circuit breaker current limiters are covered in Underwriters Laboratories Electrical Construction Materials Directors in Guide Card (DIRW). They are described as "Circuit breaker current limiters are designed to be used in conjunction with specific circuit breakers and to be directly connected to the load terminals of the circuit breakers. They contain fusible elements which function only to increase the fault current interrupting ability of the combination which is intended for use in the same manner as circuit breakers when

installed at the service and as branch circuit protection. The limiters are rated 600 volts or less." [U]

It is vital to carefully follow manufacturer's instructions when applying current-limiting fuses or circuit breakers.

In some cases, depending on the available fault current, this equipment may not be current limiting at all. This is due to the "inverse time" nature of these overcurrent devices. "Inverse time" means that as the amount of current flowing through the device increases, the operating time of the device reduces. Overcurrent devices see fault current at lower operating ranges as a load and react much slower that at their "current-limiting" range.

Many installations of fuses or circuit breakers are installed in a series configuration, that is two or more fuses or circuit breakers are in series for the circuit. These devices that have been tested for their operating compatibility may be marked by the manufacturers with a "series-combination" rating. In this case, usually, the downstream overcurrent device is rated below the fault current that is available at its line terminals while the overcurrent device closest to the source is rated at or above the fault current that is available at its line terminals.

Where a series-rated system is installed, only the equipment that has been tested to determine its suitability can be installed in this manner. Suitability is determined by reference to manufacturer's identification of components on decals located on the equipment. For additional information see "Interplay of Energies in Circuit Breaker and Fuse Combinations." [3]

The time-current characteristic curve of such fuses starting from a value of from five to six times the rating of the fuse indicates a clearing time of about one second and a time of well below one cycle for values of 50 times fuse rating. A further study indicates that such high-interrupting capacity current-limiting fuses have a pronounced current-limiting effect at values as high as fifty times the fuse rating.

Whether a fault current be a short circuit or a ground fault, the overcurrent devices ahead of the point of fault will see all the fault current. Placing an overcurrent protective device at the point the conductor receives its supply, as required by Section 240-21, will provide the short-circuit protection that is necessary. The rating of the overcurrent device is based upon Section 240-3. The basic requirement is that conductors be protected in accordance with their rated ampacity in accordance with Section 310-15.

Clearing ground faults

A ground-fault circuit is different in ways than a short circuit. In a ground fault, there can be such a high impedance of the fault circuit that the controlling factor of current flow is the impedance of the ground-fault circuit. In this case, the only part the available short-circuit capacity plays is in its ability to maintain voltage. In some circuits, the amount of current that will flow in a ground-fault circuit is not dependent on the available capacity of the system, other than its ability to maintain full voltage during a ground fault.

Where it comes to clearing a "ground fault" two things control the flow of current, that is,

(1) The impedance of the circuit, and

(2) The available short-circuit amperes of the system.

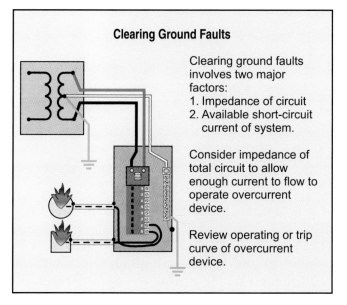

Clearing Ground Faults

Clearing ground faults involves two major factors:
1. Impedance of circuit
2. Available short-circuit current of system.

Consider impedance of total circuit to allow enough current to flow to operate overcurrent device.

Review operating or trip curve of overcurrent device.

Figure 11-6

A relatively large voltage drop can be tolerated in this grounding conductor. The impedance of the complete ground-fault circuit, from the point of fault back to the transformer, should never be higher than would permit a minimum current flow required for the overcurrent device to operate within its instantaneous range. (Review the operating characteristic chart for the overcurrent device for the correct value.) This will provide a factor of safety to allow for the variable impedance at the point of fault. Some overcurrent devices operate at their instantaneous current range at about five times the rating of the overcurrent device. For example, some 50-ampere

overcurrent devices have an instantaneous trip rating of about 250 amperes. In other cases, the instantaneous rating will be from six to ten times the rating of the overcurrent device.

Clearing faults in service equipment

The service equipment enclosure is probably the most vulnerable for ground faults. There is no overcurrent protection on the line side of service conductors, only short-circuit protection provided by the utility's transformer primary overcurrent protection. Overload protection is provided by the service overcurrent device in series with the service-entrance conductors at the load end of the service-entrance conductors. This is the reason the Code requires the service disconnecting means to be located outside the building, or, if inside, nearest the point of entrance of the conductors into the building. This limits the risk of having conductors with no short-circuit protection inside a building.

If, on the other hand, a ground fault does not develop on the load side of the service, but at a point on the line side of the service, then that fault can only be cleared by the overcurrent device ahead of the transformers. This protection is the primary fuses for the transformer which can be many times the rating of the secondary overcurrent device.

In many cases, a ground fault on the line side of the service will not clear through the primary overcurrent devices and can only clear by developing into a short circuit or by burning itself clear.

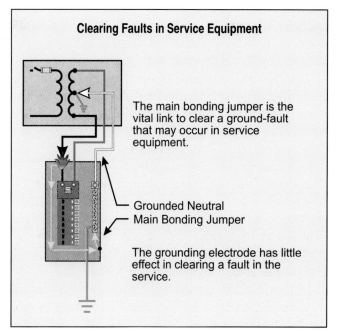

Clearing Faults in Service Equipment

The main bonding jumper is the vital link to clear a ground-fault that may occur in service equipment.

Grounded Neutral
Main Bonding Jumper

The grounding electrode has little effect in clearing a fault in the service.

Figure 11-7

If the electrical equipment on the line side of the service is not properly bonded and a properly-sized main bonding jumper installed, it is unlikely enough current will flow in the path to clear the ground fault through the overcurrent protective devices on the line side of the utility transformer.

Analysis of clearing ground-faults

The following examples illustrate a simple electrical system and analyze the conditions existing in the event of a ground fault. In the first case, the neutral is run from a grounded system where it is connected to the service equipment. In the second example, the grounded system (service) conductor has not been run from the system to the service disconnecting means.

Purposes served by the neutral on a grounded system

For maximum safety and to comply with Section 250-24(b), the neutral or grounded conductor must be run from a grounded system to all services and be bonded to each service disconnecting means enclosure even though all power may be utilized at line voltage only. This is required even though the service supplies only line-to-line loads. See Sections 250-24(b) and 250-28.

The neutral of any grounded system serves two main purposes. First, it permits utilization of power at line-to-neutral voltage, and thus will serve as a current-carrying conductor to carry any unbalanced current back to the source. Second, it plays a vital part in providing a low-impedance path for the flow of fault currents to facilitate the operation of the overcurrent devices in the circuit, as required by Section 250-2(d).

The neutral provides the lowest impedance return path for fault currents to the transformer neutral point as can be traced in Figure 11-8. If the neutral is not needed for voltage requirements, it still must be run to the service and connected to the equipment grounding conductor at the service. In that event, the neutral conductor no longer serves as a neutral but as a grounded conductor. If that is not done, it is difficult, if not impossible, to clear a fault on the system.

Grounded (neutral) conductor installed

In Figure 11-8, the neutral of the system serving as the grounded conductor is carried through to the service equipment and bonded to the equipment grounding conductor to provide a low-impedance

path directly to the transformer. Here it can be seen that a fault current is not required to go through the earth to complete the circuit, but will go through the grounded service conductor, which is a low impedance path. This will very likely permit sufficient current to flow to operate the overcurrent device. The parallel path through the grounding electrode conductor and the grounding electrode and the earth still exists. The grounded service conductor, forming a relatively very low impedance path, will carry most of the fault current, probably 90 percent or more.

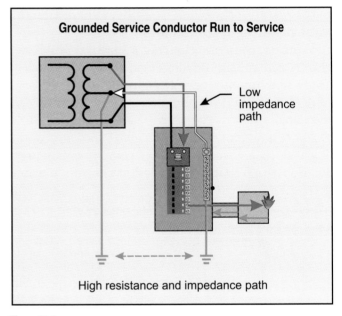

Grounded Service Conductor Run to Service

Low impedance path

High resistance and impedance path

Figure 11-8

It is obvious, then, that to get all the protection afforded by a grounded system, the grounded conductor must be run to the service and must be bonded to the equipment grounding conductor even though the neutral is not required to serve any load whatever. For the same reason, where the neutral is needed for voltage requirements, the neutral size should be based not only on the neutral load demands, but on the basis of the service overcurrent device and the amount of fault current that must flow in order to operate the overcurrent devices.

Neutral not installed

Figure 11-9 shows a 120/240-volt single-phase grounded system where all the load is supplied at 240 volts. A grounding electrode conductor is properly connected to a low-resistance grounding electrode, the water supply system, and bonded to the equipment grounding conductor, all in accordance with the requirements of the Code (prior to the 1962 Code). However, the neutral is *not* carried into the service equipment since all power utilization is at 240 volts only. For serving the load, the neutral has no useful purpose, and it would at first appear as if it could be omitted.

From the standpoint of limiting the voltage between equipment and ground under normal conditions and from the standpoint of utilization of power at 240 volts, the circuit, as shown in Figure 11-9, will work perfectly, so long as the insulation remains perfect and no ground fault occurs on the system.

If, however, a ground fault should occur on the system in the equipment, as illustrated in Figure 11-9, owing to insulation failure, the voltage between equipment and ground will rise considerably. Starting at the point of fault, first is the impedance of the fault, then the conduit itself becomes a conductor in the fault-current circuit, then the grounding electrode conductor and the grounding electrode. However, the fault current must travel from the grounding electrode through the earth itself to the point where the neutral is grounded at the transformer, then through the transformer and back to the service and, finally, through the overcurrent device to the point of fault.

A study of the fault circuit as diagrammatically represented in Figure 11-9 shows that it would be unlikely to have an impedance of less than 22 ohms as the sum of all the impedances shown. That is an optimistically low figure at best. The maximum current that can flow in this circuit is 5.5 amperes for the 120/240 volt circuit shown. (120 volts ÷ 22 ohms = 5.5 amperes) With a 100-ampere service and 20-ampere overcurrent device, it is obvious that the overcurrent device would not operate. A serious shock hazard, as well as fire hazard, will exist until the circuit is manually opened. If the circuit was opened as a result of being grounded properly, then the only period of time a shock hazard would exist would be for the duration of the fault. The fire hazard, as well as shock hazard, would be reduced to a relatively short period of time.

When a fault occurs on a system, it is only during the period while the fault exists that a potential hazard is present. It is, thus, of the utmost importance to safety that the fault clearing time be held to the shortest practical period of time.

With the wiring method as shown in Figure 11-9, the fault will not clear and may exist for minutes, hours or even days before it is recognized. Further, that recognition may come from observ-

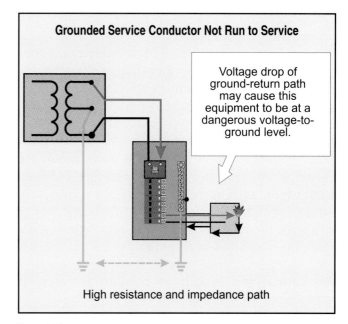

Grounded Service Conductor Not Run to Service

Voltage drop of ground-return path may cause this equipment to be at a dangerous voltage-to-ground level.

High resistance and impedance path

Figure 11-9

ing a fire, or from noting that a victim has received a shock, which sometimes may even be fatal. It is a matter of record that many serious, and sometimes fatal, accidents have resulted from faults which were not cleared promptly because the system was not properly grounded.

If the neutral at the transformer was grounded to the same water pipe system as the service (may not be too likely), then the resistance of the fault path would be appreciably decreased. Because of the wide separation between the service conductor and the water pipe, the reactance and, thus, the impedance of the fault circuit would remain high. The probabilities are that the fault current would not reach a high enough value to operate the overcurrent device. Again, fire and damage to equipment would continue until the circuit was manually opened. To improve the safety of such a system, a low-impedance path must be provided to pass enough current to clear the circuit through the overcurrent devices.

Three-phase services

Identical reasoning to that for single-phase systems may be applied to any multi-wire, multi-phase grounded system. Any 3-phase power supply taken from such a grounded transformer bank must have the neutral or grounded conductor brought into the service of the power supply to satisfy Section 250-24(b). This is true regardless of whether or not there is neutral load at the service. The grounded conductor provides a low-impedance path for fault current to return to the source.

Open system

The above statements applying to maximum ground-fault currents would not apply to an open system. If there are no metallic enclosures, the impedance of the ground-fault circuit would be much lower. Accordingly, greater ground-fault currents may be expected in actual practice in an open or nonmetallic installation. However, since most systems of 600 volts or less are metal enclosed, there may not be such high ground-fault currents in such systems. For open systems, it is vital that the equipment grounding conductor be run with the circuit conductors to maintain a low impedance ground-fault path.

Recommended length of conduit for use as equipment grounding means

The National Electrical Code® places no restriction on the size or length of rigid metal, intermediate metal conduit or electrical metallic tubing where used as an equipment grounding conductor. Independent tests have shown that consideration must be given to both size and length of conduit.

(Extensive work to determine the maximum safe length of conduit to serve as an equipment grounding conductor has recently been done by the School of Electrical and Computer Engineering at the Georgia Institute of Technology in Atlanta, Georgia.[4] Computer software has been developed which will allow the calculation of almost an unlimited combination of conductors and conduit for use in the ground-return path. This software can also be used for calculating the maximum length of equipment grounding conductors that are safe to use where not installed in metallic conduits.)

Where a metallic conductor enclosure is used as an equipment grounding conductor, it must have continuity as well as the required conductivity to pass enough current to facilitate the operation of the overcurrent devices. Some electrical inspection authorities require that an equipment grounding conductor be installed inside the conduit to account for poor workmanship of the raceway installation or to maintain continuity where fittings may be broken during use.

The engineer should examine the equipment grounding conductor (the metal enclosure) to be assured it will function properly in the event of a ground fault. Where wireways, auxiliary gutters and busways have steel enclosures, it will be found that there may be enough cross section to serve as

an equipment grounding conductor. It is questioned if electrical connections between lengths is adequate for carrying enough fault current to clear the fault. These raceways may require a supplemental equipment grounding conductor.

The conductor enclosure, conduit, raceway, etc., also may be acceptable as the equipment grounding conductor in lieu of the copper conductors given in Table 250-122. However, it may be said, in general, that if a conduit, aluminum or steel, or electrical metallic tubing meets the Code requirements for conductor fill, the enclosure will provide an acceptable equipment grounding conductor.

In the case of the average busway up to 1,500 amperes rating, there is enough steel in the enclosure to provide an acceptable equipment grounding conductor if proper conductivity is assured at the joints. For busways of higher rating, it is doubtful if the steel enclosure is heavy enough. For all sizes of busways, the matter of good electrical connection at the joints must be checked carefully. Many busways have aluminum enclosures. In such busways, the enclosure has sufficient conductivity. Good electrical connections at the joints also must be checked.

Long conduit runs

Consider an installation where a 3-inch conduit run is 1,000 feet long. Assume that the minimum fault current would be 600 percent of the overcurrent device rating, or, in this case, 2,400 amperes. The impedance of 3-inch conduit at 2,400 amperes is approximately 0.0875 ohms/1000 feet. For a 208Y/120 volt system with zero impedance at the point of fault, the maximum current that will flow will be about 1400 amperes. That value of current would operate a 400-ampere high-interrupting capacity current-limiting fuse in about two seconds. However, such ground faults are nearly always arcing faults. This adds resistance to the circuit and reduces the fault current that will flow in the circuit. Moreover, since conduit impedance increases with decrease in fault current, the conduit impedance would be 0.129 ohms/1000 feet at a current of 1,400 amperes. Allowing for the arc impedance in the circuit described and the increase in conduit impedance at 1400 amps (about 50 percent) and other variable factors including a further increase of conduit impedance at the still lower current, the ground-fault current flow is more likely to be in the order of about 300 amperes. With a 300-ampere fault current, the 400-ampere high-interrupting capacity current-limiting fuse obviously would not clear the fault.

For a long feeder such as this, it may be necessary to increase conductor sizes to account for voltage drop. Section 250-122 requires that the equipment grounding conductor also be increased in size in proportion to size of the ungrounded conductors. For example, if the 3-inch conduit run was only 100 feet long, the anticipated fault current would be about 3,000 amperes, which would satisfactorily operate the overcurrent device. Both fault-current values were based on the impedance of only that part of the circuit that would be within the conduit, not the impedance of the entire circuit. The fault-current values met in practice would likely be less than those given here.

What can be done to get maximum safety? The system must be designed to see that grounding fulfills all the requirements as set forth in Section 250-2(d) to obtain proper or effective grounding. This may mean adding an equipment-grounding conductor within the conduit in parallel with the conduit if the calculations indicate its need.

Where an equipment-grounding conductor is added inside a raceway, it is done to increase the capacity, that is, decrease the impedance of the raceway that is serving as an equipment grounding conductor. Where a metallic system is installed, the conductor enclosure also is a part of the equipment-grounding conductor. The assistance of that parallel conduit path is valuable, but proper use can be made of it only if the equipment grounding conductor and the conduit are bonded together as frequently as is practical.

If the impedance of the ground-fault circuit is higher than will pass enough current to properly operate the overcurrent devices in a reasonable time, then a lower impedance can be obtained by adding an equipment grounding conductor within the conduit in parallel with the conduit. That conductor must be run within the conduit, that is, run with the circuit conductors. The equipment grounding conductor should never be run outside the conduit or raceway through which the conductors serving the equipment are run. Where run external to the conduit, it becomes quite ineffective in the grounding circuit, for virtually all the ground-fault current will return on the conduit. Further, the equipment grounding conductor must be run as close to the phase conductors as is practical, right to the point where it attaches to the neutral conductor at the service.

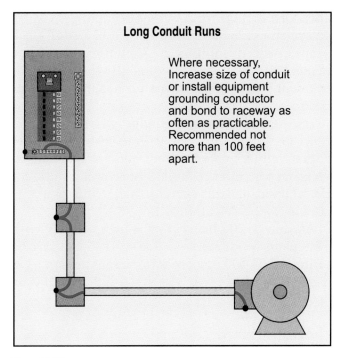

Long Conduit Runs

Where necessary, Increase size of conduit or install equipment grounding conductor and bond to raceway as often as practicable. Recommended not more than 100 feet apart.

Figure 11-10

By determining the impedance of the circuit involved when a ground fault occurs, it can be determined whether the metallic enclosure will make an acceptable equipment grounding conductor or whether it will be necessary to supplement the enclosure with an equipment grounding conductor. Where an iron conduit is a part of the electric circuit, as it will be where a ground fault occurs, there will be a large increase in both the resistance and reactance of the circuit, which will vary considerably with the amount of fault current.

Laboratory tests have shown that where a single-phase current flows in a conductor within an iron conduit, the impedance of the circuit is approximately equal to the impedance of the conduit itself. The size of the conductor within the conduit has relatively little effect on the circuit impedance. Also, despite the fact that there are many parallel paths external to the conduit, the current flowing in all the parallel paths will be very small, and under normal conditions would be under 10 percent of the total fault current.

Two other factors are to be taken into account in estimating the ground-fault current flow. They are the effects of the conduit couplings in increasing the impedance of the circuit and the voltage drop across the point of fault. If conduit couplings are installed wrenchtight, as required by Code, the increase in impedance of the conduit with couplings is about 50 percent more than the impedance for a straight run without couplings. This is where the value and importance of installing a properly-sized equipment grounding conductor inside the conduit and bonding it to the conduit at frequent intervals is proven for ensuring low impedance and safety.

Impedance values for conduit can be obtained from manufacturers. By using these impedance values, adjusting for the couplings and allowing 50 volts drop across the fault, a reasonable value for the amount of fault current that will flow in the circuit can be determined.

Assume a 200-foot run of 3-inch conduit with 500 kcmil conductors on a 208Y/120-volt circuit protected by 400 ampere overcurrent devices. On a ground fault, the current that will flow will thus be:

$$\frac{E \text{ (voltage)}}{Z \text{ (impedance)}} = I \text{ (current)}$$

With a 50-volt drop at the fault and Z (impedance) = 0.02970, the current that will flow will be about 2,350 amperes. The use of a 3-inch conduit as the equipment grounding conductor where 400 ampere overcurrent devices are used is, thus, satisfactory for this run.

A simpler method of determining if the conduit or metallic enclosure will perform satisfactorily is to first calculate the minimum desired fault current flow (5 times the overcurrent device rating or better yet to reach the instantaneous portion of the time/current curve) which, in this case, is 5 x 400 or 2,000 amperes. Then, on the basis of 70 volts available for a 120-volt-to-ground circuit, calculate Z (impedance), which will be found to be 0.035. A straight run of 3-inch conduit has about 0.099 ohms impedance per M feet at 2000 amperes. To that figure add 50 percent to include a factor of safety. That will give an impedance of 0.01485 ohms per 100 feet. The impedance value that will allow 2,000 amperes to flow in that circuit was found to be 0.035. Since 235 feet of 3-inch conduit carrying 2,000 amperes will have an impedance of 0.035, it has been determined that to have a minimum current flow of 2,000 amperes in a ground fault, up to 235 feet of 3-inch conduit can be installed where a 400-ampere overcurrent device is involved.

If 4-inch conduit was used instead of 3-inch, and the overcurrent device rating did not change but remained at 400 amperes, the maximum length of conduit can be determined by reference to the table. With some interpolation and using the same

calculations, it is found that 260 feet of 4-inch conduit could be installed and provide a satisfactory equipment grounding conductor for this circuit.

It can be determined for any circuit and any size conduit, for any size overcurrent device, the maximum safe length of conduit which will allow a fault current to pass which will be sufficient to facilitate the operation of the overcurrent device. Should the circuit length exceed the maximum safe length as calculated, then it will be necessary to add a metallic (copper or aluminum) equipment grounding conductor in parallel with the conduit or increase the size of the conduit. The equipment grounding conductor size is per Table 250-122.

It would be neither practical nor desirable to substitute a copper or aluminum conductor for the conduit. Rather, add the copper or aluminum conductor, and connect it in parallel with the conduit to form an equipment grounding conductor of two conductors in parallel. The copper or aluminum conductor and the conduit should be connected together at convenient practical intervals, about every 100 feet or less. That will reduce the length of circuit through which the ground-fault current may flow on the conduit alone.

The Code permits an equipment grounding conductor to be bare or insulated. However, if the conductor is bare, there may be some arcing between the bare conductor and the interior of the conduit at points other than the point at which the ground fault occurs. This is due to slight differences in impedance between the raceway and wire, resulting in potential differences which cause the arcing. Such arcing may damage the phase conductors without adding to the proper functioning of the ground-fault circuit. That makes a strong case for the use of insulated equipment grounding conductors where installed in a metallic enclosure.

If aluminum conduit was used instead of steel conduit for the same conditions cited above, (500 kcmil copper conductors, 3-inch conduit and a 400-ampere overcurrent device) the circuit run could be about 900 feet long, and the aluminum conduit would provide a satisfactory equipment grounding conductor. A 3-inch aluminum conduit has a dc resistance of about 0.0088 ohms/1000 feet and 500 kcmil copper cable has a dc resistance of 0.0222 ohms/1000 feet.

Flexible metal conduit is suitable as an equipment grounding conductor for not more than 6-foot lengths in the ground-fault return path and with not more than 20 ampere overcurrent protection of the contained conductors. As such, flexible metal conduit should include a metallic equipment grounding conductor. The Code requires that the various metal raceways shall be so constructed that adequate electrical and mechanical continuity of the complete system be secured. However, owing to the various joints involved, it is well for the engineer to investigate such conductor enclosures to be assured that their impedance is sufficiently low to function along the lines cited above in the event of a ground fault.

Adequate size of conductor

In general, the minimum size of equipment grounding conductor is given in Table 250-122 and is based on the rating of the overcurrent device ahead of the conductors. (An analysis of Table 250-122 is provided in Table Seven of Chapter 18.) The rule of thumb may be applied, but it should be checked by calculation that the equipment grounding conductor should not be less than 25 percent of the capacity of the phase conductors or the overcurrent device that supply the circuit. A note has been added to the table indicating that the size of equipment grounding conductor given in the table must be increased if necessary to comply with Section 250-2(d). This adds emphasis to the heading of Table 250-122 which indicates that equipment grounding conductors given in the table are the "minimum size."

An analysis of Table 250-122 of the Code shows that the relation of the equipment grounding conductor to the size of the overcurrent device (based on the continuous rating of 75°C-rated wire). It is from 50 to 125 percent of the phase conductor for overcurrent devices up to 100-ampere rating. The rating varies from 33 to 25 percent for overcurrent devices rated up to 400 amperes and is from 22 percent for 600 ampere overcurrent devices to a low of only 8 percent for an overcurrent device of 6,000 amperes.

Obviously, the equipment grounding conductor must be large enough to carry that amount of current for a given time, which is required to clear the overcurrent device with which it is associated, and not result in extensive damage. In addition, it is vital that the equipment grounding conductor be run with the circuit conductors or enclose the circuit conductors. This was covered extensively in Chapter Nine of this text.

Conductor withstand rating

Section 110-9 of the NEC® states, "Equipment intended to interrupt current at fault levels shall have an interrupting rating sufficient for the nominal circuit voltage and the current which is available at the line terminals of the equipment."[N] This includes fuses, circuit breakers, disconnect switches, contactors, and similar equipment.

Section 110-10 reads in part, "The overcurrent protective devices, the total impedance, the component short-circuit current ratings, and other characteristics of the circuit to be protected shall be selected and coordinated to permit the circuit protective devices used to clear a fault to do so without extensive damage to the electrical components of the circuit. This fault shall be assumed to be either between two or more of the circuit conductors, or between any circuit conductor and the grounding conductor or enclosing metal raceway."[N]

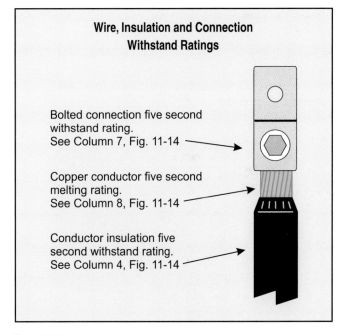

Wire, Insulation and Connection Withstand Ratings

Bolted connection five second withstand rating. See Column 7, Fig. 11-14

Copper conductor five second melting rating. See Column 8, Fig. 11-14

Conductor insulation five second withstand rating. See Column 4, Fig. 11-14

Figure 11-11

Grounding conductors, circuit conductors, bus bars, bonding jumpers, etc., are not intended to break current. These conductors must be large enough to safely carry any short-circuit and/or ground-fault current for the time that it takes the overcurrent protective device to clear the fault. This is clearly stated in Sections 110-10, 240-1 Fine Print Note (FPN), 250-1 (FPN No. 1 and 2), 250-2(d), 250-90, and 250-96(a). Section 310-10 also discusses in detail the temperature limitations for conductors.

The integrity of equipment grounding conductors, grounding electrode conductors, main bonding jumpers and other circuit conductors must be assured by sizing them properly.

Equipment-grounding conductors that are too small are of little value in clearing a fault and may, in fact, give one a false sense of security. Grounding conductors must not burn off under ground-fault conditions, leaving the equipment "live" thus, in many cases, creating a shock hazard which could be fatal.

Safety cannot be compromised. In fact, the first sentence of the National Electrical Code® states that, "the purpose of this Code is the practical safeguarding of persons and property from hazards arising from the use of electricity."*

Bolted connections

It can be shown by calculation, using values from the Insulated Cable Engineers Association Publication P 32-382 (1994), that an insulated copper conductor with a bolted connection may safely carry one ampere for every 42.25 circular mils for five seconds without destroying its validity. That will be the short-time rating or I^2t (am-

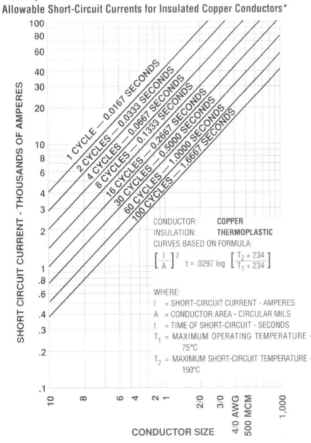

Short-Circuit Current Withstand Chart for Copper Cables with Thermoplastic Insulation
Allowable Short-Circuit Currents for Insulated Copper Conductors*

Figure 11-12. *Courtesy of Insulated Cable Engineers Association*

peres x amperes x time) value of the conductor. Then, from the time-current characteristic curves of various approved overcurrent devices, the amount of current will be necessary to clear the overcurrent device in five seconds can be found.

Using that formula, the size of equipment grounding conductors which will be proportional to those given in Table 250-122, as analyzed in Table Seven of Chapter 18 can be determined.

Grounding conductors are permitted to be bare, and, in most cases, are pulled into the same raceway as the insulated phase conductors. This presents a potential problem. Where a grounding conductor is carrying ground-fault current, an extreme rise in its temperature can cause the insulation on the adjacent phase conductors to melt, causing further damage. Again, the potential for equipment damage and electrical shock hazard to personnel is increased. It is desirable to limit the heat of the faulted circuit to reduce damage to adjacent insulated conductors.

Thus, as discussed in this chapter, for copper conductors, the clearing time and short-circuit current must be limited to:

One ampere...

for five seconds...

for every 42.25 circular mils.

This can be expressed by the term "ampere squared seconds (I^2t)."

For example, from Table 8 of Chapter 9 of the NEC®, a No. 8 AWG conductor has a cross-sectional area of 16,510 circular mils. By dividing the circular mil area of the conductor by 42.25, the conductor's five-second withstand rating can be calculated.

$$\frac{16,510}{42.25} = 391 \text{ amperes}$$

Stated another way, this conductor has an I^2t five-second withstand rating of: 391 x 391 x 5 = 764,405 ampere squared seconds.

From this five-second withstand rating value, it is easy to calculate the conductor's withstand rating for other values of time and/or for other values of current.

Example 1: How many amperes will the No. 8 AWG copper conductor be able to safely carry if the impedance of the circuit along with the operating characteristics of the overcurrent device protecting the circuit results in a 2 cycle (0.0333 seconds) opening time? Example 1 is solved as follows.

$$I^2t = 764,405 \text{ ampere squared seconds}$$
$$I^2 = \frac{764,405}{t}$$
$$I = \frac{764,405}{0.0333}$$
$$I = 4,791 \text{ amperes}$$

Example 2: How many amperes will the No. 8 AWG copper conductor be able to safely carry if the impedance of the circuit along with the operating characteristics of the overcurrent device protecting the circuit results in a ¼ cycle (0.0042) opening time? Example 2 is solved as follows:

$$I^2t = 764,405 \text{ ampere squared seconds}$$
$$I^2 = \frac{764,405}{t}$$
$$I = \frac{764,405}{0.0042}$$
$$I = 13,491 \text{ amperes}$$

Note that in the example above, because a much faster total clearing time is achieved, the allowable fault current that the conductor will be subjected to can be increased. This is a result of substituting different time values in the I^2t formula.

Generally, where current-limiting overcurrent devices are protecting the circuit, the equipment grounding conductor sizes are determined directly from Table 250-122. Where available fault currents are high, and the overcurrent protective device takes longer than ¼ cycle to clear the fault, is it suggested that the grounding conductor be sized per Figure 11-12, to be on the safe side.

Figure 11-14 provides information to assist the installer in the proper selection of equipment grounding conductors. Among other information, it includes:
(1) safe values for 75°C thermoplastic insulated conductors,
(2) safe values for bolted connections,
(3) unsafe (melting) values for the copper conductor itself.

Because the "weakest link" in any system is the insulation short-circuit withstand rating as found in columns 4, 5, and 6 of Figure 11-14, it is recommended that column 4 be the deciding factor where selecting equipment grounding conductors. The Insulated Cable Engineers Association data, Figure 11-14, Column 4 calculates out to the previously discussed:

Do not exceed one ampere...
for five seconds...
for every 42.25 circular mils of copper conductor.
This chart also shows the No. 8 AWG conductor used in the above text examples.

Where you are absolutely certain that the equipment grounding conductor will not come into contact with any of the current-carrying insulated circuit conductors, the withstand rating of the grounding conductor can be established as follows:

Do not exceed one ampere...
for five seconds...
for every 29.1 circular mils of copper conductor.
See Column 7, Figure 11-14.

This is the standard previously referred to in this text as the "Do not exceed one ampere for five seconds for every thirty circular mills of conductor area." This value is only to be used where bare equipment grounding conductors are used in such a manner that they will not come into contact with insulated conductors. In this application, the limiting element of the circuit is the bolted connection of the lug.

Column 8 of Figure 11-14 gives the current in amperes at which the melting temperature of copper conductors is reached. Of course, you never

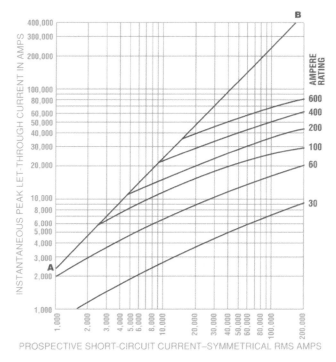

FUSETRON® Class RK5 Dual-Element Time-Delay Fuses
FRN-R

Figure 11-13. *Courtesy of Bussmann Division, Cooper Industries*

Copper 75° C. Thermoplastic Insulated Conductor, Bare Conductor, Bolted Connection Five Second Withstand Rating In Amperes and Melting of Copper Wire

1	2	3	4	5	6	7	8
Wire size	Area in circular mils	Area in square mm	ICEA Amperes	IEC Amperes	IEE Amperes	Bolted connection 250°C. Amperes	Melting of conductor 1,083°C. Amperes
14	4,110	2.080	97	107	107	141	254
12	6,530	3.310	155	170	170	224	403
10	10,380	5.261	246	271	271	357	641
8	16,510	8.367	391	430	430	567	1,020
6	26,240	13.300	621	684	684	902	1,621
4	41,740	21.150	988	1,088	1,088	1,435	2,578
3	52,620	26.670	1,245	1,372	1,372	1,808	3,251
2	66,360	33.620	1,571	1,729	1,729	2,281	4,099
1	83 690	42.410	1,981	2,181	2,181	2,876	5,170
1/0	105 600	53.490	2,499	2,751	2,751	3 629	6,523
2/0	133,100	67.430	3,150	3,468	3,468	4,574	8,222
3/0	167,800	85.010	3,972	4,372	4,372	5,767	10,366
4/0	211,600	107.200	5,009	5,513	5,513	7,272	13,071
250	250,000	126.700	5,918	6,516	6,516	8,592	15,443
300	300,000	152.000	7,101	7,818	7,818	10,310	18,532
350	350,000	177.300	8,285	9,119	9,119	12,029	21,621
400	400,000	202.700	9,467	10,425	10,425	13,747	24,709
500	500,000	253.300	11,834	13,027	13,027	17,184	30,887
600	600,000	304.000	14,201	15,636	15,636	20,621	37,064
700	700,000	354.700	16,568	18,243	18,243	24,057	43,241
750	750,000	380.000	17,752	19,544	19,544	25,776	46,330
800	800,000	405.400	18,935	20,850	20,850	27,494	49,419
900	900,000	456.000	21,302	23,453	23,453	30,931	55,596
1,000	1,000,000	506.700	23,669	26,060	26,060	34,368	61,773

Column 4 - Insulated Cable Engineers Association publication P32-382. One ampere for five seconds for every 42.25 circular mils of conductor area.

Column 5 - International Electrotechnical Commission publication 364-4-43.

Column 6 - Institute of Electrical Engineers publication 434-6.

Column 7 - Calculated from data in Electrical Engineers Handbook (75°C ambient). One ampere for five seconds for every 29.1 circular mils of conductor area.

Column 8 - Calculated from data in Electrical Engineers Handbook (75°C ambient). One ampere for five seconds for every 16.19 circular mils of conductor area.

Figure 11-14

want to reach the current shown as the equipment grounding conductor will burn off leaving the equipment ungrounded and a possible shock hazard.

Conclusion on equipment grounding conductor

For a grounded system, it is vital to safety that a low-impedance equipment grounding conductor path be provided in addition to a good grounding electrode with as low a resistance as practical. This allows for sufficient current to flow to clear a ground fault automatically in a limited time, which would be as quickly as is practical, without undue interruption of service.

The I^2t values found in Column 7 of Figure 11-14 are based on the adequacy of a copper conductor and its bolted joints to carry the current values without destroying its validity. The values are obtained from an IEEE committee report in "A Guide to Safety in AC Substation Grounding." The figure expressed in amperes per circular mil is one ampere for every 29.1 circular mils cross section. The time of five seconds was used to provide a safety factor and was considered a reasonable approach for distribution systems of 600 volts or less protected by high-interrupting-capacity current-limiting fuses and having equipment ground-fault protection.

As previously stated, where grounding conductors might come into contact with insulated phase conductors, use the values found in Column 4 of Figure 11-14. This column is based on one ampere for every 42.25 circular mils of conductor for five seconds. Figure 11-14 has all the calculations done and is much easier to use rather than performing complicated calculations.

The short-time rating of the equipment grounding conductor bears an approximately constant relation to the size of the overcurrent device. The I^2t values of the conductors given in Table 250-122 are between about 13 and 28 times their nominal continuous rating based on one ampere for every 42.25 circular mils cross section.

References

[1] Electric Power Distribution for Industrial Plants. The Institute of Electrical and Electronics Engineers, Inc., © 1954 AIEE (now IEEE), 445 Hoes Lane, PO Box 1331, Piscataway, NJ 08855-1331.

[2] Portions of "Some Fundamentals of Equipment-Grounding Circuit Design" by R. H. Kaufmann, Paper 54-244 presented at the AIEE Summer and Pacific General Meeting, Los Angeles, California, June 24-25, 1954, © 1954 AIEE, (now IEEE), 445 Hoes Lane, PO Box 1331, Piscataway, NJ 08855-1331.

[3] 1991 IEEE Industry Applications Society Annual Meeting, Volume II. The Institute of Electrical and Electronics Engineers, Inc., © 1954 AIEE (now IEEE), 445 Hoes Lane, PO Box 1331, Piscataway, NJ 08855-1331.

[4] Modeling and Testing of Steel EMT, IMC and Rigid (GRC) Conduit
School of Electrical and Computer Engineering
Georgia Institute of Technology
Atlanta, Georgia 30332

Chapter Eleven: The questions included here were developed using material included in this chapter. The answers can be found by reviewing the text. It is also important that students make use of the 1999 NEC®, where many answers can be found. See page 279 for answers.

1. A conducting connection, whether intentional or accidental, between any of the conductors of an electrical system whether it be from line to line or line to the grounded conductor, is defined as a ____.
  a. ground fault
  b. phase fault
  c. short circuit
  d. unidentified fault

2. A conducting connection, whether intentional or accidental, between any of the conductors of an electrical system and the conducting material which encloses the conductors, or any conducting material that is grounded, or that may become grounded is defined as a ____.
  a. identified fault
  b. ground fault
  c. short circuit
  d. failure of the system

3. A short circuit may be from one phase conductor to another phase conductor, or from one phase conductor to the grounded conductor or ____.
  a. unidentified conductor
  b. enclosure
  c. neutral
  d. equipment grounding conductor

4. The grounding system must provide a permanent and electrically continuous path; must have ample carrying capacity to conduct safely any fault currents likely to be imposed on it; and must have an impedance sufficiently low to limit the voltage to ground and to facilitate the operation of the circuit protective devices in the circuit. This is best described as being ____.
  a. improved
  b. effective
  c. sufficient
  d. required

5. In general, the rule of thumb may be applied, but should be checked by calculation, that the equipment grounding conductor should not be less than ____ percent of the capacity of the phase conductors that supply the circuit.
  a. 25
  b. 20
  c. 15
  d. 18

6. There is no overcurrent protection on the line side of service conductors, there is only ____ protection on the primary of the transformer.
  a. ground-fault
  b. short-circuit
  c. overload
  d. equipment ground fault

7. The grounded service conductor must be run to each service disconnecting means and bonded to the enclosure even if the service supplies only ____.
  a. 240 volt loads
  b. 208 volt loads
  c. 480 volt loads
  d. line-to-line loads

8. A No. 8 THW copper conductor has an allowable ampacity of 50 amperes according to Table 310-16. This conductor also has a safe (ICEA) five-second withstand rating of ____ amperes, but will melt if subjected to ____ amperes for five seconds.
  a. 8,367   10,000
  b. 391   1,020
  c. 30   500
  d. 4,110   254

9. When ICEA conductor short-circuit withstand rating tables are not available, it is possible to calculate the short-circuit current withstand ratings for a conductor. For example, to determine the safe withstand rating for copper conductors with 75° C insulation, use ____ ampere for every ____ seconds for every ____ circular mils of cross sectional area of the conductor.
  a. one, five, 42.25
  b. one, five, 16.9
  c. one, five, 29.1
  d. one, five, 10.2

10. Table 250-122 shows the minimum size equipment grounding conductors. Where high values of fault current are available, the grounding conductors may have to be ____ in size to be capable of safely carrying the available fault current.
  a. decreased
  b. increased

11. Referring to Figure 11-12, find the minimum conductor sizes required for the fault current values indicated.
  a. 10,000 amperes for one cycle.   Size ____
  b. 20,000 amperes for two cycles.   Size ____
  c. 4,200 amperes for one cycle.   Size ____
  d. 10,000 amperes for eight cycles.   Size ____

Grounding Separately Derived Systems

Objectives

After studying this chapter, the reader will be able to understand:

- General requirements and definitions for separately derived systems.
- Installation and sizing of bonding jumper for separately derived systems.
- Sizing and types of grounding electrode conductors for separately derived systems.
- Transformer overcurrent protection.
- Generator types and systems.
- Ground-fault systems.

Definition

Separately Derived System. "A premises wiring system whose power is derived from a battery, a solar photovoltaic system, or from a generator, transformer, or converter windings, and that has no direct electrical connection, including a solidly-connected grounded circuit conductor, to supply conductors originating in another system."[N] (NEC® Article 100.)

General

In many distribution systems of 600 volts or less for commercial or industrial occupancies, it is common to have separately derived systems at another voltage level lower or higher than the electrical system supplied by the service. Section 250-20 of the NEC® gives requirements for system configurations and voltage levels for which the system must be grounded. Where the separately derived system meets the criteria found in Section 250-20, it must be grounded according to the requirements of Section 250-30. Where the separately derived system is not required to be grounded by Section 250-20, grounding of the derived system is optional.

Each voltage level of a system that is required to be grounded must be grounded independently according to the requirements of Section 250-30 of the NEC®. The equipment grounding conductor path of each voltage level must meet the requirements of Section 250-2(d) in being electrically continuous and permanent, having ample capacity to carry the currents that may be expected and being connected to the grounded conductor (at that

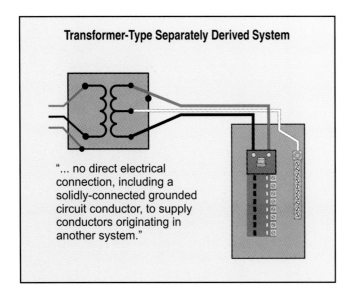

Transformer-Type Separately Derived System

"... no direct electrical connection, including a solidly-connected grounded circuit conductor, to supply conductors originating in another system."

Figure 12-1

voltage level) within the enclosure, as is done in the case of service equipment. That connection will become a common point for the grounded conductor and the equipment grounding conductor where the common conductor may be attached and run to the grounding electrode.

Transformer-type separately derived system

Figure 12-1 is a one-line diagram of a transformer-type separately derived system. (For simplicity, grounding and bonding conductors are not shown.) The supply to the transformer is at one voltage level, and the secondary is often at another voltage level, either lower or higher. As shown, there is no direct electrical connection between the transformer primary and secondary so the installation meets the definition of a separately derived system. Where 3-phase systems are employed, the primary is usually connected delta so a neutral conductor is not needed. The secondary may be connected delta or wye as desired.

Grounding primary side equipment

An equipment grounding conductor must be supplied with the primary circuit to provide a low-impedance fault-current path from the transformer case to the main service or source of supply. The equipment grounding conductor is permitted to consist of any of the means included in Section 250-118 including wires and the wiring method, where appropriate. The overcurrent device on the primary of the transformer will then clear a short circuit or ground fault up to, and including, the transformer primary windings.

The equipment grounding conductor connection to the transformer enclosure, plus the bonding jumper from the secondary does not constitute a "direct electrical connection from the primary system to the secondary system. The equipment grounding conductor is not a "system" conductor as the ungrounded and grounded conductors are.

A short circuit or ground fault on the secondary of the transformer is seen as a load by the transformer primary winding. The amount of current flowing in the fault on the secondary side will determine whether or not the primary overcurrent device will open.

Bonding jumper

Section 250-30(a)(1) requires a bonding jumper be installed for a derived system that is required to be grounded by Section 250-20.

Section 250-30(a)(1) permits the bonding jumper to be installed at any point from the source of the separately derived system to the first system disconnecting means or overcurrent device. Many transformer manufacturers supply a bonding jumper strap and terminals for connection inside the transformer enclosure. However, in some cases, this bonding strap may be for the purpose of bonding the transformer core to the enclosure and not for the derived system.

In Figure 12-2, the neutral of the separately derived system is shown connected inside the transformer case to the case through the bonding jumper. The bonding jumper is required to be sized according to Table 250-66 based on the size of the derived phase conductors on the secondary side of the transformer. Where the derived phase conductors are larger than given in Table 250-66, the bonding jumper must be not less than 12½ percent of the circular mil area of the derived phase conductors.

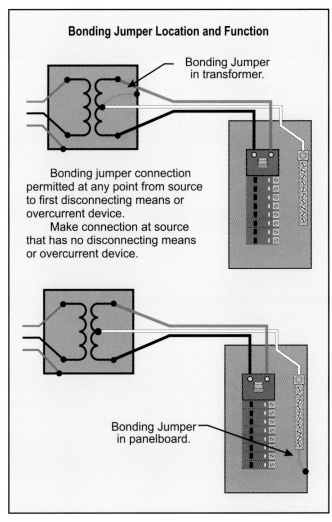

Bonding Jumper Location and Function

Bonding Jumper in transformer.

Bonding jumper connection permitted at any point from source to first disconnecting means or overcurrent device.
Make connection at source that has no disconnecting means or overcurrent device.

Bonding Jumper in panelboard.

Figure 12-2

For example, if the derived phase conductors are 3/0 copper, the bonding jumper must be no smaller that No. 4 copper or No. 2 aluminum. In this regard, requirements for separately derived systems are similar to services. Like services, the size of the bonding jumper is not based on the rating of the main overcurrent device or on the rating of the equipment being supplied but on the size of the ungrounded conductor supplied.

An alternate, and equally correct method would be to bond the derived grounded conductor or neutral to the grounded conductor (neutral) terminal bar in the distribution cabinet. As indicated in Section 250-30(a)(1), the derived grounded conductor (neutral) could be bonded at any other convenient point between the source of power and the first system disconnecting means or overcurrent device such as in a pull or junction box.

The bonding jumper is a most critical part of the separately derived system. Without it, there will not be a path for ground faults on the derived system to return to the source. All ground-fault current downstream from the derived system will pass over the bonding jumper as it returns to its source. This ground-fault current from the derived system will not attempt to return to the service equipment as that is not its source.

The general rule for the bonding jumper is that it be installed at one location only and that the grounding electrode conductor for the separately derived system be connected at the same location. In addition, the grounded conductor of the derived system is permitted to be bonded at both the source and first overcurrent device or disconnecting means only where doing so does not create a parallel path for the grounded circuit conductor. See Section 250-30(a)(1) Exception No. 1. An example of this is where a nonmetallic raceway is installed between a transformer and panelboard and there are no other paths for grounded conductor current including through building steel, water pipes, etc.

Bonding jumper location exceptions

Section 250-30(a)(1) recognizes the provision for connecting the bonding jumper for a separately derived system according to the requirements for a high-impedance grounded system. These rules are given in Sections 250-24(a)(1), 250-36 and 250-186 of the NEC®. These systems are grounded only through an impedance device (often a resistor) usually at the panelboard or motor control center and not at the transformer. These systems are

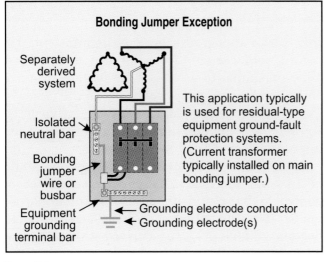

Bonding Jumper Exception

Separately derived system

Isolated neutral bar

Bonding jumper wire or busbar

Equipment grounding terminal bar

This application typically is used for residual-type equipment ground-fault protection systems. (Current transformer typically installed on main bonding jumper.)

← Grounding electrode conductor
← Grounding electrode(s)

Figure 12-3

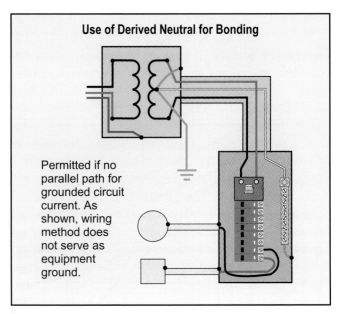

Use of Derived Neutral for Bonding

Permitted if no parallel path for grounded circuit current. As shown, wiring method does not serve as equipment ground.

Figure 12-4

designed to limit the fault current of the first ground fault to a predetermined value and often incorporate an indication or alarm system. Additional information on high-impedance grounded systems is provided in Chapter 15 of this text.

Section 250-30(a)(1) permits the grounding electrode connection to be made to an equipment grounding terminal bar, rather than to the neutral bar where the main bonding jumper is a wire or busbar, as permitted by Section 250-24(a)(4). This provides for residual-type equipment ground-fault protection systems where, often, a current transformer is located on the bonding jumper to measure ground-fault currents flowing back to the source. Another common application of this exception is in switchboards where the main bonding jumper is a wire or busbar.

Use of derived neutral for bonding

Section 250-142(a)(3) permits the grounded circuit conductor to be used for grounding equipment "On the supply side or within the enclosure of the main disconnecting means or overcurrent devices of a separately derived system where permitted by Section 250-30(a)(1)."[N] As used in Section 250-142(a)(3), the term, "on the supply side" no doubt means "up to or within the enclosure for the disconnecting means or overcurrent devices" as it is widely accepted to make bonding connections of the neutral or grounded circuit conductor within the disconnecting means enclosure for the separately derived system and upstream to the transformer or generator.

As shown in Figure 12-4, the grounded circuit conductor serves two purposes. First, it allows

line-to-neutral or grounded conductor loads to be supplied, and the conductor carries the unbalanced loads from ungrounded conductors. Secondly, it serves as the equipment grounding conductor and will carry line-to-ground fault currents back to the source. This grounding scheme parallels that for services and for supply to additional buildings on the premises served from a common service.

Where this scheme is used, it is not necessary to install an equipment grounding conductor between the source and the disconnecting means or overcurrent protection enclosure. This will allow nonmetallic raceways to be used. Where nonmetallic raceways are used, a parallel path for neutral current does not exist.

Keep in mind that this method of bonding the grounded conductor at both ends is permitted only where doing so does not create a parallel path for grounded conductor current. A parallel path for neutral current exists where a metal raceway is used between the source of the separately derived system and the metal enclosure for the disconnecting means or overcurrent device. A parallel path can also be established by other means such as through metal pipes, cable trays and structural steel. The neutral current will divide between the available paths depending upon the impedance of the paths. The lowest impedance path will obviously carry the most current.

Grounding electrode conductor

Next, a grounding electrode conductor must be

installed to comply with Section 250-30(a)(2). The size of this conductor is determined by Table 250-66, based upon the size of the derived phase conductors. For example, if the derived phase conductors are 500 kcmil aluminum, the minimum size grounding electrode conductor determined by reference to Table 250-66 is No. 2 copper or 1/0 aluminum. Unlike the bonding jumper, the maximum size of grounding electrode conductor is 3/0 copper or 250 kcmil aluminum.

The connection of the grounding electrode conductor is generally required to be made at the same point the bonding jumper is installed. Similar to the provision for installing the bonding jumper, the grounding electrode conductor is permitted to also be installed:

(1) Where separately derived systems supply a dual fed (double ended) switchboard, panelboard or motor control center (or similar) the grounding electrode conductor is permitted to be made to the tie point.

(2) The grounding electrode is permitted to be

connected to the equipment grounding terminal bar rather than to the terminal bar for the grounded system conductor where the system supplies a bonding-jumper type equipment ground fault protection system.

An exception is also provided for Class 1, Class 2 or Class 3 transformer derived circuits (usually control circuits) where the transformer is rated not more than 1000 volt-amperes in accordance with the requirements contained in NEC® Section 250-30(a)(2) Exception.

Grounding electrode for separately derived system

Section 250-30(a)(3) of the NEC® is very specific about the grounding electrode required to be used for separately derived systems. The grounding electrode is to be the nearest one of the following:

(a) An effectively grounded structural metal member of the structure; or

(b) An effectively grounded metal water pipe within 5 ft. from the point of entrance of the water pipe within the building; or

(c) Other grounding electrodes as specified in Sections 250-50 and 250-52 where electrodes indicated in (a) or (b) are not available.

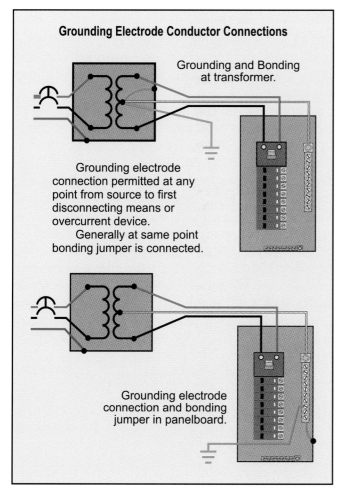

Grounding Electrode Conductor Connections

Grounding and Bonding at transformer.

Grounding electrode connection permitted at any point from source to first disconnecting means or overcurrent device.
Generally at same point bonding jumper is connected.

Grounding electrode connection and bonding jumper in panelboard.

Figure 12-5

Grounding electrode for separately derived systems.

- **Near as practicable.**

- **Nearest effectively grounded structural metal member.**

- **Effectively grounded water pipe within 5 ft. of entrance into building.**

- **Other electrodes where the above are not available.**

Note that the term "effectively grounded" is defined in Article 100 of the NEC® and means, "Intentionally connected to earth through a ground connection or connections of sufficiently low impedance and having sufficient current-carrying capacity to prevent the buildup of voltages that may result in undue hazards to connected equipment or to persons."[N]

An exception to the general requirement in Section 250-30(a)(3)(b) is provided for industrial and commercial buildings where conditions of

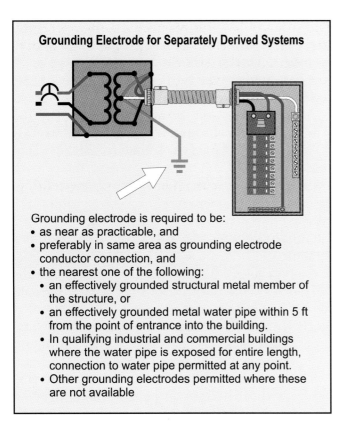

Grounding Electrode for Separately Derived Systems

Grounding electrode is required to be:
• as near as practicable, and
• preferably in same area as grounding electrode conductor connection, and
• the nearest one of the following:
  • an effectively grounded structural metal member of the structure, or
  • an effectively grounded metal water pipe within 5 ft from the point of entrance into the building.
  • In qualifying industrial and commercial buildings where the water pipe is exposed for entire length, connection to water pipe permitted at any point.
  • Other grounding electrodes permitted where these are not available

Figure 12-6

(Note that is some cases, the grounding electrode conductor is required to be larger for the separately derived system than for the service based on the size of the ungrounded phase conductor of the systems.)

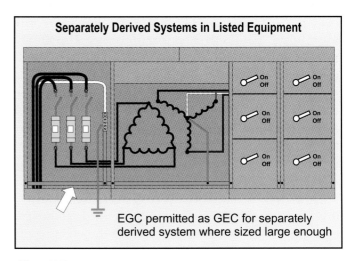

Separately Derived Systems in Listed Equipment

EGC permitted as GEC for separately derived system where sized large enough

Figure 12-7

maintenance and supervision ensure that only qualified persons will service the installation and the entire length of the interior metal water pipe that is being used for the grounding electrode is exposed. This exception permits the connection to the water pipe to be located at any point on the water pipe system rather than within 5 ft. of the point where the water pipe enters the building.

An exception to Sections 250-30(a), (b), and (c) permits the service equipment grounding electrode conductor to serve the separately derived system or the internal bussing to serve as the grounding electrode conductor for separately derived systems under the following conditions:

(1) The separately derived system originates in listed equipment also used as service equipment. (This equipment may be a switchboard, motor control center or unit substation.)

(2) The grounding electrode conductor for the service is large enough for the separately derived system.

(3) Where the equipment ground bus in the service equipment is large enough, the bus is permitted to serve as the grounding electrode conductor for the separately derived system.

The Code goes on to say that in all other respects, grounding methods shall comply with other requirements in the Code. This includes requirements for installation methods and protection of conductors from damage and connection of the grounding electrode conductor to grounding electrodes.

Section 250-58 requires that the same grounding electrode be used for grounding all ac systems in or at the building. This applies to services as covered in Section 250-24 and to two or more buildings or structures in Section 250-32, but does not directly apply to separately derived systems. Where more than one grounding electrode is installed, they then must be bonded together and are considered to be one grounding electrode. See Section 250-58. Since the building service is grounded to effectively grounded building structural steel, and the interior water pipe is required to be bonded, the separately derived system and the service neutral or grounded conductor will be electrically connected together.

In some installations, the same grounding electrode will be used for both the utility supplied electrical system (service) and the separately derived system. This occurs where an effectively grounded structural metal member of the structure or effectively grounded water pipe, as directed in Section 250-50, is available and used for the grounding electrode. In other cases where these "system" grounding electrodes are not available, other electrodes such as concrete encased electrodes, ground rings or made electrodes, such as metal pipes, tanks, rods and plates, are required to be used.

Note that Section 250-50 requires that all grounding electrodes at the building or structure served be bonded together to form the grounding electrode system. So, if an additional grounding electrode is installed for the separately derived system, it must be bonded to the other grounding electrodes that are present and available at the building or structure.

Grounding of ungrounded separately derived systems

Rules are provided in Section 250-30(b) of the NEC® for grounding ungrounded separately derived systems. Basically, a grounding electrode conductor connects the metal equipment of the separately derived system to a grounding electrode similar to the rules for grounded systems. However, the system itself is not grounded.

Ground-fault return path

As with all electrical systems, it is vital to be certain that, for separately derived systems, an effective ground-fault return path exists from the furthermost point on the electrical system back to the source.

Fault current downstream from the separately derived system follows the same laws of physics as does the fault current on the line side. It attempts to return to its source, which is the secondary windings of the transformer, and seeks the lowest impedance path.

As can be seen in the figure, fault current downstream from the separately derived system will return to the secondary windings of the transformer, not to the primary or to the service. The transformer primary sees the fault current on the secondary as a load, and the overcurrent devices on the line side of the transformer will respond according to their rating.

As illustrated in Figure 12-7, be certain that a path for ground-fault current is permanent and electrically continuous, of adequate capacity, and low impedance from the equipment supplied back to the separately derived system. Often, it is best to make a one-line diagram of the installed system to be certain a complete and low-impedance path is provided.

One fairly common error in installing these systems is to use flexible metal conduit between the transformer and panelboard without installing an equipment grounding conductor through the flexible conduit. As fully explored in Chapter 5 of this textbook, flexible conduit is suitable as an equipment grounding conductor only where the contained conductors have overcurrent on the line side not greater than 20 amperes. This requirement practically prevents flexible metal conduit from being used between the transformer and the panelboard or fused switch unless an equipment grounding means is installed between the transformer and the overcurrent device.

If the bonding jumper is installed in the panelboard or fusible switch rather than in the transformer, and flexible metal conduit is used as the wiring method, an equipment grounding conductor must be installed between the transformer enclosure and the overcurrent device enclosure. In this case, a parallel path will exist through the equipment grounding conductor and the metal wiring method.

Equipment grounding conductors

The equipment grounding conductors, of both the primary and the separately derived system,

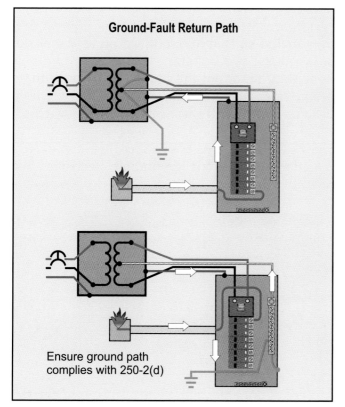

Ground-Fault Return Path

Ensure ground path complies with 250-2(d)

Figure 12-8

may be obtained by using the metal conductor enclosures, such as conduit or electrical metallic tubing, where permitted by Code. An equipment grounding conductor may be used, if necessary, to be certain the circuit is of low enough impedance to be effective. In both the primary and the separately derived system, a low-impedance path to the grounded conductor or neutral of each system respectively must be provided. This is vital to ensure that overcurrent devices will function to remove ground-faults that may occur.

The important connections here require an equipment grounding conductor for both the primary and separately derived systems independent of each other and terminating at the respective points of supply at their grounded conductor or neutral. Both equipment grounding conductors will meet at some common point, as at the point where the bonding jumper is connected, and they both may connect to the common grounding electrode.

Interconnecting the equipment grounding conductor of the two systems does not violate the definition of separately derived system. This is due to the fact that the equipment grounding conductor is not a system or power conductor nor a grounded conductor. Also, the equipment grounding conductor usually does not make a direct electrical connection of the neutral to ground. By following, on a one-line diagram, the bonding and equipment grounding conductors of the two systems, it can be seen that the systems are, in fact, connected together by these conductors.

Bonding metal water piping

Water pipe in the area served by the separately derived system must be bonded to the derived grounded conductor at the nearest available point. This will ensure that this metal water piping is not electrically isolated and, should it become energized, would not become a shock hazard.

The connection to the separately derived system must be made at the same point where the grounding electrode conductor is connected.

The bonding conductor must be sized in accordance with Table 250-66 based on the size of the derived phase conductors.

Transformer overcurrent protection

Overcurrent protection of the primary as well as the secondary disconnect for the transformer must comply with Section 450-3. In addition, Section 384-16(e) generally requires that where a trans-

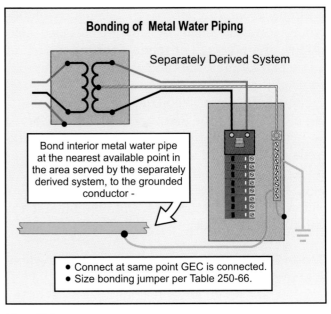

Bonding of Metal Water Piping

Separately Derived System

Bond interior metal water pipe at the nearest available point in the area served by the separately derived system, to the grounded conductor -

• Connect at same point GEC is connected.
• Size bonding jumper per Table 250-66.

Figure 12-9

former supplies a panelboard, overcurrent protection, not exceeding the rating of the panelboard, be provided on the secondary of the transformer.

An exception that the supply conductors terminate in an overcurrent protective device in a panelboard is provided for certain transformers installed in compliance with Section 240-21(c)(1).

Generator-type separately derived system

Figure 12-9 illustrates a separately derived system from a generator. Note that there is no connection, including that of a solidly grounded neutral, between the two systems. An easy way to determine whether or not the generator is a separately derived system is to examine the transfer switch. If the neutral and all phase conductors are switched,

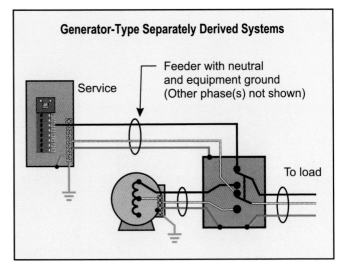

Generator-Type Separately Derived Systems

Service

Feeder with neutral and equipment ground (Other phase(s) not shown)

To load

Figure 12-10

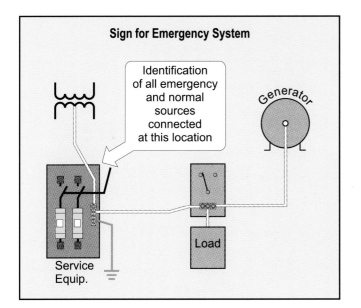

Figure 12-11

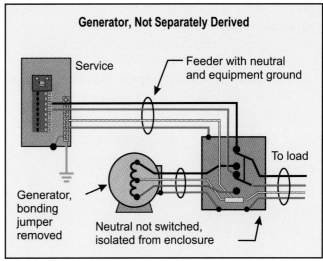

Figure 12-12

it is a separately derived system. If the neutral is not switched, but is solidly connected, then it is not a separately derived system.

Where the generator is a separately derived system, the neutral bonding jumper must be installed, either at the generator or at the first disconnecting means or overcurrent device or any point between. In addition, a grounding electrode conductor must be installed between the neutral and a grounding electrode.

Section 700-8(b), regarding grounding of emergency systems, requires that where the grounded circuit conductor connected to the emergency source is connected to a grounding electrode conductor at a location remote from the emergency source, a sign must be placed at the grounding location that shall identify all emergency and normal sources connected at that location.

Identical requirements are in Section 701-9(b) for legally required standby systems and in Section 702-8(b) for optional standby systems.

Alternate power supply systems that have an unswitched neutral, that are grounded at the service entrance, present a safety concern, as electricians that are unfamiliar with the system grounding scheme may inadvertently disconnect the grounded (neutral) conductor when working on the normal source. If the emergency system is operating, the neutral conductor from the transfer switch to the location where it is grounded will function as a grounding electrode conductor, and disconnecting it may present a serious shock hazard.

Generator not separately derived

Where the generator is not a separately derived system, the neutral bonding jumper must be removed from the generator, and the neutral must not be grounded at the generator or at any point between the generator up to the service.

In this case, the system is grounded by its solid connection to the neutral of the premises wiring system. An equipment grounding conductor is installed throughout the system with the circuit conductors between the service, transfer switch, and other noncurrent carrying parts of the installation.

Ground-fault systems

A word of caution. It is imperative that care be exercised where generators supply systems that have equipment ground-fault protection. In most cases, the engineer will specify that the neutral be switched by the transfer switch to avoid grounding the neutral downstream from the ground-fault protection equipment. This is vital to prevent desensitizing the protection system. In this case, the generator is a separately derived system, and it must be grounded and bonded separately.

Section 700-26 indicates that equipment ground-fault protection with automatic disconnecting means is not required for emergency systems. However, indication of a ground fault on the emergency system must be given in accordance with the rules of Section 700-7(d). It is required that, where practicable, audible and visual signal devices be provided that will indicate a ground fault in a solidly-grounded wye connected emergency system of more than 150 volts to ground and circuit protective devices rated 1,000 amperes or more. A similar exception from the requirement for ground-fault protection of equipment is provided in Section 701-17 for legally required standby systems.

Chapter Twelve: The questions included here were developed using material included in this chapter. The answers can be found by reviewing the text. It is also important that students make use of the 1999 NEC®, where many answers can be found. See page 279 for answers.

1. A premises wiring system whose power is derived from generator, transformer, or converter windings without any direct electrical connection, including a solidly-connected grounded circuit conductor to supply conductors originating in another system, describes some ____.
   a. medium voltage systems
   b. utility supplied systems
   c. separately derived systems
   d. high voltage systems

2. Phase conductors used for a separately derived system sized at 500 kcmil aluminum, require a grounding electrode conductor not smaller than No. ____.
   a. 4 copper or 2 aluminum
   b. 2 copper or 1/0 aluminum
   c. 6 copper or 4 aluminum
   d. 8 copper or 6 aluminum

3. The connection of the grounding electrode conductor is permitted to be made at the transformer, or the ____ disconnecting means or overcurrent device or any point between.
   a. first
   b. second
   c. any
   d. line-side

4. The grounding electrode for a separately derived system is required to be located as near as practicable to, and preferably in, the ____ as the separately derived system.
   a. same enclosure
   b. same cabinet
   c. same area
   d. same pull box

5. A grounding electrode conductor for a separately derived system is required to be connected to the nearest available, effectively grounded ____ in the structure.
   a. underground metal gas pipe
   b. structural metal member
   c. service equipment enclosure
   d. metal raceway

6. Where the grounding electrode for a separately derived system, as in question No. 5, is not available, the grounding electrode is required to be ____.
   a. the nearest available effectively grounded metal enclosure
   b. the nearest available effectively grounded metallic gas pipe
   c. an effectively grounded metal water pipe within 5 ft. of entry.
   d. the nearest available effectively grounded plate electrode

7. Where installed in or at the building, the same ____ is required to be used for the grounding of all ac systems.
   a. grounding electrode
   b. grounded raceway
   c. electrode enclosure
   d. bonding jumper

8. The ____ grounding conductors of both the primary and the separately derived system are permitted to be obtained by using the conductor enclosures, where permitted by Code.
   a. isolated
   b. identified
   c. system
   d. equipment

9. If the neutral and all phase conductors are switched, it ____ a separately derived system.
   a. is
   b. is not
   c. may be
   d. cannot be considered as

10. Where a generator is a separately derived system, the ____ bonding jumper must be installed, either at the generator or at the first disconnecting means or overcurrent device.
   a. equipment
   b. neutral
   c. service
   d. No. 10

11. Where the generator is not a separately derived system, the ____ bonding jumper must be removed from the generator and the neutral must not be grounded. In this case, the system is grounded by its solid connection to the neutral of the premises wiring system.
   a. No. 10
   b. equipment
   c. service
   d. neutral

Notes:

Chapter 13

Grounding at More Than One Building

Objectives

After studying this chapter, the reader will be able to understand:

- General requirements for grounding at more than one building.
- Sizing of grounding conductors.
- Bonding grounding electrodes together.
- Disconnecting means for separate buildings.
- Objectionable currents.
- Requirements for mobile homes, recreational vehicles, and agricultural buildings.

Definition

"**Grounding Conductor:** A conductor used to connect equipment or the grounded circuit of a wiring system to a grounding electrode or electrodes."[N] (See Article 100 of the NEC®)

As indicated in the definition, this conductor includes both the grounding electrode conductor used at services and separately derived systems as well as the conductor used to connect the equipment or the grounded circuit conductor to a grounding electrode(s) at a second or additional building or structure.

General

Section 250-32 provides requirements for grounding of electrical systems and equipment at buildings or structures that are supplied from a common service.

Included are rules for grounding electrodes, for grounded systems, ungrounded systems and grounding at buildings where the disconnecting means are located in a separate building or structure on the same premises.

Grounding electrode required

The general rule in Section 250-32(a) is that at each building or structure served from a common service by one or more feeders or branch circuits, a grounding electrode meeting the requirements of Article 250 Part C must be connected in a manner specified in Section 250-32(b) or (c). Where there are no existing grounding electrodes at the building or structure, the grounding electrode(s) required in Part C of Article 250 must be installed.

Grounding electrodes required in Part C of Article 250 and thus must be installed or used as required in that part include:

(1)   A metal underground water pipe – 250-50(a).
(2)   Metal frame of the building – 250-50(b).
(3)   Concrete-encased electrode – 250-50(c) .
(4)   Ground ring – 250-50(d)
(5)   Made electrodes of the following type where the above electrodes are not available at the building or structure:
  (a) Metal underground gas piping systems are not permitted to be used as a grounding electrode – 250-52(a)
  (b) Local metal underground systems or structures such as piping systems and underground tanks – 250-52(b)
  (c) Rod and pipe electrodes – 250-52(c)

(d) Plate electrodes – 250-52(d)

The exception to Section 250-32(a) of the NEC® makes it clear that a grounding electrode is not required at the separate building or structure where only one branch circuit supplies the building or structure and the branch circuit includes an equipment grounding conductor for grounding equipment supplied by the branch circuit.

See Chapter Six of this text for additional information on installing grounding electrodes.

Grounding of grounded systems

Two general methods are provided in Section 250-32(b) of the National Electrical Code® for grounding electrical systems and/or equipment at additional buildings or structures on the premises. Buildings or structures supplied by a feeder(s) or more than one branch circuit that contain an equipment grounding conductor must comply with the rules in Section 250-32(b)(1). Buildings or structures supplied by a feeder(s) or more than one branch circuit that do not contain an equipment grounding conductor and meet the additional requirements must comply with the rules in Section 250-32(b)(2).

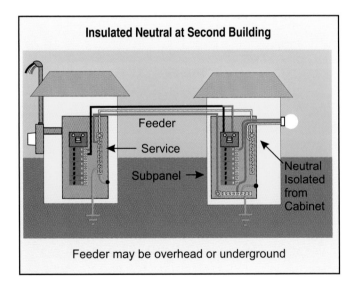

Insulated Neutral at Second Building

Feeder

Service

Subpanel →

Neutral Isolated from Cabinet

Feeder may be overhead or underground

Figure 13-1

Grounding with an equipment grounding conductor

In this method, an equipment grounding conductor is run with the feeder(s) or branch circuit(s), along with the ungrounded and grounded conductors to the additional building or structure. See Section 250-32(b). In this method, the grounded (neutral) conductor cannot be connected to the equipment grounding conductor or to a grounding electrode at the additional building or structure but is insulated from the equipment enclosure. This method is similar to or identical to installing a feeder to a feeder panel in the same building or service where the service is located.

The equipment grounding conductor, that is run from the building or structure where the service is located, is connected to a terminal bar on the supply side of or inside the building or structure disconnecting means where it is grounded by connecting the grounding electrode to the equipment grounding terminal bar. Equipment grounding conductors of the type included in Section 250-118 are acceptable and include wires as well as some conduits or other raceways.

The number of conductors for various systems that must be installed is summarized in the following table.

Table 13-1
Equipment Grounding Conductor
Installed

System	Un-grounded	Grounded	Equip. Ground.
120V	1	1	1
120/240V	2	1	1
208/120V	3	1	1
480/277V	3	1	1

The grounded (often a neutral) conductor, installed as a part of the feeder or branch circuit from the building or structure where the service is located to the second or additional building or structure, must be an insulated conductor. See Section 310-2(a).

By using this method, the grounded (often a neutral) conductor(s) must be connected to the terminal bar for the grounded conductors, and the equipment grounding conductor(s) must be connected to the equipment grounding terminal bar where the grounding (electrode) is connected inside the building or structure disconnecting means enclosure.

The equipment grounding conductor that is run to the additional building or structure must be sized from Table 250-122 of the NEC® based on the rating of the overcurrent device at the beginning of the feeder or branch circuit.

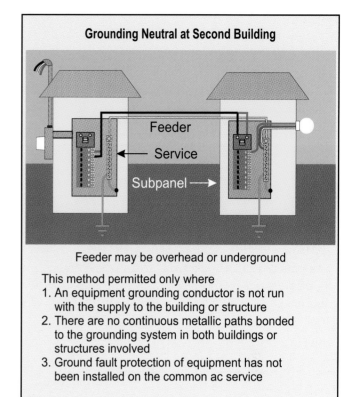

Grounding Neutral at Second Building

Feeder may be overhead or underground

This method permitted only where
1. An equipment grounding conductor is not run with the supply to the building or structure
2. There are no continuous metallic paths bonded to the grounding system in both buildings or structures involved
3. Ground fault protection of equipment has not been installed on the common ac service

Figure 13-2

Grounding without an equipment grounding conductor

Section 250-32(b)(2) permits the grounded circuit conductor (often a neutral) to be grounded again at the disconnecting means for the building or structure only where all the following conditions are complied with:

(1) An equipment grounding conductor is not run with the supply to the building or structure, and

(2) There are no continuous metallic paths bonded to the grounding system in both buildings or structures involved, and

(3) Ground fault protection of equipment has not been installed on the common ac service.

In this case, the electrical system at the additional building or structure is treated like a service for grounding purposes, although the building or structure is supplied from a service in another building or structure by a feeder or branch circuit. The grounded circuit conductor is bonded to the disconnecting means enclosure for the building or structure. The grounding electrode connection to the grounded (often a neutral) conductor must be made on the supply side of or within the building or structure disconnecting means.

In this method, both the grounded circuit con-

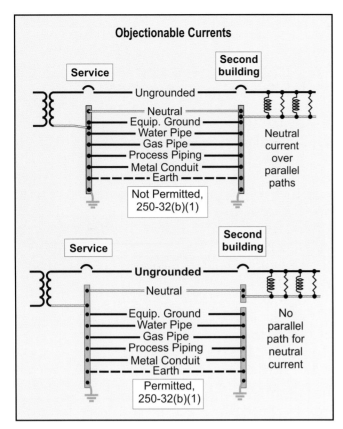

Figure 13-3

ductor (usually a neutral) and the noncurrent-carrying equipment are connected to a grounding electrode (system) at the additional building or structure. The number and type of conductors that must be taken from the building or structure where the service is located to the second building or structure is summarized in the following table.

Table 13-2
Equipment Grounding Conductor
Not Installed

System	Un-grounded	Grounded	Equip. Ground.
120V	1	1	0
120/240V	2	1	0
208/120V	3	1	0
480/277V	3	1	0

Both the grounded (usually a neutral) conductor(s) and the equipment grounding conductor(s) are permitted to be connected to the terminal bar where the grounding electrode conductor connects to the grounded conductor.

Note that the grounded conductor from the building or structure where the service is located to the second or additional building or structure must generally be an insulated conductor. See Section 310-2(a). Section 338-3(b) permits a Type SE cable without insulation on the grounded circuit conductor to be used as a feeder to supply other buildings on the same premises.

Size of grounded conductor

The grounded circuit conductor must be sized no smaller than an equipment grounding conductor from Table 250-122 where an equipment grounding conductor is not run from the service to the additional building or structure. In this case, the grounded circuit conductor serves three purposes: first, to permit line-to-neutral loads to be utilized; second, to carry unbalanced loads back to the source; and, third, as an equipment grounding conductor. As such it must be sized for the calculated load according to Section 220-22 and not smaller than an equipment grounding conductor from Table 250-122.

Grounding conductor

A grounding conductor (a grounding electrode conductor at services) is used to connect the grounded circuit conductor or equipment grounding conductor and equipment to the grounding electrode system that exists or is installed at the additional building or structure. As covered above, where no grounding electrodes exist at the additional building or structure, one or more grounding electrodes must be installed in compliance with Part C of Article 250.

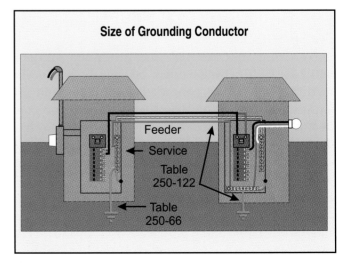

Figure 13-4

This connection must be made inside the building or structure disconnecting means to the equipment grounding terminal bar or to the terminal bar for the grounded conductor as appropriate.

Sizing grounding conductor

The grounding conductor (grounding electrode conductor at services) is sized from Table 250-122 based upon the ampere rating of the overcurrent protective device ahead of the feeder or circuit. See Section 250-32(f).

Bonding grounding electrodes together

Section 250-58 does not require that grounding electrodes at separate buildings or structures be bonded back to the service. A study of the system will show that there is not really two isolated grounding electrode systems. The grounding electrode system for the electrical system at the individual buildings or structures are connected together either by the grounded conductor or by the equipment grounding conductor. This is illustrated by both Figures 13-1 and 13-2.

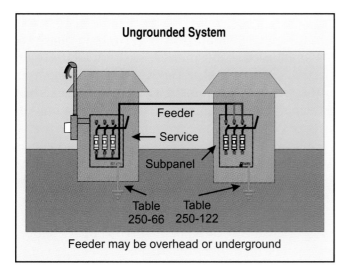

Ungrounded System

Feeder

Service

Subpanel

Table 250-66

Table 250-122

Feeder may be overhead or underground

Figure 13-5

Ungrounded systems

Where more than one building or structure is supplied with an ungrounded electrical system, a grounding conductor must connect the electrical equipment enclosure to the grounding electrode that exists at the building or structure. A grounding electrode system in compliance with Part C of Article 250 must be installed at the building or structure where none exists.

The exception in Section 250-32(a) excludes buildings or structures supplied by only one branch circuit that includes an equipment grounding conductor from the requirement to install a grounding electrode system at the building or structure.

Disconnecting means located in separate building or structure on the same premises

Special rules have been provided for large capacity, multibuilding industrial installations under single management in Section 250-32(d). These occupancies have trained and qualified personnel and have established procedures for safe switching of electrical feeders or circuits. As a result, the disconnecting means are permitted to be located at other locations on the premises rather than at the building served. Often, the switching is managed by automatic or manual means from a control room or station.

The special rules for grounding the electrical system at these separate buildings or structures are as follows:

1. The grounded circuit conductor cannot be grounded at the additional building or structure.

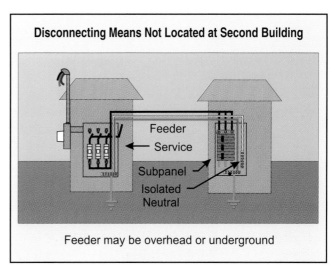

Disconnecting Means Not Located at Second Building

Feeder

Service

Subpanel

Isolated Neutral

Feeder may be overhead or underground

Figure 13-6

2. An equipment grounding conductor must be run with the circuit conductors to the additional building or structure. There, it must be connected to noncurrent-carrying electrical equipment, interior metal piping systems and building or structural metal frames and to existing grounding electrodes. Where no grounding electrodes exist at the building or structure, a grounding electrode in compliance with Part C of Article 250 must be installed where more than one branch circuit is supplied.

3. The equipment grounding conductor run to

the building or structure must be bonded to the noncurrent-carrying electrical equipment and to the grounding electrode in a junction box, panelboard or similar enclosure such as a switchboard or motor control center.

Mobile homes and recreational vehicles

The basic requirement in Section 550-23(a) is that mobile home service equipment be located remote from the structure but within sight from and not more than 30 feet from the mobile home. There, the grounded (usually a neutral) conductor is grounded to a grounding electrode. Four insulated and color coded conductors, one of which is an equipment grounding conductor, must be run to the mobile home distribution panelboard. See Section 550-24(a).

Section 550-23(a) permits the service equipment to be located remote from the mobile home site, if a disconnecting means suitable for use as service equipment is installed within sight from and not more than 30 feet from the mobile home it serves, and grounding at the disconnecting means complies with Section 250-32. This exception permits the two options for grounding to be used as described earlier in this chapter.

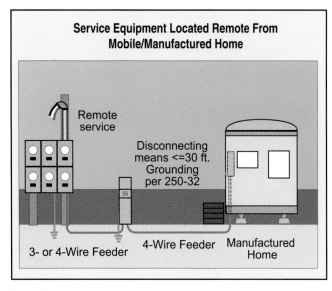

Service Equipment Located Remote From Mobile/Manufactured Home

Remote service

Disconnecting means <=30 ft. Grounding per 250-32

3- or 4-Wire Feeder

4-Wire Feeder

Manufactured Home

Figure 13-7

A three-conductor feeder is permitted to be run from the service located remote from the mobile home disconnecting means where the neutral is grounded, and four insulated conductors in compliance with Section 550-24(a) are run from the mobile home disconnecting means to the mobile

home. This permits the mobile home service equipment to be located at a common location, as is required by some serving electric utilities. The grounding of the grounded circuit conductor at the disconnecting means must comply with Section 250-32(b) as covered above.

Section 550-23(b) permits service equipment to be installed directly on the manufactured home (not mobile home – see the definitions in Section 550-2) under four conditions.

(1) The manufactured home is secured to a permanent foundation that complies with the applicable building code, and

(2) Service equipment is installed in a manner acceptable to the authority having jurisdiction, and

(3) The service equipment installation complies with Article 230, and

(4) Means are provided for the connection of a grounding electrode conductor to the service equipment and routing it outside the structure.

Where this concept is chosen, the service is grounded to the grounding electrode at the manufactured home, and no service disconnecting means remote from the manufactured home is required.

For additional information, see Part 3280, Manufactured Home Construction and Safety Standards, of the federal Department of Housing and Urban Development for requirements related to manufactured homes.

Agricultural buildings

The grounding of agricultural buildings is covered in Article 547 of the NEC®. Section 250-32(e) requires that where the building or structure houses livestock, an insulated or covered copper conductor must be installed where the equipment grounding conductor is run underground from one building or structure to another. See Chapter Sixteen of this text for additional information on the subject.

Two buildings in one structure

Figure 13-8 shows two buildings in one structure that are separated by a fire wall. The two buildings may be served by a single service or by multiple services in accordance with Sections 230-2(a), (b), (c) or (d).

As previously discussed, Section 250-58 requires that all grounding electrodes at a building be bonded together. If two or more grounding electrodes are used, they become, in effect, a single

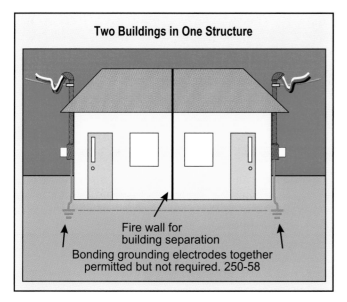

Figure 13-8

electrode if they are connected together with a conductor properly sized per Section 250-66.

Since the fire wall creates two buildings, bonding of the electrodes together is desirable but not required by Code. The electrodes can be bonded together by connecting them together by a properly sized bonding jumper.

Chapter Thirteen: The questions included here were developed using material included in this chapter. The answers can be found by reviewing the text. It is also important that students make use of the 1999 NEC®, where many answers can be found. See page 279 for answers.

1. A conductor used to connect equipment or the grounded circuit of a wiring system to a grounding electrode or electrodes is defined as a ____.
    a. main bonding jumper
    b. grounding conductor
    c. grounded conductor
    d. neutral conductor

2. A grounding electrode is required at a separate building or structure under which of the following conditions?
    a. one grounded system feeder is installed
    b. one ungrounded system feeder is installed
    c. more than one circuit is installed
    d. all of the above

3. Which of the following methods is NOT acceptable for grounding the electrical system at separate buildings or structures?
    a. grounding the grounded system conductor
    b. grounding the equipment grounding conductor
    c. grounding the ungrounded conductor

4. Where a grounding electrode(s) is not installed at the second building or structure, ___.
    a. one or more must be installed.
    b. installing one or more is optional.
    c. an equipment ground must be installed
    d. a grounded conductor must be installed

5. An accepted method for grounding at separate buildings recognizes the installation of an ____ where it is run with the feeder or circuit ungrounded (hot) conductors.
    a. approved conductor
    b. identified bonding jumper
    c. equipment grounding conductor
    d. acceptable conductor

6. All of the following are acceptable as a grounding electrode at a separate building or structure EXCEPT
    a. ½-inch trade size conduit
    b. ⅝-inch iron or steel not less than 8 feet long
    c. ¾-inch galvanized conduit
    d. plate electrode

7. Where an equipment grounding conductor is run with the feeder or circuit from a first building to a separate building, the size is based upon the ampere rating of the overcurrent protective device ____ the feeder or circuit.
    a. on the load side of
    b. in the transformer of
    c. on the line side of
    d. in the service disconnect of

8. A grounding conductor is used to connect the ____ circuit conductor, or equipment grounding conductor and equipment, to the grounding electrode system at the separate building(s).
    a. feeder
    b. main bonding
    c. grounded
    d. branch

9. Under certain circumstances, the grounded circuit conductor ____ to be used for grounding equipment on the line (supply) side of the main disconnecting means for separate buildings.
    a. is permitted
    b. is not permitted
    c. is permitted by special permission
    d. is required

10. Currents that introduce noise or data errors in electronic equipment, such as data processing equipment, are ____ considered objectionable currents.
    a. not to be
    b. considered to be
    c. always to be
    d. sometimes

11. Where the additional building or structure houses livestock, an insulated or covered ____ conductor must be installed where run underground.
    a. bonding
    b. copper clad
    c. aluminum
    d. copper

Ground-Fault Circuit-Interrupters

Objectives

After studying this chapter, the reader will be able to understand:

- Requirements for replacement of ungrounded receptacles.
- Requirements for locations of GFCI-protected receptacles.
- Ground-fault circuit-interrupter principles of operation
- Ratings of GFCI devices.
- Markings for GFCI devices.
- GFCI application and consideration.

Shock Hazard Protection. [1]

The Underwriters Laboratories (UL) requirement for Class A Ground-Fault Circuit Interrupters (GFCIs) is that tripping shall occur when the continuous 60 - hertz differential current exceeds 6 mA but shall not trip at less than 4 milliamperes (5 mA ± 1 mA). Some people contend that 5 mA is too low and should be increased to 10 mA or higher.

Several eminent investigators, including C.F. Dalziel, F.P. Kouwenhoven, O.R. Langworthy and others, have prepared papers on the dangers of electric shock hazards. They define "let-go" current as the maximum current at which a person is able to release a conductor by commanding those muscles directly stimulated by the shock. Currents over the "let-go" levels are said to "freeze" the victim to the circuit.

Appended to this report is one of Dalziel's many papers, titled "Electric Shock Hazard" and published in the Institute of Electrical and Electronics Engineers (IEEE) Spectrum, Vol. 9, February 1972. It summarizes the studies that estimate shock currents based on the effective impedance of the body under various conditions. According to Dalziel, the reasonably safe electric current for normal healthy adults is the let-go current from which 99.5 percent of the population can extricate themselves from the circuit by releasing the conductor. On page 44, Dalziel states, "[s]o far, it has been impossible to obtain reliable [let-go] values for children; they just cry at the higher values." However, the IIT Research Institute report cites a value for children evidently based on engineering judgment by Dalziel and others. The following summarizes the let-go currents for 99.5 percent of the population:

For children	4.5 milliamperes (mA)
For women	6.0 mA
For men	9.0 mA

Based on information provided in the report by IIT Research Institute, 4.5 milliamperes for children may be the appropriate GFCI trip level relative to the 6 mA and 9 mA thresholds for men and women, but the rationale for this selection is not very evident in Dalziel's or other publications.

Based on Dalziel's data, UL's estimates of the percentage of the population that would be protected against inability to let-go at various current levels is shown in the following table.

For general purposes, a "let-go" limit for GFCIs higher than 6 mA appears inappropriate because too large a fraction of the population would be left unprotected. UL has decided to continue to designate 5 mA (±1 mA) as the limit based on tolerated reaction and physiological effects. Moreover, UL retains the 5 mA limit because the 4.5 mA level is only an estimate that is not based on data and because the 5 mA limit has withstood the test of time with no evidence of being inadequate and has appeared in a number of standards for many years.[2]

Table 14-1
Percentage of the Population Estimated to Be Protected Against Inability to Let Go for Several Levels of Shock Current

Level of Shock Current	6mA(rms)	10mA (rms)	20mA (rms)	30mA (rms)
Men	100%	98.5%	7.5%	0%
Women	99.5%	60%	0%	0%
Children*	92.5%	7.5%	0%	0%

*half of let-go threshold for men

The first sensation of electricity can be felt by most people at currents considerably less than 0.5 mA, 60 Hz (frequency in Hertz). Current near 0.5 mA may produce an involuntary startled reaction such as to cause a person to drop a skillet of hot grease or cause a workman to fall from a ladder. As the current increases, involuntary muscular contractions increase, accompanied by current-generated heat.

Higher currents of longer duration than one second can cause the heart to go into ventricular fibrillation. Ventricular fibrillation is considered the most dangerous effect of electric shock. Once fibrillation begins, it practically never stops spontaneously. Death is almost certain within minutes. The rhythmic contractions of the heart become disordered, its pumping action stops, and the pulse soon ceases altogether. Fibrillation in adults can occur at 52 mA and in children at 23 mA. According to V.G. Biegelmeier, the onset of fibrillation in a 50-kilogram (110 pound) adult occurs within the range of 50 mA to 200 mA when the duration of the shock exceeds two seconds.[3] The table on the following page shows implied safe voltages based on these values.

The table's 3,000-Ω (ohm) body resistance column, for example, indicates that 15 volts would result in a shock current of 52 milliamperes.

Table 14-2
Implied Safe Voltage [4] **Based on Several Published Values of Body Resistance and Selected Body Current Safety Criteria as Published by Dalziel**

Criterion	Body Resistance 300 ohms	Body Resistance 500 ohms	Body Resistance 1500 ohms	Body Resistance 3000 ohms
"let-go" 4.5 mA for children	1.35 volts	2.25 volts	6.75 volts	13.5 volts
"let-go" 9 mA for adult males	2.7 volts	4.5 volts	13.5 volts	27 volts
fibrillation at 23 mA 5 sec. pulse of 60-Hz current for 18 kg children	6.9 volts	11.5 volts	34.5 volts	69 volts
fibrillation at 52 mA 5 sec. pulse of 60-Hz current for 50 kg adult	15.6 volts	26 volts	78 volts	156 volts

Extensive work on this subject has been reported in "Effects of Current on Human Beings and Livestock"[5] IEC 479-1 by the International Electrotechnical Commission. Several charts and graphs with background material are provided. Measurements were made on 50 living persons at a touch voltage of 15 volts and on 100 persons at a touch voltage of 25 volts. The total body impedance of one living person was measured with touch voltages of up to 200 volts. Measurements were also made on a large number of corpses.

Three-Wire Grounded System vs. GFCI. [6]

For many years, grounding was emphasized as the primary means for the protection of the electrical system, equipment and personnel from fires and injury, as well as for operating and maintenance advantages. The grounded neutral as a protective element was recognized more and more with each succeeding edition of the National Electrical Code® (NEC®). The Code emphasizes "effective" grounding in order to ensure that wiring faults to ground became "overcurrents" as required to activate the overcurrent device. The wiring and equipment were designed to be protected, and the thought that the grounded system also provided adequate protection for people became ingrained and accepted.

The trend toward grounding equipment and appliances has been gradual, characterized as deliberate but cautious. The belief that grounding provided adequate protection against electric shock and fire hazards became so ingrained that consumers generally have not recognized its limitations, and they find it difficult to accept other more effective means of protection from electric shock and electric fire hazards.

Ground faults occur when electrical current flowing in a circuit returns:

- Through the equipment grounding wire,
- Through conductive material other than the electrical system ground (metal, water, plumbing pipes, etc.),
- Through a person, or
- Through a combination of these ground return paths.

When a person becomes a pathway for electricity flowing to ground, the person will incur an electrical shock, may be seriously injured or may be electrocuted depending on:

- The amount of current or amperes flowing, and
- The duration or time the current flows, and
- The size of the person, and
- The pathway the current follows through the body.

On the other hand, arcing ground faults can occur just about anywhere on an electrical system, resulting in a fire. In either case, the ground faults may be of too low a magnitude to activate the overcurrent device and interrupt the flow of electricity.

A person may become a pathway to ground in one of two ways; in series contact or in parallel contact. In series contact, the person is the only current path to ground and the ground wire is not involved in the circuit. There are many ways in which contact can occur. One example of series contact was that of an infant that stuck a hairpin into a receptacle slot while sitting on a floor heating vent. The infant was electrocuted. Another case involved a man operating a metal-encased electric drill that incorporated a 3-wire grounded cord. He used a "two-to-three" wire adapter but did not connect the adapter pigtail to the wall plate grounding screw. Inadvertently, the pigtail touched the plug blade, thus energizing the drill case and electrocuting the man. In this incident, the equipment ground contributed to the electrocution by providing a current path from the drill case through the body to ground. In both cases, the 3-wire grounded system was totally ineffective since the ground wire was not involved in the current path, and the current through the body was not large enough to trip the overcurrent protective device.

In parallel contact, the victim becomes a path to ground in parallel with the equipment grounding conductor. One scenario of parallel contact occurs when the metal case of an appliance becomes charged electrically by some internal fault condition resulting in current leakage to ground via the grounding conductor. The leakage to ground, however, may not be sufficient to activate the branch-circuit overcurrent protective device. A person who touches the charged case and, at the same time, contacts a grounded surface such as a water faucet or pipe will be subjected to an electric shock. In such parallel contact situations, the effectiveness of the grounding conductor in preventing electrocution of the victim is dependent on several variables, including the following:

- On whether or not the ground-fault current reaches the instantaneous trip level of the overcurrent protective device (which is relatively high-over 15 amperes),
- On how fast the overcurrent device reacts,
- On the voltage level from faulted enclosure to ground, and
- On the impedance of the grounding path (composed of connections, contacts and the ground wire).

An effective grounding system depends upon the integrity of many series connections which must be properly made and maintained. The higher the resistance or impedance of the ground path, the less effectual will be the protection provided by the 3-wire grounded system. Higher ground impedances may be due to long wire lengths, small wire sizes, loose and/or corroded ground wire clamps and connections and other causes. A detailed analysis of the relationship of the impedance ratio of the line circuit to ground circuit to shock current levels is described in the previously cited Consumer Product Safety Commission paper titled "Three Wire Ground System vs. GFCI" and the cited IITRI report.

A similar study by Mr. R.H. Lee of Dupont Company[7] corroborates IITRI's analysis of circuit impedances. His paper deals with the hazard vs. safeguard of a 3-phase grounded power distribution system.

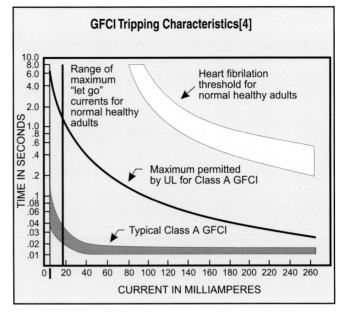

Figure 14-1

An assessment of the effectiveness of the 3-wire grounded system for providing protection against electric shock hazards was conducted by Mr. A.W. Smoot of the Underwriters Laboratories.[8] He analyzed 164 fatal electric shock accidents occurring over a 3¾-year period in and around homes. His study indicated the limitations of the 3-wire system and was submitted to the National Electrical Code® Committee to support proposed GFCI amendments to the 1971 NEC®.

Principles of operation. [9]

"The GFCI sensing system continuously monitors the current balance in the ungrounded 'hot' conductor and the neutral conductor. If the current in the neutral wire becomes less than the current in the 'hot' wire, a ground fault could exist. A portion of the current returns to the supply by

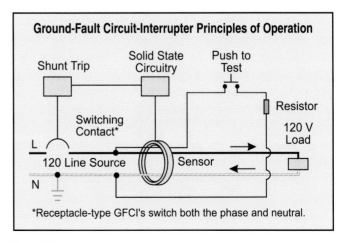

Figure 14-2

some path other than the neutral wire. With a current imbalance as low as 6 milliamperes, the GFCI will interrupt the circuit and this will be shown by a trip or 'off' indicator on the device." See Figure 14-2.

"The GFCI does not limit the magnitude of the ground-fault current. It limits the time that a current of given magnitude can flow. The trip level-time combinations are based on physiological data established for avoiding injury to normal healthy persons. These trip level-time combinations may be too high for persons with heart problems, such as those wearing a pacemaker or under treatment in health care facilities."

The principle of operation of ground-fault circuit-interrupters provides a significant advancement in safety for both equipment that is grounded by an equipment grounding conductor as well as for equipment that is ungrounded. Since the GFCI detects an imbalance of current in both the supply and return paths, it protects equipment supplied by both a 2-wire circuit and a 2-wire with ground circuit. This is the reason some NEC® sections will allow a grounding-type receptacle to be supplied on a 2-wire circuit that has GFCI protection.

Several kitchen appliances, as well as portable heaters, are manufactured with a 2-wire supply cord and, thus, are ungrounded. A significant advancement in safety is realized where these appliances are supplied from receptacle outlets that have GFCI protection.

GFCI Required for Replacement Receptacles

The general requirement in Section 210-7 is that receptacles installed on 15- and 20-ampere branch circuits be of the grounding type and have their grounding contacts effectively grounded by an equipment grounding conductor of the supply branch circuit.

Ground-fault circuit-interrupter protection of receptacles is required for receptacles that are replaced in outlets or at locations that are now required by the Code to be GFCI protected. [Section 210-7(d)]

This requires that installers be aware of the rules that call for GFCI protected receptacles in areas covered by Section 210-8 for dwellings, as well as many other locations in the Code for other facilities. This includes on 15- and 20-ampere, 125-volt receptacles installed in dwelling unit kitchens, bathrooms, garages, outdoor receptacles, etc. Also, this replacement requirement extends to other than dwelling units such as commercial repair garages, elevators and elevator pits, health care facilities and bathrooms in commercial and industrial facilities, as provided in Section 210-8(b).

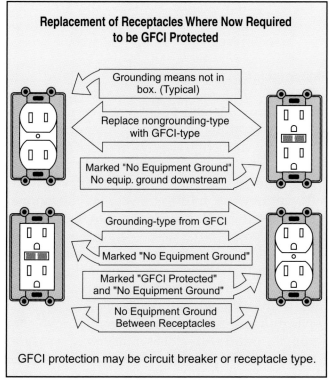

Replacement of Receptacles Where Now Required to be GFCI Protected

Grounding means not in box. (Typical)

Replace nongrounding-type with GFCI-type

Marked "No Equipment Ground" No equip. ground downstream

Grounding-type from GFCI

Marked "No Equipment Ground"

Marked "GFCI Protected" and "No Equipment Ground"

No Equipment Ground Between Receptacles

GFCI protection may be circuit breaker or receptacle type.

Figure 14-3

Table 14-3
GFCI Requirement Locations

Location	NEC® Section
Accessory buildings	210-8(a)(2)
Agricultural buildings	547-9(c)
Basements, dwelling units	210-8(a)(5)
Bathrooms, dwellings	210-8(a)(1)
Bathrooms, nondwellings	210-8(b)(1)
Boathouses	555-3
Carnivals, Circuses, etc.	525-18
Construction sites	305-6
Crawl spaces, dwellings	210-8(a)(4)
Electric Vehicle Charging	625-22
Elevators, etc.	620-85
Existing	210-7(d)
Feeders	215-9
Fountains	680-51
Garages, dwellings	210-8(a)(2)
Garages, commercial	511-10
Health care facilities	517-20(a)
	517-21

Ground-fault circuit-interrupters of various types, the most common of which are the circuit-breaker and receptacle-type, are permitted to be used to provide the protection required in some locations, the particular type of device is specified. For example, Section 620-85 requires that the ground-fault circuit-interrupter protection "in pits, elevator car tops and in escalator and moving walk wellways: be provided by receptacle-type GFCIs.

Specific marking requirements exist where nongrounding-type receptacles are replaced with grounding-type receptacles where an equipment ground does not exist at the outlet. These receptacles must be marked "No Equipment Ground" so the user will be informed that even though the receptacle has an equipment-ground slot, an equipment grounding conductor is not connected to the device. An equipment grounding conductor is not permitted to be connected from the GFCI-type receptacle to any outlet supplied from the GFCI receptacle.

Where grounding-type receptacles are not on a circuit that includes an equipment grounding conductor but are protected on the line side by a ground-fault circuit-interrupter device, the receptacle(s) must be marked "GFCI Protected" and "No Equipment Ground." This will inform the user that even though the receptacle has a grounding terminal, it is not in fact grounded but is provided with GFCI protection. The grounding-type receptacle is protected by the GFCI device, but an equipment grounding conductor is not present as may be required for some equipment such as computers. An equipment grounding conductor is not permitted to be connected between the grounding-type receptacles as this would present incorrect information to the users. As explained earlier, the GFCI device will protect the user from line-to-ground faults even though an equipment grounding conductor is not connected to the devices.

Section 250-130(c) provides a method for installing an equipment grounding conductor for receptacles that are being replaced at a location where an equipment grounding conductor is not present in the box. An equipment grounding conductor is permitted to be connected from the receptacle to:

(1)   Any accessible point on the grounding electrode system as described in Section 250-50.

(2)   Any accessible point on the grounding electrode conductor.

(3)   The equipment grounding terminal bar within the enclosure where the branch circuit for the receptacle or branch circuit originates.

(4)   For grounded systems, the grounded service conductor within the service equipment enclosure.

(5)   For ungrounded systems, the grounding terminal bar within the service equipment enclosure.

Receptacles in Bathrooms

Section 210-8(a)(1) requires that all 125-volt, single-phase, 15- and 20-ampere receptacles installed in dwelling unit bathrooms have ground-fault circuit-interrupter protection for personnel.

The word "bathroom" is defined in Article 100 as "An area including a basin, with one or more of the following: a toilet, a tub, or a shower." It is important to recognize that the definition does not use the word "room" but rather "an area" to describe the location where these receptacles must be provided with GFCI protection. This makes it clear that GFCI protection of these receptacles is required even though the receptacle(s) may be in separate rooms but are in the same "area." For example, a wall with a door or doorway separates a basin from the area where the toilet and tub or shower are installed. No distance measurement is given in the Code beyond which GFCI protection of

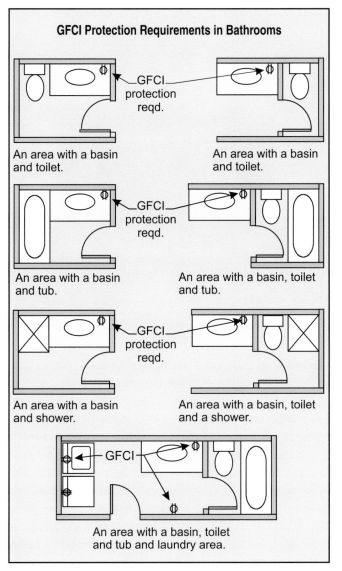

Figure 14-4

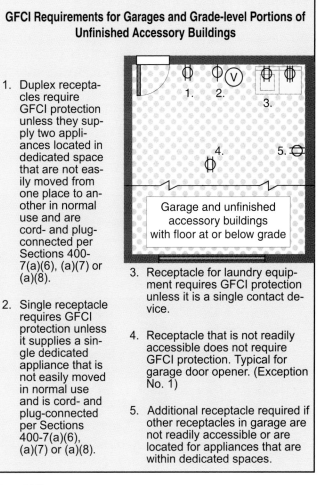

Figure 14-5

these receptacles is not required. This definition applies to all bathrooms regardless of type of facility, due to the requirement in Section 210-8(b) for other than dwelling units.

See Figure 14-4 for a graphic illustration of bathroom "areas."

Receptacles in dwelling-unit garages and unfinished accessory buildings

Section 210-8(a)(2) generally requires GFCI protection for all 125-volt, single-phase, 15- and 20-ampere receptacles installed in dwelling unit garages and in accessory buildings that have a floor located at or below grade level, that are not intended as habitable rooms but are used for storage or work areas.

Exception No. 1 excludes those receptacles that are not readily accessible. "Readily Accessible" is

defined in Article 100 as "Capable of being reached quickly for operation, renewal, or inspections, without requiring those to whom ready access is requisite to climb over or remove obstacles or to resort to portable ladders, etc." No specific height above the garage floor is given as a qualifier. In most cases, receptacles that are located on the ceiling for garage door operators are not considered "readily accessible."

Exception No. 2 to Section 210-8(a)(2) excludes from GFCI requirements a single receptacle for a single appliance or a duplex receptacle for two appliances that occupies dedicated space for each appliance, that in normal use is not easily moved from one location to another and that is cord- and plug-connected. This will clarify that duplex or multi-receptacles installed at outlets in a dwelling unit garage must have ground-fault circuit-interrupter protection for personnel, unless a duplex receptacle is used for the supply to two appliances that meets all the conditions.

Dwelling unit outdoor receptacles

Section 210-8(a)(3) generally requires all 125-volt, single-phase, 15- and 20-ampere receptacles installed outdoors to have GFCI protection for personnel. This requirement applies to these receptacles that are installed outdoors, for any purpose and at any height above grade or platform. The receptacles do not have to be accessible from grade level before the requirement applies.

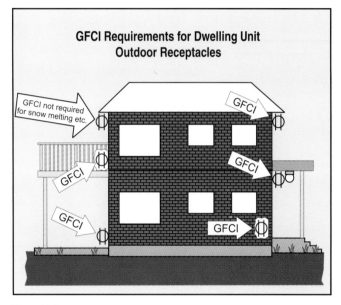

GFCI Requirements for Dwelling Unit Outdoor Receptacles

Figure 14-6

An exception excludes from the GFCI requirement those receptacles that are not readily accessible (defined in Article 100), are on a dedicated branch circuit and are for electric snow-melting or deicing equipment as covered in Article 426. Section 426-28 requires ground-fault protection of equipment (not GFCI protection for personnel) for resistance heating elements used for electric snow-melting or deicing unless the cable is mineral-insulated, metal-sheathed cable that is embedded in a noncombustible medium.

Dwelling unit crawl space receptacles

Section 210-8(a)(4) requires that receptacles that are of the 125-volt, single-phase, 15- and 20-ampere type and located in dwelling unit crawl spaces that are at or below grade level have GFCI protection for personnel.

These receptacles are often installed for servicing equipment such as heating, air-conditioning or refrigerating equipment that is located in crawl spaces.

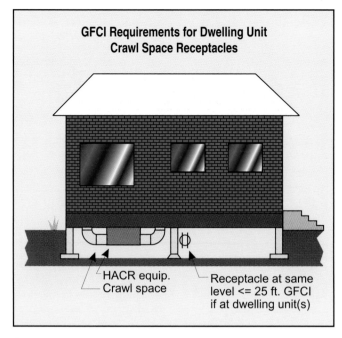

GFCI Requirements for Dwelling Unit Crawl Space Receptacles

Figure 14-7

Unfinished dwelling unit basements

Section 210-8(a)(5) requires that receptacles that are of the 125-volt, single-phase, 15- and 20-ampere type and located in dwelling unit unfinished basements have GFCI protection for personnel.

A definition for "Unfinished Basements" has been included as follows: "Portions or areas of the basement not intended as habitable rooms and limited to storage areas, work areas and the like." It is likely that building codes will be consulted for determining whether rooms in basements are permitted to be classified as habitable rooms.

Exception No. 1 excludes from the GFCI requirement those receptacles that are not readily accessible. "Readily Accessible" is defined in Article 100 as "Capable of being reached quickly for operation, renewal, or inspections, without requiring those to whom ready access is requisite to climb over or remove obstacles or to resort to portable ladders, etc."

Exception No. 2 to Section 210-8(a)(5) excludes from GFCI requirements a single receptacle for a single appliance or a duplex receptacle for two appliances that occupy dedicated space for each appliance, that in normal use are not easily moved from one location to another and that are cord- and plug-connected. This clarifies that duplex or multi-receptacles installed at outlets in a dwelling unit unfinished basement must have ground-fault circuit-interrupter protection for personnel, unless a duplex receptacle is used for the supply to two appliances that meets all the conditions.

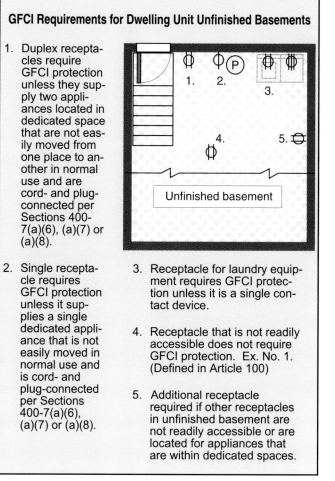

GFCI Requirements for Dwelling Unit Unfinished Basements

1. Duplex receptacles require GFCI protection unless they supply two appliances located in dedicated space that are not easily moved from one place to another in normal use and are cord- and plug-connected per Sections 400-7(a)(6), (a)(7) or (a)(8).

2. Single receptacle requires GFCI protection unless it supplies a single dedicated appliance that is not easily moved in normal use and is cord- and plug-connected per Sections 400-7(a)(6), (a)(7) or (a)(8).

3. Receptacle for laundry equipment requires GFCI protection unless it is a single contact device.

4. Receptacle that is not readily accessible does not require GFCI protection. Ex. No. 1. (Defined in Article 100)

5. Additional receptacle required if other receptacles in unfinished basement are not readily accessible or are located for appliances that are within dedicated spaces.

Figure 14-8

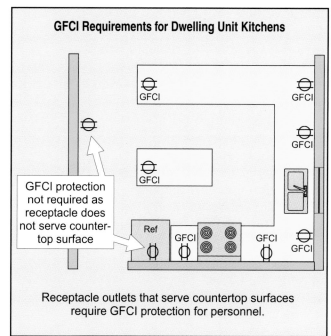

GFCI Requirements for Dwelling Unit Kitchens

GFCI protection not required as receptacle does not serve countertop surface

Receptacle outlets that serve countertop surfaces require GFCI protection for personnel.

Figure 14-9

Kitchen counter receptacles

All 125-volt, single-phase, 15- and 20-ampere receptacles that serve countertop surfaces in dwelling unit kitchens are required by Section 210-8(a)(6) to have ground-fault circuit-interrupter protection for personnel.

This requirement does not apply to those receptacles that do not serve countertop surfaces but are installed for other purposes such as for a waste disposal, a trash compactor or a dishwasher. Also excluded from this requirement are wall-mounted receptacles that are installed for equipment such as electric clocks or receptacles that are installed for appliances such as refrigerators where the receptacle is located behind the appliance and does not serve the counter surface.

Wet bar sinks

Section 210-8(a)(7) requires that all 125-volt, single-phase, 15- and 20-ampere receptacles installed to serve the countertop surfaces and are located within 6 feet of the outside edge of a wet bar sink in a dwelling unit.

Receptacles that serve wall spaces rather than countertop surfaces and are located in the vicinity of the wet bar sink but are not located at or within six feet of the wet bar sink countertop are not required to have GFCI protection.

Rooftop heating, air-conditioning and refrigeration equipment receptacle outlet

Receptacles installed on rooftops of dwelling units and other than dwelling units must have ground-fault circuit interrupter protection for personnel in accordance with Sections 210-8(a)(3) and (b)(2). Note that a receptacle outlet for servicing this equipment is required for other

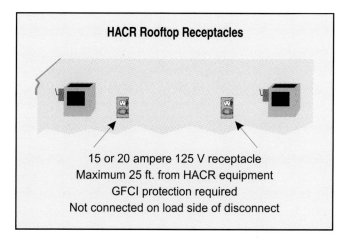

HACR Rooftop Receptacles

15 or 20 ampere 125 V receptacle
Maximum 25 ft. from HACR equipment
GFCI protection required
Not connected on load side of disconnect

Figure 14-10

than dwelling units but not for dwelling units. However, the receptacle must have GFCI protection if installed at these locations.

This receptacle is for the purpose of servicing heating, air conditioning and refrigeration equipment on rooftops.

For other than dwelling units, Section 210-8(b)2. Exception excludes from the GFCI requirement those receptacles that are not readily accessible (defined in Article 100), are on a dedicated branch circuit and are for electric snow-melting or deicing equipment as covered in Article 426. Section 426-28 requires ground-fault protection of equipment (not GFCI protection for personnel) for resistance heating elements used for electric snow-melting or deicing unless the cable is mineral-insulated, metal-sheathed cable that is embedded in a noncombustible medium.

GFCI requirements on construction sites

Section 305-6 contains requirements for ground-fault protection for personnel using temporary power at buildings, structures or on equipment or similar activities while performing any of the following tasks:

- Construction,
- Remodeling,
- Maintenance,
- Repair,
- Demolition, or
- Similar activities.

Ground-fault circuit-interrupter protection is generally required for all 125-volt, single-phase, 15-, 20-, and 30-ampere receptacles used for these purposes. The rule is the same regardless of whether the power is from a temporary construction service, from an existing permanent power source or from one or more receptacles installed as a part of the permanent wiring of the building or structure.

A cord set incorporating listed ground-fault circuit-interrupter protection is permitted to be used by personnel in lieu of permanently or temporarily installed receptacle outlets. This may be the most practical means of complying with this requirement, especially for maintenance and repair activities.

An exception from the requirement for ground-fault circuit-interrupter protection of 15-, 20-, and 30-ampere receptacles is provided for industrial establishments where conditions of maintenance and supervision ensure that only qualified persons are involved. The exception permits compliance

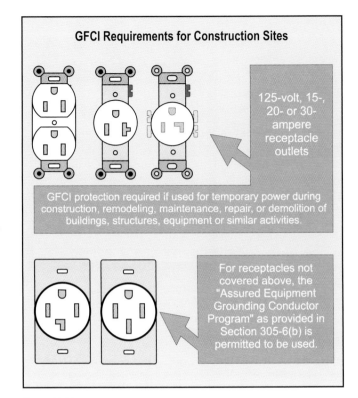

GFCI Requirements for Construction Sites

125-volt, 15-, 20- or 30-ampere receptacle outlets

GFCI protection required if used for temporary power during construction, remodeling, maintenance, repair, or demolition of buildings, structures, equipment or similar activities.

For receptacles not covered above, the "Assured Equipment Grounding Conductor Program" as provided in Section 305-6(b) is permitted to be used.

Figure 14-11

with the assured equipment grounding conductor program outlined in Section 305-6(b).

For receptacles like 250-volt, 15- or 20-ampere or 125/250-volt, 30- or 50-ampere, the assured equipment grounding conductor program, as provided for in Section 305-6(b), is permitted. This program requires a written procedure enforced by one or more individuals at the construction site. Detailed requirements for testing of the equipment grounding conductor continuity on a specific time schedule are provided. A record of this testing program must be maintained and provided to the authority having jurisdiction.

Application guide for ground-fault circuit-interrupters. [9]
Definitions

Ambient temperature - The temperature of the surrounding medium that comes in contact with the device or equipment. (For an enclosed device, it is the temperature of the medium outside the enclosure.)

Current interrupting rating - The value of the available RMS symmetrical current at a specified voltage that the device is capable of interrupting under prescribed test conditions.

Electrical shock hazard - A condition, which could result in injury or electrocution to a person, caused by the flow of electrical current through the body.

Endurance test - A test made to determine compliance with specified mechanical and electrical life requirements.

Ground fault - An unintentional electrical path between a part operating normally at some potential to ground, and ground.

Ground-fault circuit-interrupter (GFCI) - A device intended for the protection of personnel. It deenergizes a circuit or portion thereof within an established period of time when a current to ground exceeds some predetermined value that is less than that required to operate the overcurrent protective device of the supply circuit.

1. Class A GFCI - A ground-fault circuit-interrupter that is designed to trip when fault current to ground is 6 milliamperes or more.

2. Class B GFCI - A ground-fault circuit-interrupter that is designed to trip when fault current to ground is 20 milliamperes or more.

GFCI test - The integral part of a GFCI which checks the ground fault sensing and tripping function of the device.

Ground-fault current - Current that flows through a ground fault.

Ground-fault trip time - the elapsed interval between the time when the ground-fault current is first applied and the time when the circuit is interrupted.

Nuisance trip - Tripping caused by conditions other than those for which the device is intended to respond.

Rated Continuous Current - The maximum RMS current which a device or an assembly is designed to carry.

Rated frequency - The frequency for which the device is designed.

Rated maximum ambient temperature - The maximum ambient temperature at which the device is designed to carry continuous current.

Rated voltage - The nominal RMS voltage for which the device is designed to be used.

Service conditions - The conditions under which a device is designed to be used.

Trip free device - A device wherein the tripping operation cannot be overridden manually.

Trip threshold - The ground-fault current which, when it is reached, will cause tripping of a GFCI.

Tripping - The automatic opening of a device.

Double Grounded Neutral Protection

A Class A GFCI is designed so that it will automatically trip if the neutral conductor is grounded on the load side of the sensor.

If a load-side neutral fault to ground and a ground fault occurred simultaneously, some of the fault current would flow through the neutral wire to the sensor, and some would flow through the inadvertent ground path. If such a ground connection occurred, it would be possible for a person to contact a hot wire and ground, having the ground-fault current flow through the inadvertent neutral ground and the neutral to the service entrance. Under this condition, there may not be enough imbalance in current through the sensor to cause the GFCI to trip.

The double grounded neutral protection feature prevents this condition from occurring.

This protective feature demands proper load-side building wiring. It continuously monitors the load-side wiring to detect a double-grounded neutral condition.

Test means

Each GFCI is equipped with a means of testing the ground-fault trip capability. A simulated ground fault through a test resistor is applied to the sensor through the activation of the test means. It is important that user testing procedures be followed.

GFCI tripping characteristics

Typical trip characteristics of GFCIs are shown in Figure 14-1.

Types of GFCIs.

Basic GFCI.

The basic GFCI consists of a ground-fault detecting means coupled with a circuit-interrupting means.

The detecting means measures the ground-fault current as the difference between outgoing and incoming load current on the protected conductors. The current difference is suitably processed to activate the circuit-interrupting means.

The circuitry module and associated interrupting means which make up the basic GFCI should be mounted in an enclosure or combined with some other device in order to be suitable for installation on a wiring system. The basic GFCI is combined into a complete device to make up the various types described below.

Blank-face type GFCI

The blank face type GFCI has no receptacle openings. It provides both personnel ground-fault protection and switching capabilities. It may be rated for motor switching if so marked. It is not intended to be used as a substitute for a wall-mounted switch.

This type of device may have flexible insulated leads or binding screw terminals or push-in terminals and is to be mounted in an enclosure such as an outlet box. Means for five field connections are provided: line, line neutral, load, load neutral and equipment grounding.

This device may be mounted in listed industrial control panels (UL 508, Industrial Control Equipment) and may be used as a motor controller if so marked. It may be mounted in other equipment and be listed by UL in combination with the equipment.

Permanently-mounted type of GFCI

The permanently-mounted type of GFCI has an integral enclosure as a part of the total unit and is designed to be permanently wired into the supply circuit.

Circuit breaker type

The circuit breaker-type GFCI incorporates within one device the ability to provide all the functions of a circuit breaker, as well as ground-fault protection for personnel for the entire circuit.

It is intended to be mounted in an enclosure and frequently may be installed as a direct replacement for a circuit breaker in a load center or a panelboard.

Circuit breaker-type GFCIs are available as single-pole and two-pole units with various plug-in and lug type electrical connections. Line neutral connections are usually made by means of a wire lead, and care should be taken to install the device according to the manufacturer's instructions.

Receptacle-type GFCIs

The receptacle-type GFCI incorporates within one device one or more receptacle outlets, protected by the GFCI. It is available as a "non-feed-through" unit which provides GFCI protection only to the outlets in the unit, and as a "feed-through" unit which provides GFCI protection to loadside receptacles as well. The feed-through unit may also be used as a non-feed-through unit.

Receptacle-type GFCIs are available with vari-

ous terminating means, including flexible insulated leads and binding screw terminals or push-in terminals. The feed-through unit provides for five field connections: line, line neutral, load, load neutral and equipment grounding. The non-feed-through unit provides connections for line, line neutral and equipment grounding.

Receptacle-type GFCIs are intended to be mounted in an enclosure such as an outlet box.

Portable type GFCI

The portable-type GFCI is an easily transportable, self-contained, enclosed GFCI with one or more integral receptacle outlets protected by the GFCI module. In lieu of an integral receptacle outlet, the load conductors may be attached by a cord and connector. Portable type GFCIs are normally connected to a supply receptacle by an integral attachment plug or a plug and cord. Receptacle cord connectors and plugs are available in various general purpose and locking configurations.

The enclosures for these units may be of the indoor type or the outdoor type. Indoor types usually have nonmetallic enclosures. The outdoor types may have metallic or nonmetallic enclosures suitable for the environment in which they are used. (See NEMA Standards Publication No. 250, *Enclosures for Electrical Equipment (1000 Volts Maximum)*, for enclosure information.)

The portable-type GFCI has a no-voltage release feature which will automatically interrupt the circuit when any of the supply conductors are open.

Cord-connected type

A cord-connected type GFCI consists of an attachment plug containing a GFCI, and a connected cord. It provides GFCI protection on the loadside for the cord and equipment to which the cord is attached.

A cord-connected type may be part of other equipment and is listed by UL in conjunction with the equipment.

The cord-connected type GFCI also has a no-voltage release feature which will automatically interrupt the circuit when any of the supply conductors are open.

Rating information

The rating of a GFCI includes the following: class, rated voltage, rated frequency, rated continuous current and, where applicable, current-interrupter rating.

Class

The GFCI designated "Class A" will trip when there is a fault current to ground of 6 milliamperes or more. The GFCI designated "Class B" will trip when there is a fault current to ground of 20 milliamperes or more "Class B" GFCIs should be used only for underwater swimming pool lighting installed prior to May 1965 (prior to the adoption of the 1965 National Electrical Code®).

Rated voltage

The common alternating-current voltage ratings of GFCIs are 120, 120/240, and 240 volts, single phase; and 208Y/120 and 240 volts, 3-phase.

Rated frequency

The rated frequency of GFCIs is 60 Hz.

Rated continuous current

The common continuous current ratings of GFCIs are 15, 20, 25, 30, 40, 50 and 60 amperes.

Rated withstand current

Receptacle, blank-face, portable and cord-connected types have current-withstand ratings. Typically, they are 2,000 amperes, RMS symmetrical. Permanently-mounted types normally have current-withstand ratings of 5,000 amperes, RMS symmetrical.

Rated interrupting current

Circuit breaker-types normally have current-interrupting ratings of 5,000, 10,000, or 22,000 amperes, RMS symmetrical.

Markings
General

GFCIs are marked with the manufacturer's name or trademark, voltage rating and load capacity in amperes. Additional information may also be marked on the GFCI or included with the manufacturer's literature.

Types which are permanently connected into circuit

For types which are permanently connected into a circuit, which include the switch-type, permanently-mounted-type, circuit breaker-type, and receptacle-type, the following additional informa-

tion should be provided:

1. Installation instructions,
2. Test instructions to operate upon installation and at least monthly,
3. Test reminder with provisions for checking off monthly tests,
4. Warning against use in wet locations,
5. Warning that GFCI does not protect against shock hazard when both conductors are touched,
6. Statement- "For Use on Grounded Circuits Only,"
7. Class (A) or (B), and
8. Proper conductor termination instructions.

Marking on the device or literature packaged with the device should include the following or equivalent statement: "To avoid accidental by-passing of protection, the load-circuit wiring should be separated from other wiring by suitable insulation, barriers, or restraints."

Class A devices

Marking on a Class A GFCI or literature packaged with the device should include the following or equivalent statement:

"To minimize false tripping do not connect to swimming pool equipment installed before adoption of the 1965 National Electrical Code®."

Class B devices

Marking on a Class B GFCI should include the following or equivalent statements: These instructions may appear in the literature packed with the GFCI.

"Use only with underwater swimming pool lighting fixtures:

CAUTION - To avoid possible shock, disconnect power before servicing fixture. To minimize false tripping, do not connect to more than __ feet of load conductor for the total one-way run."

Portable and cord-connected types

For portable and cord-connected types which are cord connected, the following additional information should be provided:

1. Load capacity in watts.
2. The word WARNING and a statement to the effect that, to ensure protection against electric shock the test instructions should be followed before an appliance is plugged into any receptacle on the device. The marking describes the significance of the test

and informs the user that, in the event of an indication of improper functioning, the cause of the malfunction should be corrected before further use of the device.

3. Unless a cord-connected GFCI has been found suitable for use in wet locations, it should be marked "Do not use where water is likely to enter case" or the equivalent.

4. A portable or cord-connected device should be marked with a statement to the effect that the device does not guard against electric shock resulting from (1) some possible defects or faults in an extension cord or other wiring supplying the GFCI, or (2) contact with both circuit conductors.

5. Unless otherwise marked, a cord-connected GFCI not provided with a permanently attached cord 6 feet or longer in length should be marked "This product should be used only with a three-phase conductor, 120 volts, ___*** ampere supply cord set employing Type ST, SO, STO, SJT, SJO, or SJTO cord. In the event of cord set damage, it should be replaced only with an equivalent cord set" or similar wording.

***15, 20, or 30 ampere, to be specified by manufacturer.

GFCI application

In most cases tripping of a GFCI is an indication of a dangerous condition that needs correction. GFCI tripping under such conditions is not "nuisance" tripping. The cause of the tripping should be identified and corrected.

Unless the device is a circuit breaker type GFCI, it will not protect the circuit conductors against overcurrent. Separate overcurrent protection must be provided.

Grounding and GFCI protection are used to complement one another, not to replace one another.

Periodic tripping tests should be made in accordance with the manufacturer's recommendations, using the test means on the unit. When the test button is pushed, a predetermined value of ground-fault current is supplied to the GFCI, tripping the unit and thereby testing the GFCI.

To avoid damage to the GFCI, it must be disconnected before feeders or branch circuits are subjected to a megger, high voltage, or hi-pot test. Disconnect the electrical power supply and load conductors to isolate the GFCI.

Carefully read the manufacturer's instructions before installation of the GFCI.

A GFCI does not protect a person who comes in contact with two "hot" wires or any "hot" wire and the neutral wire.

A GFCI does not protect a person from feeling and reacting to a shock.

When using a feed-through receptacle-type GFCI, be certain that outlets to be protected are all on the load terminals or wires of the GFCI.

When a feed-through receptacle-type GFCI is used only to protect its own receptacle, the unused load wire connections, if any, should be properly insulated to prevent electrical hazard.

When a feed-through receptacle-type GFCI is used to protect a complete load circuit or only a portion of a load circuit, the "load" wires should be connected to the remaining receptacles in the branch. THE LINE AND LOAD CONNECTIONS MUST NOT BE REVERSED; OTHERWISE THE GFCI'S OWN RECEPTACLE WILL BE UNPROTECTED AND WILL REMAIN ENERGIZED AFTER THE UNIT HAS TRIPPED.

Depending upon where it is installed on a branch circuit, a feed-through receptacle-type GFCI can protect all of the receptacles on the circuit, a portion of the receptacles on the circuit, or only its integral receptacle.

Suitable enclosures should be used for GFCIs or receptacles protected by GFCIs. This is especially important in the presence of moisture or a corrosive atmosphere.

The GFCI contains an electronic circuit which is designed to be resistant to electrical interference. In the same way that a television set or radio suffers from momentary "static" in the presence of interference, so may the GFCI. However, in the television or radio, the interference is merely a temporary nuisance. The GFCI, as a safety device, may interpret this interference as a ground fault and trip. Therefore, the GFCI should not be used in an application where continuity of service takes precedence over the personnel protection feature, such as life-sustaining equipment in health care facilities.

The trip characteristics established for the GFCI ground fault sensor are based on physiological data taken for normal healthy persons. These levels may be too high for persons with heart problems such as those wearing pacemakers or those under treatment in health care facilities.

Portable or cord-connected GFCI types will not

protect the receptacle or any portion of the branch circuit into which it is plugged. It will protect only those conductors and devices connected to its load side.

Equipment with high temperature-sheathed heating elements, such as dishwashers and electrical space heaters, tend to have high electrical leakage which may cause tripping when used with GFCIs. This type of equipment should be effectively grounded.

Permanently-connected lights, without receptacles, need not be protected by GFCIs.

Stationary appliances, such as sump pumps and clothes washers, should be effectively grounded. These types of appliances may have high electrical leakage caused by moisture. Refrigerators and freezers, particularly older units, may also have this type of leakage. When these appliances are on a GFCI-protected circuit, the leakage currents could exceed the GFCI trip threshold and cause the GFCI to trip.

High moisture conditions may result in electrical leakage of portable tools. This will cause the GFCI to trip in some instances even if the tool is turned off, if the leakage exceeds the GFCI trip value.

High moisture conditions may result in leakage of unprotected extension cord connectors and aged cord insulation. Replace or repair any aged wiring, extension cord, or portable tool which, under high moisture, causes the GFCI to trip.

GFCI protection cannot be used for ranges, ovens, cooking units, and clothes dryers when the frame of the appliance is connected to the grounded circuit conductor.

If a GFCI is used with a small portable generator, the GFCI circuit should be turned off when starting and until the generator has attained constant speed and stabilized output voltage.

A GFCI feeding an isolation transformer, which supplies circuits such as underwater swimming pool fixtures, will not detect a ground fault on the secondary side of the transformer.

GFCIs should be used only on systems with a grounded circuit conductor. A GFCI can be used on a 2-wire circuit without a ground (green) wire as long as the circuit has a grounded circuit conductor. It should be noted that the equipment ground conductor "green wire" is completely independent of the GFCI circuitry.

The National Electrical Code® requires that the white (or grounded) conductor be grounded at the service entrance, at certain separately derived systems and is permitted to be grounded at additional buildings or structures supplied from a common service under specific conditions. If this conductor is grounded again on the load side of the GFCI, tripping will occur.

A 120/240V two-pole circuit breaker-type GFCI need not have the load side neutral connected for 240V rated loads.

Certain portable voltage testers or test lamps, when connected between line and ground on the load side of the GFCI, may cause the GFCI to trip.

Some office and laboratory equipment (e.g., personal computers) exhibit ground leakage currents due to designed-in features such as RFI filtering and transient voltage surge suppression. While this current is permitted by applicable product standards, it can result in tripping of a GFCI. This possibility is increased when two or more pieces of equipment, each with some leakage current, are connected to the same branch circuits.

After a GFCI has tripped and the cause has been determined to be a result of this accumulated permitted leakage current, the number of such devices on one branch circuit must be reduced. It may be necessary to provide dedicated circuits or individual GFCI receptacles if this inherent leakage is high.

The equipment manufacturer should be contacted if there is any question pertaining to the expected leakage current level.

GFCIs are specifically tested for immunity to radio frequency noise. See NEMA PP 1, *Procedure for Evaluating Ground-Fault Circuit-Interrupters for Response to Conducted Radio Frequency Energy.*

Circuit breaker or receptacle-type GFCIs are not intended to be field wired to an attachment plug to create a portable GFCI. Such an arrangement will not provide the full protection afforded by a portable or cord-connected GFCI.

References:

[1] A. Albert Biss, *Ground-Fault Circuit-Interrupter (GFCI) Technical Report,* (Washington, D.C.: U.S. Consumer Product Safety Commission, February 28, 1992).

[2] Walter Skuggevig, *5-Milliampere Trip Level for GFCIs,* (Northbrook, IL: Underwriters Laboratories, March 1989).

[3] *Electric Shock Prevention, p. 25.*

[4] *From the content of the IITRI report, the*

voltage and current values are assumed to be "root-mean-square" (RMS) values.

[5] Technical Report IEC 479-1, Effects of current on human beings and livestock. International Electrotechnical Commission. Available from Global Engineering Documents, 15 Inverness Way East, Englewood, Colorado 80112.

[6] A. Albert Biss, Three-Wire Grounding System vs. GFCI, (Washington, D.C.: U.S. Consumer Product Safety Commission, December 1985).

[7] Ralph H. Lee, Electrical Grounding: Safe or Hazardous?," Chemical Engineering, July 28, 1969.

[8] A.W. Smoot, Analysis of Accidents, (Northbrook, IL: Underwriters Laboratories, 1971).

[9] Application Guide for Ground-Fault Circuit-Interrupters, Standards Publication No. 280-1990, National Electrical Manufacturers Association, 1300 North 17th Street, Suite 1847, Rosslyn, VA 22209. Reprinted with permission.

Chapter Fourteen: The questions included here were developed using material included in this chapter. It is also important that students make use of the 1999 NEC®, where many answers can be found. See page 279 for answers.

1. Where existing 125 volt, single-phase, 15- and 20-ampere rated receptacles are replaced, or where new receptacles are installed in locations that are required by the Code to be GFCI protected, they must be ___.
   a. of the grounding type
   b. GFCI protected
   c. on dedicated circuit
   d. polarized

2. Circuit breaker or receptacle-type ground-fault circuit-interrupters ____ be used to protect the receptacle, where nongrounding types of receptacles are replaced with grounding type receptacles where an equipment ground does not exist in the receptacle enclosure.
   a. are permitted to
   b. cannot
   c. must always
   d. by special permission can

3. Where used for a single cord- and plug-connected appliance that occupies a dedicated space in a residential garage, GFCI protection is not required where a ____ receptacle is installed and dedicated for use by that appliance.
   a. duplex
   b. single
   c. isolated
   d. 15-ampere rated

4. Portions or areas of a dwelling unit basement that are not intended to be used as habitable rooms and limited to storage areas, work areas and the like, defines:
   a. a basement
   b. a finished basement
   c. an unfinished basement
   d. a storage space

5. Where installed in dwelling unit kitchens, all 125-volt, single-phase, 15- and 20-ampere receptacles that are installed ____ must be ground-fault circuit-interrupter protected.
   a. within 6 feet of the sink
   b. on the dishwasher circuit
   c. to serve counter surfaces
   d. to serve refrigeration equipment

6. The distance of the required receptacle on a rooftop, other than at one- and two-family dwellings, must be located to be not further than ____ feet from the heating, air-conditioning and refrigeration equipment.
   a. 10
   b. 15
   c. 25
   d. 50

7. Where installed in ____ of commercial, industrial and all other nondwelling-type occupancies, all 125-volt, single-phase, 15- and 20-ampere receptacles must be ground-fault circuit-interrupter protected.
   a. kitchens
   b. basements
   c. closets
   d. bathrooms

8. Where installed on roofs of which of the followings locations, all 125-volt, single-phase, 15- and 20-ampere receptacles must be ground-fault circuit-interrupter protected.
   a. one family dwelling units
   b. two family dwelling units
   c. dwelling unit garages
   d. all the above

9. With a current imbalance as low as ____ milliamperes, a GFCI will interrupt the circuit and this will be shown by a trip or "off" indicator on the device.
   a. 3
   b. 4
   c. 5
   d. 6

10. A GFCI of the Class ___ type is designed so that it will automatically trip if the neutral conductor is grounded on the load side of the sensor.
   a. A
   b. B
   c. C
   d. D

11. The basic GFCI consists of a ground-fault detecting means that is coupled with a circuit-interrupting means. This detecting means measures the ground-fault current as the difference between outgoing and incoming load current on the ____ conductors.
   a. protected
   b. neutral
   c. bonded
   d. service

12. A condition which could result in injury or electrocution to a person, caused by the flow of electrical current through the body, is defined as:

    a. a high resistance ground

    b. an electrical shock hazard

    c. a short circuit

    d. ground fault

13. To determine compliance with specified mechanical and electrical life requirements, ___ is made.

    a. an endurance test

    b. a ground fault test

    c. an operational test

    d. an overcurrent test

Chapter 15

Equipment Ground-Fault Protection Systems

Objectives

After studying this chapter, the reader will be able to understand:

- Requirements for static and lightning protection.
- Bonding in hazardous (classified) locations.
- Requirements for agricultural buildings.
- Requirements for health care facilities.
- Requirements for swimming pools, hot tubs and spas.
- Requirements for electric signs.

Definition

Ground-fault protection of equipment: "A system intended to provide protection of equipment from damaging line-to-ground fault currents by operating to cause a disconnecting means to open all ungrounded conductors of the faulted circuit. This protection is provided at current levels less than those required to protect conductors from damage through the operation of a supply circuit overcurrent device."[N] (NEC® Article 100)

Protection required

Section 230-95 of the NEC® requires ground-fault protection of all solidly-grounded wye electrical services of more than 150 volts to ground, but not exceeding 600 volts phase-to-phase for each service disconnect rated 1,000 amperes or more. As can be seen, this protection is required for nominal 480/277-volt, three-phase, 4-wire wye connected systems. These provisions do not apply to fire pumps or continuous industrial processes where a nonorderly shutdown would introduce additional or increased hazards.

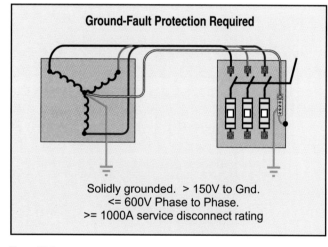

Ground-Fault Protection Required

Solidly grounded. > 150V to Gnd.
<= 600V Phase to Phase.
>= 1000A service disconnect rating

Figure 15-1

The maximum setting of the devices is 1,200 amperes, and the maximum time delay is one second for ground-fault currents equal to or greater than 3,000 amperes.

The magnitude of current which will flow in the event of a line-to-ground fault on a grounded system is usually determined by the reactance of the grounded apparatus, the reactance of the lines or cables leading to the fault and the resistance and reactance of the ground return path including any intentional grounding resistance or reactance. For interconnected systems, calculation of the current

may be rather complicated. For simpler cases, an approximation of the available fault current may be obtained.

This equipment protection is required due to a history of destructive burndowns of electrical equipment operating at this voltage level. An electric arc, which generates a tremendous amount of heat, is readily maintained at 277 volts to ground. Since it is not a solid ground (bolted) connection, and the current is limited by the resistance of the arc, often, not enough current will flow in the circuit to allow the overcurrent device ahead of the fault to open. A great deal of damage is done to the electrical equipment while the arc is burning.

There are basically two types of equipment ground-fault protection systems in use although these systems may have different names in the industry. The most common types are zero-sequence which may have more than one form and residual type which is sometimes referred to as ground-strap or ground-return type. Both types are designed to protect downstream equipment from destructive arcing burndowns. Note that this equipment will not protect equipment or the system on its line side from line-to-ground faults as this fault current will not pass through the ground-fault equipment.

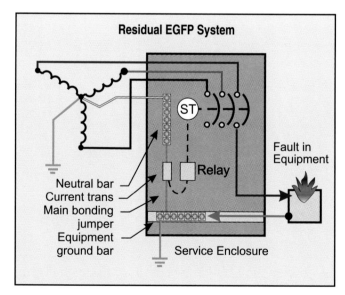

Residual EGFP System

Neutral bar
Current trans
Main bonding jumper
Equipment ground bar
Relay
Service Enclosure
Fault in Equipment

Figure 15-2

Residual-type system

The residual or ground-strap type of equipment ground-fault protection system consists of a current transformer, control relay and, usually, a shunt trip circuit breaker. The main bonding jumper passes through the current transformer as shown in Figure 15-2.

The advantage of this system is it is probably the least expensive, but it is limited to application at the main service. It cannot be used downstream of the main bonding jumper.

With the connections as shown, the maximum ground-fault current will flow through the main bonding jumper and will, thus, be recognized by the current transformer. Although there are many parallel bypass circuits for ground-fault current to flow, the relative impedance of the main circuit is so low compared to all the parallel bypass circuits that a minimum of 90 percent of the total ground-fault current will flow in the main bonding jumper and, thus, be seen by the current transformer.

If a current transformer of properly-selected value is placed in the main bonding jumper and has its secondary feeding a magnetic current re-lay, the relay may be adjusted to close a contact at any desired value of fault. The current trans-former value is selected on the basis of the smallest current transformer whose short-time rating will not exceed the maximum value of ground-fault current that may be anticipated.

Examination of the diagram Figure 15-2, de-scribing the ground-fault protective device, will show that the main ground-fault current path is from the transformer to the service, through the feeder to the fault, back to the service over the equipment grounding conductor through the main bonding jumper where it returns to the trans-former by the neutral conductor. The parallel cir-cuit from the grounding electrode at the service to the grounding electrode is both a high-resistance and a high-reactance circuit. As a result, little current will return through the earth.

Some small ground-fault current will be carried by the building steel if it is in the circuit. It is preferable to adequately ground the building steel using the same grounding electrode as used for the service. This prevents the building steel from ris-ing to a dangerous potential above ground and often serves as a grounding electrode as required by Section 250-50(b). Even though the building steel represents a parallel path for fault current to flow, most of the current will flow back to the transformer through the neutral because of the lower reactance of that path as compared to the reactance of the other parallel paths.

Zero-sequence transformer-type system

Probably the most popular and common type of

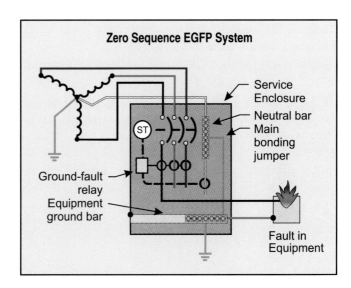

Zero Sequence EGFP System

Service Enclosure
Neutral bar
Main bonding jumper
Ground-fault relay
Equipment ground bar
Fault in Equipment

Figure 15-3

equipment ground-fault protection system is the zero-sequence type. This system is shown in Fig-ure 15-3 and Figure 15-4. It consists of a current transformer that is placed around all the conduc-tors of the circuit, including the grounded (neutral) conductor, a control relay and usually a shunt trip circuit breaker. As shown, the current transformer must be placed downstream from the main bond-ing jumper. The equipment grounding conductor does not pass through the window.

The general difference between the types of zero sequence equipment shown in Figures 15-3 and 15-4 is the current transformers are generally built into the circuit breaker shown in Figure 15-4 with an external current transformer through which the neutral passes that may be field in-stalled. Often, these current transformers are used

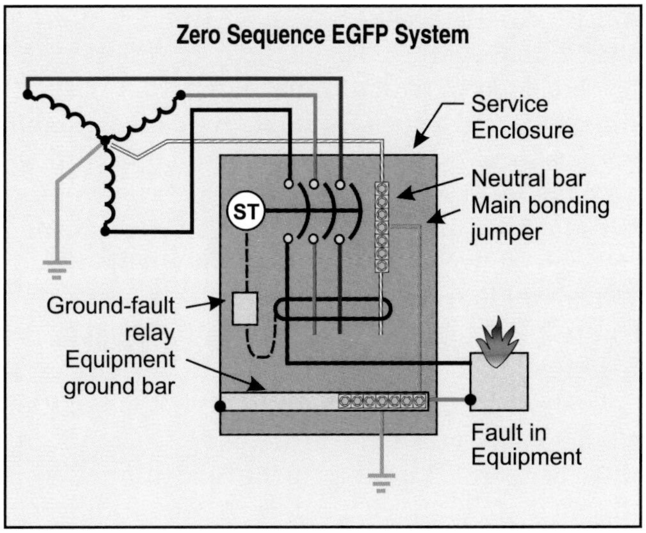

Zero Sequence EGFP System

Service Enclosure
Neutral bar
Main bonding jumper
Ground-fault relay
Equipment ground bar
Fault in Equipment

Figure 15-4

by the circuit breaker as a part of its internal operational system. The equipment ground-fault protection system shown in Figure 15-4 is sometimes referred to as a residual type as the ground-fault relay or system sums up the current through all four coils and considers any excess current as "residual." Generally, the current transformer through which all conductors of the circuit pass as shown in Figure 15-3 is installed by the manufacturer of the switchboard. Cables for feeders are field installed.

Under normal operation, the vector summation of all phase and neutral (if used) currents approaches zero. This is due to the cancelling effect of the currents in the conductors. A sensor (differential current transformer) around the phase and neutral conductors detects the current imbalance when a ground fault occurs. This is due to the fault current passing outside the current transformer window(s) which sets up the imbalance which is detected by the system.

The output of the sensor is proportional to the magnitude of the ground-fault current. This output is fed to a ground-fault relay. The relays are usually field adjustable. Pickup ranges of from 4 to 1,200 amperes are common. Time delay settings are available from instantaneous (1.5 cycles) to 36 cycles delay. When the ground-fault current exceeds a pre-selected level, the relay will activate the circuit-interrupting device, which usually is a shunt trip circuit breaker, to open the circuit.

Ground-fault protection not required

Ground-fault protection of equipment is not required for the service for a continuous industrial process where a nonorderly shutdown will introduce additional or increased hazards. (See NEC® Section 230-95 Exception No. 1)

Likewise, ground-fault protection of equipment is not required for fire pump services. (NEC® Section 230-95 Exception No. 2) For an emergency system, Section 700-26 excludes the alternate source of power from the requirement to have ground-fault protection of equipment with automatic disconnection means. This applies to fire pumps that are classified as an emergency system in accordance with Section 700-1. Indication of a ground fault of the emergency source is required to comply with Section 700-7(d). The indication of the ground fault is required for a solidly grounded wye emergency system of more than 150 volts to ground and circuit protective devices rated 1000 amperes

or more. The sensor is required to be located at, or ahead of, the main system disconnecting means for the emergency source. Instructions on the course of action to be taken must be located at or near the sensor location.

A similar exclusion from the ground-fault protection of equipment requirement is provided for legally required standby systems in Section 701-17. No such exclusion is provided for optional standby systems installed in accordance with Article 702.

System coordination

System coordination with circuit breaker systems is easily accomplished by use of ground-fault protective devices. These ground-fault protective devices may be cascaded where the economics of the design warrant doing so. The time-delay settings may become lower and lower as the device gets further from the service, so that the last ground-fault protective device in the system may even be set for instantaneous trip. Similar coordination also can be obtained by locking out the relay or relays upstream from the sensor that sees the fault first.

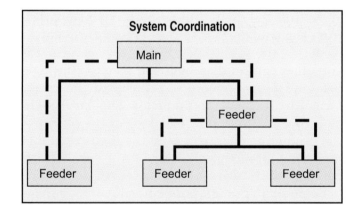

Figure 15-5

Since it cannot be know where a fault-to-ground or enclosure may originate in the system, and since it is desirable to coordinate the protection and clear the fault nearest to its point of origin, it will be necessary to delay the action of tripping the main overcurrent device. The magnetic current relay is thus used to energize a time delay relay to provide proper coordination and maximum continuity of service consistent with safety.

When a fault-to-ground or enclosure occurs, the possibility should be considered that the fault may reach a high value. If so, the normal main overcurrent device should be set to a time value to allow it to function and clear the fault through that means. Even though the ground-fault protective relay should be set at as low a value as possible, enough time delay should be set to permit an overcurrent device near the point of origin of the fault to clear first. This ground-fault protective device is an adjunct to, but does not replace, the main overcurrent device. It gives protection below the rating of the main overcurrent device. Being affected only by ground-fault currents, it does not interfere with normal operation of the main overcurrent device. In effect then, the ground-fault protective device will keep all ground faults to a limited time, selected on the basis of experience, to give maximum safety without undue interruption of service.

Second level required in health care facilities

Special rules for ground-fault protection of electrical systems apply to health care facilities. Section 517-17 requires that where ground-fault protection is provided on the service or feeder disconnecting means in accordance with NEC® Sections 230-95 or 215-10, a second level of ground-fault protection be provided in the next level of feeder downstream.

This rule clarifies that regardless of whether or not the health care facility is supplied by a service or feeder, where equipment ground-fault protection is provided at the service or feeder disconnecting means, a second level is required downstream.

The additional level(s) of ground-fault protection is not permitted to be installed:

(1) On the load side of an essential system transfer switch, or

(2) Between the on-site generating unit(s) described in Section 517-35(b) and the essential system transfer switch(es), or

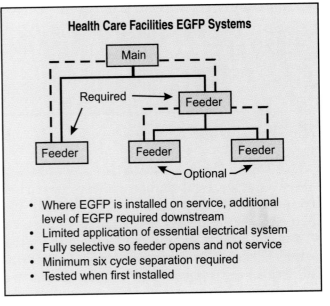

Figure 15-7

(3) On electrical systems that are not solidly grounded wye systems with greater than 150 volts to ground, but not exceeding 600 volts phase-to-phase.

This feeder protection is required to be 100 percent selective so that where a ground fault occurs downstream from the feeder overcurrent device, only the feeder overcurrent device will open and the service or feeder main will remain closed. This, of course, is to prevent the blackout of a facility caused when a main would open for a fault that can be isolated to a single feeder. To achieve this coordination, a six cycle, or greater, separation between the service and feeder tripping times is required. Each level of the ground-fault protection system must be tested when first installed.

Feeder ground-fault protection

Figure 15-8 illustrates a situation where equipment ground-fault protection is required for feeders. Section 215-10 requires this protection for feeders that are of the same voltage and current rating as similar services.

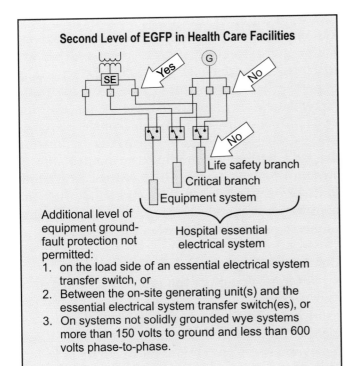

Figure 15-6

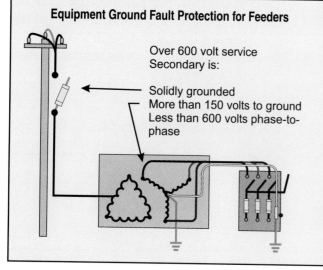

Figure 15-8

This covers the situation where the electrical system is delivered at more than 600 volts and has the service disconnect at that level. A transformer then reduces the voltage of the feeder to the level where equipment ground-fault protection is required. The risk of destructive arcing burndowns is the same regardless of whether the system is a service or feeder.

Three exceptions to the general requirement are:

(1) For continuous industrial process where a nonorderly shutdown will introduce additional or increased risks. See Article 685 – Integrated Electrical Systems for more information on these systems.

(2) For fire pumps. See Article 695 and NFPA-20 for information on electrical requirements for fire pumps.

(3) For feeders that have ground-fault protection on the line side of the feeder. However, care must be exercised to be certain that each new system created by a separately derived system has the protection required.

Ground-fault protection of building or structure main disconnecting means

For each building or structure disconnecting means meeting the criteria, Section 240-13 requires ground-fault protection of equipment to be installed in accordance with Section 230-95. This requirement applies for systems of identical voltage and amperage rating to that for services. Three exceptions to the general requirement are provided.

(1) Excludes a continuous industrial process where a nonorderly shutdown will introduce additional or increased risks. See Article 685 for additional information.

(2) Excludes the disconnecting means installation from the ground-fault protection requirements where the service or feeder provides the necessary protection.

(3) Excludes fire pumps installed in accordance with Article 695 from the requirements.

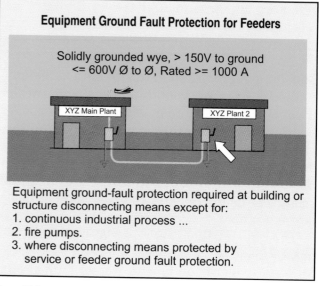

Figure 15-9

It is most important that an overall engineered system approach is taken where more than one building or structure is on the premises that are fed from a common service. This is especially important where transfer switches and alternate power sources are installed in the premises wiring system. In most cases, four-pole transfer switches are installed so as to allow the system to function properly without grounding the neutral downstream from the service. It is critical to ground-fault protection systems functionality that the neutral be completely isolated from any grounding connections downstream from the service.

Combination systems

An improvement in the protection of distribution systems of 600 volts or less may be obtained by using a ground-fault protective device as a supplement to high-interrupting capacity current-limiting overcurrent protection. Fault currents in low ranges may be recognized and virtually all fault currents can be held to a limited time duration with the resultant increase to the safety of such systems.

If circuit breakers are used as the disconnecting means and overcurrent protection on such secondary distribution systems and a ground-fault reaches a high enough value, the circuit breaker will open and clear all three phases, thus preventing a feedback into the ground fault from the two unaffected phases. When a circuit breaker is used, a ground-fault protection device also may be applied to automatically open the circuit breaker in a reasonable and limited time if a ground fault should develop and not be cleared by some overcurrent device near the point of fault.

If a manually-operated, fused load-break switch is used as the disconnecting means and overcurrent protection, and if a ground fault reaches a high enough value to clear one fuse, an alarm should be provided to indicate a blown fuse so all three phases may be cleared manually.

If an electrically-operated, fused load-break switch is used, it can be provided with an automatic blown-fuse indicator which will immediately open the switch and, thus, all three phases, if any one or more fuses should blow. An electrically-operated switch also permits the application of a ground-fault protective device which will automatically open the switch in a reasonable and limited time if a ground fault should develop and not be cleared by some overcurrent device near the point of fault.

For systems with large available short-circuit current, and which require the best degree of protection, the electrically-operated switch or circuit breaker can provide full protection where it is equipped with high-interrupting capacity current-limiting fuses, a ground-fault protective device and an automatic blown-fuse indicator.

Testing of system

Section 230-95(c) requires that ground-fault protection systems be performance-tested when first installed on site to ensure that they will operate properly. Experience has shown that the majority of these systems do not operate properly, or at all, when first installed. This is often due to improper field wiring of the system.

The test must be performed in full compliance with the manufacturer's written instructions. These instructions must be furnished with the equipment. A vital part of the test is to remove the neutral disconnect link and megger the neutral to be certain that it is clear from any ground down-stream from the service. Accidental or intentional grounding of the neutral past the service will desensitize the ground-fault protection system and render it ineffective.

A written record of these tests must be made and provided to the inspection agency. Safety demands that the equipment be properly installed, as does the threat of lawsuit where negligence of the installer can be proven.

Optional protection

Figure 15-10 illustrates an installation where equipment ground-fault protection is not required even though the system voltage is the same as that where it is required. The difference is the rating of the disconnecting means is not 1,000 amperes or greater, even though the combined rating exceeds 1,000 amperes. For example, a service with a single 2000 ampere main requires equipment ground-fault protection at the stated voltage and ampere rating. If a service with four, six-hundred ampere overcurrent devices is installed rather than the single main, equipment ground-fault protection is not required. Obviously, the system can benefit from equipment ground-fault protection even though it is not required by the Code.

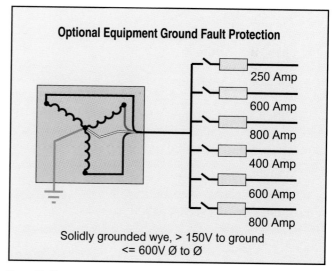

Optional Equipment Ground Fault Protection

250 Amp
600 Amp
800 Amp
400 Amp
600 Amp
800 Amp

Solidly grounded wye, > 150V to ground
<= 600V Ø to Ø

Figure 15-10

In the case where a 900-ampere fuse is installed in a 1,200-ampere switch, the requirements are different. Equipment ground-fault protection is required since the fuse holder will accept a 1,000-ampere fuse. Where adjustable circuit breakers are used, the rating, for purposes of Section 230-95, is the actual overcurrent device installed in the circuit breaker or the maximum rating that it can be adjusted to.

Chapter Fifteen: The questions included here were developed using material included in this chapter. The answers can be found by reviewing the text. It is also important that students make use of the 1999 NEC®, where many answers can be found. See page 279 for answers.

1. A system intended to provide protection of equipment from damaging line-to-ground fault currents, by operating to cause a disconnecting means to open all ungrounded conductors of the faulted circuit, is defined as ____.
   a. ground-fault circuit-interrupter
   b. ground-fault protection of equipment
   c. leakage current detector
   d. saturable core reactor

2. The NEC® requires ground-fault protection of all solidly-grounded wye electrical services of more than 150 volts to ground, but not exceeding 600 volts phase-to-phase, for each service disconnect rated ____ amperes or more.
   a. 800
   b. 1,000
   c. 600
   d. 400

3. Ground-fault protection of equipment is not applicable to ____ or continuous industrial processes where a nonorderly shutdown would introduce additional or increased hazards.
   a. electronically-actuated fuses
   b. phase converters
   c. fire pumps
   d. industrial buildings

4. Where a service is protected by a system of GFPE, the maximum setting of the GFPE devices is ____ amperes and the maximum time delay is one second for ground-fault currents equal to or greater than ____.
   a. 1,300 - 2,000
   b. 1,200 - 3,000
   c. 1,400 - 1,500
   d. 1,600 - 1,000

5. For feeders rated 1,000 amperes or more in a solidly-grounded wye system with greater than 150 volts to ground, but not exceeding 600 volts phase-to-phase, ground-fault protection ____ required.
   a. is
   b. is not
   c. sensors are
   d. indication is

6. The most common types of equipment ground-fault protection equipment is:
   a. three-phase, three wire
   b. residual and bypass relaying
   c. zero-sequence and residual
   d. isolation transformer and relay

7. One of the most critical elements of a ground-fault protection system is:
   a. the neutral must be isolated downstream
   b. a choke coil is installed downstream
   c. capacitors are installed downstream
   d. all the above

8. The equipment ground-fault protection system must be tested:
   a. by pushing the "push to test" button
   b. on an annual basis
   c. within 30 days of installation
   d. when first installed on site

9. A second level of equipment ground-fault protection is required
   a. for fire pump installations
   b. for health care facilities
   c. for emergency systems
   d. for optional standby systems

10. Equipment ground-fault protection is required for electrical systems of more than 150 volts to ground and less than 600 volts phase-to-phase:
   a. for feeders of more than 800 amperes
   b. at a building disconnecting means rated at 1200 amperes or more
   c. for feeders rated more than 100 amperes
   d. for a continuous industrial process

Special Location Grounding and Bonding

Objectives

After studying this chapter, the reader will be able to understand:

- Requirements for static and lightning protection.
- Bonding in hazardous (classified) locations.
- Requirements for agricultural buildings.
- Requirements for health care facilities.
- Requirements for swimming pools, hot tubs and spas.
- Requirements for electric signs.

Static protection

Protection against static electricity is a requirement of a number of industries and establishments. The grounding of equipment is not necessarily a solution to static problems. Each static problem requires its own study and solution. Humidity plays an important part. The higher the humidity the less the chances of static. In some industries, increasing the humidity in the area of a static discharge has been found very effective. One example is in the printing industry.

Lightning protection systems

Lightning protection is an important factor at an outdoor substation and at locations where thunderstorms are prevalent. Lightning discharges usually consist of very large currents of extremely short duration. Protection is accomplished by deliberately providing a path of low resistance to earth, compared to other paths. There is no guarantee that lightning will necessarily follow the lower resistance path that has been provided, but at least the low resistance path will reduce damage.

The steel fence around an outdoor substation, when properly grounded, offers good lightning protection for the equipment, especially if the fence is higher than the equipment within the station. Lightning conductors will enhance the safety. Lightning rods (air terminals) should project at least one foot above any part of the structure, and equipment, and the path to ground should be as direct as possible.

Each air terminal should have at least two connections to ground with air terminals placed on each side of the structure. Surge arresters also should be used on the power lines and the equipment. The lightning down conductor system should be connected to earth through driven electrodes as well as being bonded to the station ground bus. Lightning conductors should not be smaller than No. 4, increasing in size as the primary voltage of the system increases.

See NFPA-780, *Standard for the Installation of Lightning Protection Systems* for additional information on the subject.

Bonding of lightning protection systems

The ground terminals (grounding electrodes) for lightning protection systems are required to be bonded to the building or structure grounding electrode system. See Section 250-106 of the National Electrical Code®. This section no longer requires that metallic parts of electrical wiring system be bonded to the lightning protection system conductors where there is less than six feet of separation. However, specific requirements for bonding the systems together are found in the *Lightning Protection Code.*

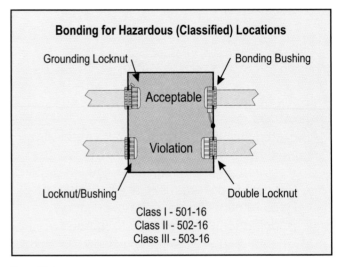

Figure 16-1

Hazardous (classified) locations

Bonding: The Code places some special requirements for grounding and bonding in hazardous (classified) locations. These requirements can be found in Section 501-16 for Class I locations, Section 502-16 for Class II locations and Section 503-16 for Class III locations. For Class I, Zone 0, 1 and 2 hazardous (classified) locations, Section 505-25 requires compliance with Section 501-16.

By these special requirements, an effort is made to provide assured grounding and bonding to reduce the likelihood that a line-to-ground fault will cause arcing and sparking at connection points of

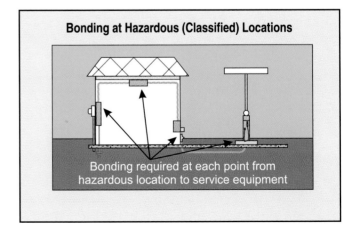

Figure 16-2

metallic raceways and boxes or other enclosures. If such arcing or sparking were to occur in a hazardous (classified) location while a flammable gas is present in its explosive range, it is likely that the flammable atmosphere would be ignited.

Generally, these requirements are that locknuts on each side of the enclosure, or a locknut on the outside and a bushing on the inside, cannot be used for bonding. Bonding locknuts or bonding bushings with bonding jumpers must be used to ensure the integrity of the bond and its capability of carrying the fault current that may be imposed, hopefully, without arcing or sparking at the connections.

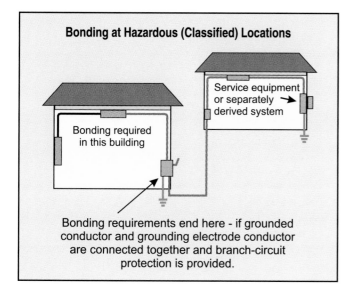

Bonding at Hazardous (Classified) Locations

Service equipment or separately derived system

Bonding required in this building

Bonding requirements end here - if grounded conductor and grounding electrode conductor are connected together and branch-circuit protection is provided.

Figure 16-3

The bonding means required here must generally be installed from the hazardous (classified) location to the service equipment, or point of grounding of a separately derived system that is the source of the circuit. This includes all raceways, fittings, junction boxes, enclosures, controllers and panelboards between the hazardous location and the service or separately derived system.

An exception to this general requirement is provided. It clarifies that the bonding is required to be taken no further than the point where the grounded circuit conductor (may be a neutral) and the grounding electrode conductor are connected together on the line side of the building or structure disconnecting means that is grounded in accordance with Section 250-32(a), (b) or (c), and that branch-circuit overcurrent protection is located on the load side of the disconnecting means.

Section 250-32 provides grounding requirements where more than one building or structure are on the same premises and are supplied from a common service. Where the grounded circuit conductor is not grounded at the building or structure, the rule in Sections 501-16(a), 502-16(a) and 503-16(a) requires that the bonding extend from the hazardous location back to the service even if it is in another building. See Chapter 13 of this text for additional information on grounding electrical systems at additional buildings or structures on the premises.

Grounding in hazardous locations. Where flexible metal conduit or liquidtight flexible metal conduit is permitted and used in Class I, Division 2 hazardous (classified) locations, they must have an internal or external bonding jumper installed to supplement the conduit. If installed outside the conduit, the bonding jumper is limited to 6 feet in length by Section 250-102(e). See Section 501-16(b).

In these Class I, Division 2 areas, the bonding jumper is permitted to be omitted under the following conditions:

(a) Listed liquidtight flexible metal conduit not more than 6 feet long with fittings listed for grounding is used;

(b) Overcurrent protection of the circuit in the raceway does not exceed 10 amperes; and

(c) The load is not a power utilization load.

See Section 502-16(b) for similar rules for Class II areas and Section 503-16(b) for Class III locations.

Agricultural buildings

Sections 547-4(f) and 547-9 contain some specific requirements for grounding and bonding in agricultural buildings to which the article applies. The major concern is twofold. The first is for the integrity of the grounding path due to corrosive conditions that exist in these buildings. The second concern is for neutral-to-earth and stray voltage which, if excessive, cause behavior responses in livestock. These behavior responses in dairy cattle can lead to loss of production and health problems.

Section 250-32(e) requires that the portion of an equipment grounding conductor that is run underground to an agricultural building where livestock is housed be insulated or covered copper.

Equipment grounding

Where equipment that must be grounded is installed in an agricultural building, a copper equipment grounding conductor must be installed to ground the equipment. This applies regardless of the type of wiring method employed. Where installed underground, the equipment grounding conductor must be either insulated or covered copper. See Section 547-4(f).

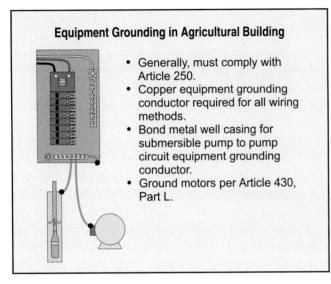

Equipment Grounding in Agricultural Building

- Generally, must comply with Article 250.
- Copper equipment grounding conductor required for all wiring methods.
- Bond metal well casing for submersible pump to pump circuit equipment grounding conductor.
- Ground motors per Article 430, Part L.

Figure 16-4

The frame of water pumps, including the submersible type, must be grounded. See Section 250-112(l). Where submersible pumps are used in metal well casings, the equipment grounding conductor must be bonded to the casing. See Section 250-112(m).

Isolated neutral or grounded conductor

One of the most important elements of a farm wiring system, especially where dairy cattle are involved, is to isolate the system neutral at barns, milking parlors, etc. Separation of the neutral conductor from the equipment grounding conductors will prevent voltage drop on the feeder neutral from becoming stray voltage in the building.

Isolation of the neutral (not re-grounding it) at an additional building or structure on the same premises as the electrical service is permitted by Section 250-32(b)(1). In the case of agricultural buildings, the conditions of Section 547-8 must be satisfied as well. Here, two general methods are provided.

(1)   Where the (service) disconnecting means and overcurrent protection are located at the distribution point, feeders to buildings must meet the requirements of Section

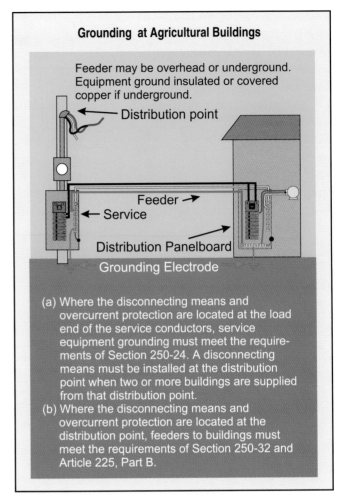

Grounding at Agricultural Buildings

Feeder may be overhead or underground. Equipment ground insulated or covered copper if underground.

Distribution point

Feeder →

← Service

Distribution Panelboard

Grounding Electrode

(a) Where the disconnecting means and overcurrent protection are located at the load end of the service conductors, service equipment grounding must meet the requirements of Section 250-24. A disconnecting means must be installed at the distribution point when two or more buildings are supplied from that distribution point.

(b) Where the disconnecting means and overcurrent protection are located at the distribution point, feeders to buildings must meet the requirements of Section 250-32 and Article 225, Part B.

Figure 16-5

250-32 so far as grounding is concerned. This means that either:

(a) The neutral is not permitted to be grounded at additional buildings if an equipment grounding conductor is run with the feeder to the buildings or structures, or

(b) The neutral is permitted to be grounded at the additional buildings or structures if (1) an equipment grounding conductor is not run with the feeder, (2) there are no continuous metallic paths bonded to the electrical systems in the buildings or structures involved, and (3) ground-fault protection of equipment has not been installed on the common ac service.

(2)   Where the (service) disconnecting means that is located at the distribution point does not have overcurrent protection and overcurrent protection is located at the disconnecting means at the building(s),

the grounded circuit conductor (often a neutral) is not permitted to be grounded at the building disconnecting means and all the following conditions must be complied with:

(a) All buildings and premises wiring are under the same ownership.

(b) Disconnecting means suitable for use as service equipment is provided at the distribution point.

(c) An equipment grounding conductor is run with the supply conductors and is of the same size as the largest supply conductor, if of the same material, or is adjusted in size in accordance with the equivalent size columns of Table 250-122 if of different materials.

(d) The equipment grounding conductor is bonded to the grounded circuit conductor at the service equipment or the source of a separately derived system.

(e) A grounding electrode system is provided and connected to the equipment grounding conductor at the building disconnecting means.

For the purposes of Article 547, a definition of "Distribution Point" is provided which reads as

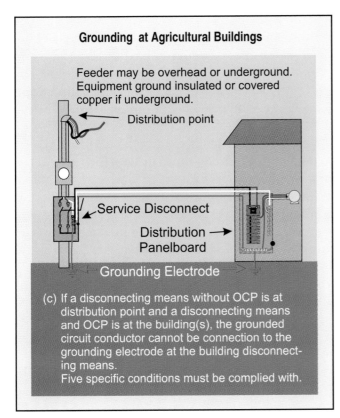

Grounding at Agricultural Buildings

Feeder may be overhead or underground. Equipment ground insulated or covered copper if underground.

Distribution point

Service Disconnect

Distribution → Panelboard

Grounding Electrode

(c) If a disconnecting means without OCP is at distribution point and a disconnecting means and OCP is at the building(s), the grounded circuit conductor cannot be connection to the grounding electrode at the building disconnecting means.
Five specific conditions must be complied with.

Figure 16-6

follows; "A centrally located electrical supply structure from which services or feeders to agricultural buildings and other buildings, including the associated farm dwelling, are normally supplied."[N]

Where this provision is employed, service equipment is often installed at a location remote from the agricultural building, and feeders are run to the agricultural buildings. Service switches are installed at the metering point, often referred to as a "May Pole" as feeders radiate from the service switch location to the various buildings. The neutral and equipment grounding conductor are bonded together in the service equipment.

It is not uncommon to find one or more transfer switches at this location for connection of portable or tractor-mounted generators for standby power.

Stray (Tingle) voltage. Voltage drop on a neutral conductor supplying a building or structure housing livestock can result in elevated levels of stray voltage. It is important to balance 120-volt loads to minimize neutral loads, operate motors at 240 volts wherever practical and size neutral conductors in service drops and feeders as large as practical.

Livestock behavioral responses, including production and health problems, can be caused by stray voltage (also referred to as "tingle-voltages"). These voltages appear between various portions of grounded or ungrounded metallic systems such as electrical equipment or piping systems and the earth or floor such that livestock can come between two different potentials. It is not uncommon to find voltage differences between adjacent concrete slabs or between a concrete slab and the adjacent earth.

Common causes of neutral-to-earth voltages (stay voltage if it is at livestock contact points) are currents on primary distribution systems, farm secondary neutral conductors and faulty wiring on the farm. Some electric utilities have installed primary-to-secondary neutral conductor isolators in an attempt to solve the stray voltage problem. Neutral conductors that are too small for the load and length of run and loose or corroded splices and terminations will frequently elevate the neutral-to-earth voltage to a level that can affect livestock.

Another common cause of stray voltage is ground faults where the fault current gets into the earth, concrete slabs or on metal equipment that may be contacted by livestock. Ground faults that can cause stray voltage may occur in water pumps, underground wires, sump pumps, manure pumps, electrically heated livestock watering fountains, electrically operated feeders and similar equipment. Proper grounding along with bonding this equipment together usually prevents the stray voltage from occurring even where there is a high impedance fault.

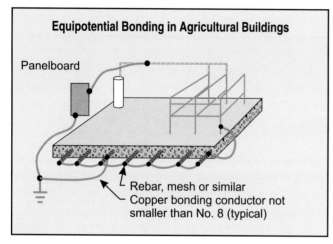

Equipotential Bonding in Agricultural Buildings

Panelboard

Rebar, mesh or similar
Copper bonding conductor not smaller than No. 8 (typical)

Figure 16-7

Equipotential plane. Section 547-9(b) requires the installation of an equipotential plane in the concrete floor of livestock confinement areas. This is the best method of controlling the effects of stray voltages on livestock. This bonding does not correct or remove the faults that are causing the problem but keeps everything in the livestock area at the same potential. This, in essence, prevents the livestock from being aware of and being affected by the stray voltage.

There are two exceptions to the basic requirement for providing an equipotential plane in an agricultural building. First, the equipotential plane is not required where there is no electrical service including branch circuits or feeders to the building and there is no metal equipment that is likely to become energized that is accessible to livestock. Second, slatted floor sections that are supported by structures that are a part of an equipotential plane are not required to be bonded. These floor sections are typically pre-cast concrete sections, which by their size and mass effectively become a part of the equipotential plane when supported by the rest of the structure. In addition, the installation of bonding conductors to these "floating" sections has proven difficult as these sections are

removed for periodic wash downs and other cleaning.

The term, "equipotential plane" is defined in Section 547-9(a) as, "An area accessible to livestock where a wire mesh or other conductive elements are embedded in concrete, are bonded to all metal structures and fixed nonelectrical metal equipment that may become energized and are connected to the electrical grounding system to prevent a difference in voltage from developing within the plane. For this section, livestock does not include poultry."[N]

Wire mesh or other conductive elements, such as reinforcing steel rods, must be installed in the concrete floors in these areas and be bonded together, as well as to the building grounding electrode system. This is accomplished by bonding the reinforcing rods and wire mesh in the concrete together with a copper conductor, not smaller than No. 8, and then connecting these bonding conductors to metal piping systems, stanchions and the building grounding electrode systems. This creates the required equipotential plane which really means that everything in the area is at the same potential. This prevents voltage differences between conductive bodies livestock can make contact with.

Obviously, it is best to install this bonding system during the original construction of the building or portion of a building that houses livestock or serves as a milking parlor. Equipotential planes have been installed in existing buildings by sawing grooves in the concrete floors, installing bonding conductors and bonding all conductive elements together.

It is recommended in Section 547-9(b) that there be a voltage gradient ramp at livestock entrances and exits that are traversed daily by the same livestock. The decision to install a voltage gradient ramp can be determined by actual measurement of the voltage gradient that exists at the edge of the equipotential plane. The purpose served by a gradient ramp is explained by Fine Print Note No. 1 which reads, "A natural voltage gradient exists at the edge of an equipotential plane. Typically, voltage gradients exceeding 1 volt per foot at the edge of the equipotential plane will require a voltage gradient ramp."[N]

Fine Print Note No. 2 following Section 547-9(b) provides information on methods on installing equipotential planes and voltage gradient ramps and reads, "Methods to establish equipotential

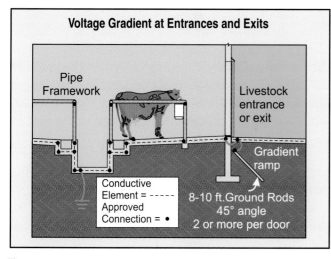

Figure 16-8

planes and voltage gradient ramps are described in *Equipotential Planes in Animal Containment Areas*, American Society of Agricultural Engineers (ASAE) EP473-1997."[N]

The method indicated for producing a voltage gradient is installing two or more 8- to 10-foot long ground rods at each livestock exit and entrance. The ground rods are installed at a 45 degree angle with the highest end at the building. The rods are connected by means of a bonding conductor to the equipotential plane of the building.

Receptacles in agricultural buildings

All 125-volt, single-phase, 15- and 20-ampere general purpose receptacles in areas having an equipotential plane are required to have ground-fault circuit-interrupter protection for personnel. While the term "general purpose receptacles" is not defined in the NEC®, it no doubt refers to those receptacles that would be used for portable tools and portable power cords and not to receptacles installed for a specific purpose such as for fixed equipment like brooders, incubators, etc.

Health care facilities

Electrical systems in health care facilities must comply with at least two safety standards, the *National Electrical Code*®, NFPA 70 and the *Health Care Facilities Code*® NFPA-99. These codes provide the minimum requirements for installation as well as maintenance and testing of electrical systems in health care facilities. Depending on the type of health care facility involved and the scope of the project, several other electrical codes and standards may be involved. These include *Centrifugal Fire Pumps*®, NFPA-20; *National Fire Alarm Code*®, NFPA-72; *Emergency and Standby Power Systems*®, NFPA-110; and *Installation of Lightning Protection Systems*®, NFPA-780.

Article 517 provides special requirements for grounding of equipment in certain health care facilities, particularly in patient care areas. Reasons for the extra or specialized requirements are given in the Fine Print Note following Section 517-11. It reads' "In a health care facility, it is difficult to prevent the occurrence of a conductive or capacitive path from the patient's body to some grounded object, because that path may be established accidentally or through instrumentation directly connected to the patient. Other electrically conductive surfaces that may make an additional contact with the patient, or instruments that may be connected to the patient, then become possible sources of electric currents that can traverse the patient's body. The hazard is increased as more apparatus is associated with the patient, and, therefore, more intensive precautions are needed. Control of electric shock hazard requires the limitation of electric current that might flow in an electric circuit involving the patient's body by raising the resistance of the conductive circuit that includes the patient, or by insulating exposed surfaces that might become energized, in addition to reducing the potential difference that can appear between exposed conductive surfaces in the patient vicinity, or by combinations of these methods. A special problem is presented by the patient with an externalized direct conductive path to the heart muscle. The patient may be electrocuted at current levels so low that additional protection in the design of appliances, insulation of the catheter, and control of medical practice are required."[N]

Grounding in patient care areas

Section 517-13(a) generally requires that, "In an area used for patient care, the grounding terminals of all receptacles and all noncurrent-carrying conductive surfaces of fixed electric equipment likely to become energized that are subject to personal contact, operating at over 100 volts, shall be grounded by an insulated copper conductor. The grounding conductor shall be sized in accordance with Table 250-122 and installed in metal raceways with the branch-circuit conductors supplying these receptacles or fixed equipment."[N]

Exception No. 1 provides that, "Metal raceways shall not be required where listed Types MI, MC, or AC cables are used, provided the outer metal

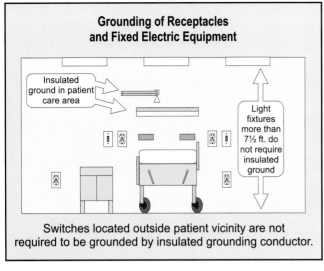

Grounding of Receptacles and Fixed Electric Equipment

Insulated ground in patient care area

Light fixtures more than 7½ ft. do not require insulated ground

Switches located outside patient vicinity are not required to be grounded by insulated grounding conductor.

Figure 16-9

armor or sheath of the cable is identified as an acceptable grounding return path."

Exception No. 2 provides that, "Metal faceplates shall be permitted to be grounded by means of a metal mounting screw(s) securing the faceplate to a grounded outlet box or grounded wiring device."

Finally, Exception No. 3 provides that, "Light fixtures more than 7½ feet (2.29 m) above the floor and switches located outside of the patient vicinity shall not be required to be grounded by an insulated grounding conductor."[N]

Both a primary and secondary, or redundant

Health Care Facilities Patient Care Area Wiring Methods

1. Rigid metal conduit.
2. Intermediate metal conduit.
3. Electrical metallic tubing.
4. Listed flexible metal conduit, max. 6 feet long, max. 20A OC protection, fittings listed for grounding.
5. Armor of listed Type AC cable where the outer metal jacket is approved as a grounding means.
6. Listed Type MI cable and MC cable where the outer metal jacket is acceptable as a ground-return path.
7. Listed liquidtight flexible metal conduit maximum 6 feet long, and
   Max. 20 A OC protection ³⁄₈ to ½ inch or
   Max. 60 A OC protection ¾ - 1¹⁄₄ inch

All wiring methods must contain an insulated, copper equipment grounding conductor.

---

means of grounding is required in patient care areas of health care facilities. See Section 517-13. The metal raceway or outer metal jacket of a listed cable, that is suitable for grounding, provides the primary means of equipment grounding. An insulated copper equipment grounding conductor (the secondary, redundant equipment grounding means) must also be installed or be a part of listed cable for grounding receptacles and noncurrent-carrying metal portions of fixed electric equipment in the patient care areas that are subject to personal contact.

The wiring methods that are acceptable for providing the primary equipment grounding fault return path are the metal raceways or cable assembly included in Section 250-118. (See the following table.) Of course, where these wiring methods are installed, it is critical that good work practices be followed to ensure that the wiring method provides a reliable, continuous and low impedance path for fault current.

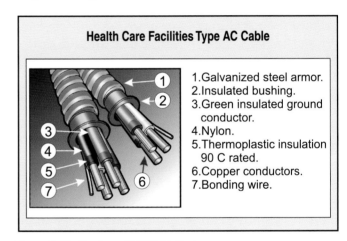

Health Care Facilities Type AC Cable

1. Galvanized steel armor.
2. Insulated bushing.
3. Green insulated ground conductor.
4. Nylon.
5. Thermoplastic insulation 90 C rated.
6. Copper conductors.
7. Bonding wire.

Figure 16-10. *Courtesy of AFC Cable Systems*

Sections 517-13(a) and (b) place additional restrictions on the use of cable wiring methods that are used for wiring in patient care areas. The outer jacket of the cable is required to qualify as an equipment grounding return path in accordance with Section 250-118.

Types MC and MI cables are required in Section 517-13(b) to have an outer armor that itself qualifies "as an equipment grounding return path in accordance with Section 250-118." These requirements do not permit the combination of the cable outer jacket, which alone does not qualify as an equipment grounding conductor and an internal equipment grounding conductor, to meet the requirements for grounding.

Generally, the outer jacket of Type MC cable of the spiral-interlocking type is not suitable as an equipment grounding conductor and cannot be used in a patient care area of a health care facility that is required to comply with Section 517-13. Type MC cable with a smooth or corrugated continuous tube is suitable for these patient care areas if it contains a green-insulated equipment grounding conductor with a yellow stripe or surface marking or both to indicate that it is an additional equipment (or isolated) grounding conductor. This acceptable type of MC cable would not contain a supplemental bare or unstriped green-insulated grounding conductor. A supplemental bare or unstriped green-insulated equipment grounding conductor in Type MC cable of the smooth or corrugated continuous type is an indication that the outer armor by itself does not qualify as an equipment ground.

By its construction and listing by a qualified electrical testing laboratory, Type AC cable with a spiral interlocking metal jacket and a bonding strip or wire in intimate contact with the outer jacket is suitable as an equipment grounding conductor.

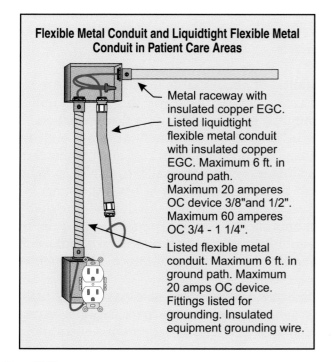

Flexible Metal Conduit and Liquidtight Flexible Metal Conduit in Patient Care Areas

Metal raceway with insulated copper EGC.
Listed liquidtight flexible metal conduit with insulated copper EGC. Maximum 6 ft. in ground path. Maximum 20 amperes OC device 3/8"and 1/2". Maximum 60 amperes OC 3/4 - 1 1/4".

Listed flexible metal conduit. Maximum 6 ft. in ground path. Maximum 20 amps OC device. Fittings listed for grounding. Insulated equipment grounding wire.

Figure 16-11

Copper, Type MI cable, by nature of its construction, has an outer jacket that is suitable as an equipment grounding conductor. This cable is manufactured with bare conductors that are physically separated from each other by the mineral compound inside the metal sheath. The individual

conductors are insulated by slip-on insulation at the time the cable is terminated.

While recognized in Section 250-118 as equipment grounding conductors, severe restrictions are placed on the use of flexible metal conduit and liquidtight flexible metal conduit due to their limited capabilities of carrying current and functioning as a ground-fault return path. These restrictions are illustrated in Figure 16-11.

Additional restrictions are placed on the wiring of the emergency system in hospitals. While these requirements relate to mechanical protection of the circuits and not directly to grounding, the installer should be aware of these rules. See Section 517-30(b)(2) for designation of these branches, Section 517-3 for definitions and Section 517-30(c)(3) for wiring requirements.

Testing of grounding system in patient care areas

The *Standard for Health Care Facilities*®, NFPA 99, requires that the integrity of the grounding path provided by the wiring method in patient care areas of health care facilities be tested before acceptance of the initial installation and after any alterations or replacement of the electrical system is made. See Section 3-3.3.2.1. Both voltage and impedance measurements must be made of exposed conductive surfaces including the grounding contacts of receptacles in the patient-care vicinity.

Excluded from the testing requirements are small, wall-mounted conductive surfaces, not likely to become energized, such as surface-mounted towel and soap dispensers, mirrors and the like.

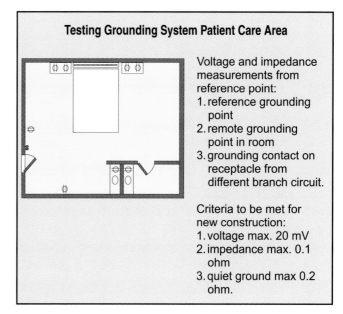

Testing Grounding System Patient Care Area

Voltage and impedance measurements from reference point:
1. reference grounding point
2. remote grounding point in room
3. grounding contact on receptacle from different branch circuit.

Criteria to be met for new construction:
1. voltage max. 20 mV
2. impedance max. 0.1 ohm
3. quiet ground max 0.2 ohm.

Figure 16-12

Also exempt are large, metal conductive surfaces not likely to become energized, such as window frames door frames and drains. A note following this section indicates that the grounding system, including both the metallic raceway and equipment grounding conductor, is to be tested as an integral system. Removing equipment grounds from receptacles or equipment is not required or recommended.

The voltage and impedance measurements are required to be measured against a reference point. The reference point is permitted to be one of the following:

(1) A reference grounding point,
(2) A grounding point in the room under test that is electrically remote from the equipment under test, such as a metal water pipe,
(3) The grounding contact of a receptacle that is powered from a different branch circuit from the receptacle being tested.

The criteria for new construction that must be met are as follows:

(1) The voltage limit is 20 mV.
(2) The impedance limit is 0.1 ohm.
(3) For "quiet ground systems" the impedance limit is 0.2 ohm.

Receptacle testing in patient care areas

In wet locations, fixed receptacles, equipment connected by cord and plug, and fixed electrical equipment are required to be tested:

(1) when first installed,
(2) where there is evidence of damage,
(3) after any repairs, or
(4) at intervals not exceeding 6 months.

Section 3-3.4.2.3 gives requirements for the testing interval for receptacles in patient care areas. It is required that:

(1) Testing be performed after initial installation, replacement, or servicing of the device.
(2) Additional testing be performed at intervals defined by documented performance data.

Receptacles that are not listed as hospital-grade are required to be tested at intervals not exceeding 12 months.

Recordkeeping requirements are found in Section 3-3.4.3. A record must be maintained of the tests required by the chapter and associated repairs or modification. At a minimum, this record must contain the date, the rooms or areas tested, and an indication of which items have met or have failed to meet the performance requirements of this chapter.

Where an isolated power system is installed, Section 3-3.4.3.2 requires that a permanent record be kept of the results of each of the tests.

Tests that must be performed for each receptacle include:

(1) Visual inspection to confirm physical integrity.
(2) Continuity of the grounding circuit.
(3) Correct polarity.
(4) Retention force (except locking type) to be not less than 115 g (4 oz).

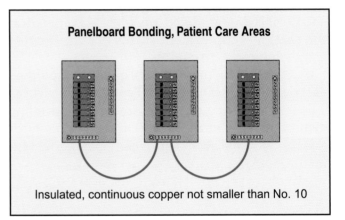

Figure 16-13

Bonding of panelboards in patient care areas

Section 517-14 requires that the equipment grounding terminal busses of the normal and essential branch-circuit panelboards be bonded together. The bonding conductor must be insulated continuous copper not smaller than No. 10. Where more than two panelboards serve the same location, it is permitted to install the bonding conductor unbroken from one panelboard equipment grounding terminal bar to another without looping an unbroken conductor between all panelboard enclosures.

This bonding is intended to ensure that little if any potential difference exists between exposed noncurrent-carrying metal portions of equipment in the patient-care area. It is not required that this bonding conductor be installed in a raceway between panelboards where it is protected by normal building construction such as studs and gypsum board.

Patient equipment grounding point

The "Patient Equipment Grounding Point" is an option for improving electrical safety in the patient care area. Though an optional feature [Section 517-19(c)], a patient equipment grounding point is provided by many hospital equipment manufacturers and specified by consulting engineers. Where installed, it consists of a grounding terminal bus and may include one or more grounding jacks. An equipment bonding jumper not smaller than No. 10 is then used to bond the grounding terminals of all receptacles in the patient vicinity to the equipment grounding point. Again, this is done to reduce the potential difference between conductive surfaces in the patient vicinity.

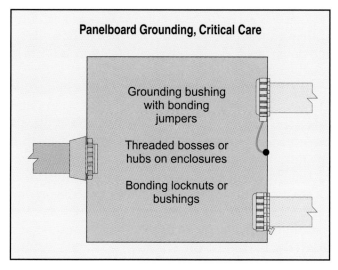

Panelboard Grounding, Critical Care

Grounding bushing with bonding jumpers

Threaded bosses or hubs on enclosures

Bonding locknuts or bushings

Figure 16-14

Panelboard grounding, critical care areas

Section 517-19(d) requires that where metal raceways or Type MC or MI cables are used for feeders, grounding of panelboards and switchboards serving critical care areas must be assured. Acceptable methods include as follows:

(1) A grounding bushing and a continuous copper bonding jumper sized in accordance with Section 250-122, with the bonding jumper connected to the junction enclosure or the ground bus of the panel, or

(2) Connection to threaded bosses or hubs on enclosures, or

(3) Other approved devices such as bonding locknuts or bushings.

Isolated power systems permitted

Isolated power systems are permitted be used for wiring in critical care areas. If used, the isolated power system equipment is required to be listed for the purpose and the system designed and installed so it meets the provisions of Section 517-

160. The audible and visual indicators of the line isolation monitor are permitted to be located at the nursing station for the area being served.

An isolated power system usually consists of a one-to-one transformer that has an ungrounded secondary. A protective shield is placed between the primary and secondary windings. A line isolation monitor with visual and audible alarms is provided to warn of excessive line to ground leakage as well as to indicate that a line-to-ground fault has occurred. The primary purpose of an isolated power system is to allow equipment to function with the first line-to-ground fault without opening an overcurrent device. Faulty equipment can be repaired or replaced at the earliest opportunity but will remain in service until a second line-to-ground fault occurs which then becomes a line-to-line fault. This would cause the overcurrent device to operate and take the faulted equipment off the line.

Other specific requirements for isolated power systems are in Section 517-160.

Special purpose receptacle grounding

Where an isolated ungrounded power source is used and limits the first-fault current to a low magnitude, the grounding conductor associated with the secondary circuit is permitted to be run outside of the enclosure of the power conductors in the same circuit.

The Fine Print Note points out that "Although it is permitted to run the grounding conductor outside of the conduit, it is safer to run it with the power conductors to provide better protection in case of a second ground fault."[N]

Wet locations in health care facilities

The term, "Wet locations" is defined in Section 517-3 as "Wet locations are those patient care areas that are normally subject to wet conditions while patients are present. These include standing fluids on the floor or drenching of the work area, either of which condition is intimate to the patient or staff. Routine housekeeping procedures and incidental spillage of liquids do not define a wet location." It is the responsibility of the governing body of the health care facility to designate areas of the health care facility in accordance with the type of patient care anticipated as well as the special conditions such as "wet locations."

The general requirement in Section 517-20 is that "Receptacles and fixed equipment within the area of the wet location shall have ground-fault circuit-interrupter protection for personnel if interruption of power under fault conditions can be tolerated, or be served by an isolated power system if such interruption cannot be tolerated."[N]

An exception provides that branch circuits supplying only listed, fixed, therapeutic and diagnostic equipment are permitted to be supplied from a normal grounded service, single- or 3-phase system, provided:

a. Wiring for grounded and isolated circuits does not occupy the same raceway, and

b. All conductive surfaces of the equipment are grounded.

It is common that operating rooms and some special procedures rooms are designated as "wet locations" particularly in hospitals. Many complex and lengthy procedures are performed in these rooms which cannot be interrupted by the operation of a ground-fault circuit-interrupter so isolated power systems are often installed.

Where an isolated power system is utilized, the equipment is required to be listed for the purpose and be installed so that it meets the provisions of, and is in accordance with, Section 517-160. This section provides the rules for isolated power systems.

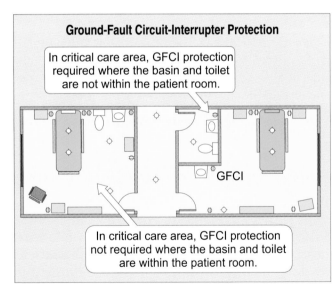

Figure 16-15

Ground-fault circuit-interrupter protection for personnel in health care facilities

The general requirement for providing ground-fault circuit-interrupters of all 125-volt, single-phase, 15- and 20-ampere receptacles in bathrooms are contained in Section 210-8(b)(1). GFCI protection of these receptacles is generally required. Note that the definition of the term "bathroom" is contained in Article 100, Part A.

NEC® Section 517-21 provides that. "Ground-fault circuit-interrupter protection for personnel shall not be required for receptacles installed in those critical care areas where the toilet and basin are installed within the patient room."[N] Toilets of the traditional configuration are not generally installed in critical care patient rooms as these patients are often too ill to get out of bed. Some special configurations of patient care equipment may include a patient toilet in the critical care room. GFCI protection of the receptacles is not required where, for example, only a basin is installed in the patient room. In this configuration, the room does not meet the definition of "Bathroom" in Article 100.

GFCI protection of the receptacles identified above is required in other patient care rooms that meet the definition of "Bathroom," and typically have a basin and a toilet and/or a shower. Note that the definition of "Bathroom" uses the term "area" rather than "room." As a result, GFCI protection is typically required for receptacles at the basin and in the room where the toilet and/or shower are installed and not for the receptacles that are not located adjacent to the basin.

Swimming pools, general

It is important to understand the organization of Article 680 for proper application of the requirements. Article 680 is divided in several parts, namely Part A through Part G. Part A is general and applies to all equipment and installations covered in Article 680. Part B covers permanently installed pools. Other parts cover the equipment or installation identified in the title of the part and do not necessarily apply to all equipment included in Article 680. For example, the rules in Part B for grounding panelboards do not apply to the wiring of a hydromassage bathtub covered in Part G.

However, be aware of the requirements for wiring spas and hot tubs. Section 680-40 reads, "A spa or hot tub installed outdoors shall comply with the provisions of Parts A and B of this article except as permitted in (a) and (b)." Generally, a spa or hot tub installed outdoors must meet all the rules for swimming pools, while a spa or hot tub installed indoors must meet the rules in Section 680-41 which are different from those for swimming pools.

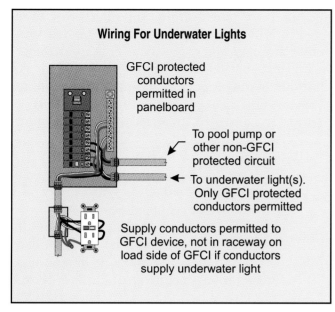

Wiring For Underwater Lights

GFCI protected conductors permitted in panelboard

To pool pump or other non-GFCI protected circuit

To underwater light(s). Only GFCI protected conductors permitted

Supply conductors permitted to GFCI device, not in raceway on load side of GFCI if conductors supply underwater light

Figure 16-16

GFCI requirements for swimming pools

Sections 680-5(b) and (c) contain specific requirements relative to ground-fault circuit-interrupters for swimming pools. They are permitted to be self-contained units, circuit-breaker types, receptacle types, or other approved types.

Conductors on the load side of a ground-fault circuit-interrupter used to provide protection for underwater lighting fixtures are generally not permitted to occupy raceways, boxes or enclosures containing other conductors, even if the other conductors are for pool-related equipment such as for a water pump. Grounding conductors are permitted in these raceways and enclosures.

Ground-fault circuit-interrupters are permitted in a panelboard that contains circuits protected by other than ground-fault circuit-interrupters.

Table 16-1
GFCI Requirements in Article 680

Fountains	680-51(a)
Fountains, cord- and plug-connected .	680-56(a)
Hydromassage bathtubs	680-70
Lighting fixtures, ceiling (paddle) fans, pools	680-6(b)(2),(b)(3) and (b)(4)
Motors in other than dwelling units ..	680-6(d)
Pool cover	680-26(b)
Receptacles, pools	680-6(a)(1) and (a)(3)
Signs with fountains	680-57(b)
Spa & hot tub lighting fixtures & ceiling (paddle) fans	680-41(b)(1) and (b)(2)
Spa & hot tub receptacles	680-41(a)(2)
Spa or hot tub, outlet	680-42
Storable pool equipment	680-31, 32(b)(3)
Therapeutic pools and tubs	680-62(a)
Underwater lighting fixtures	680-20(a)(1)

Supply conductors to a feed-through-type ground-fault circuit-interrupter are permitted in the same enclosure with the conductors supplied by the GFCI device.

GFCI requirements for receptacles

Section 680-6(a)(1) generally requires that receptacles on the property be located at least 10 feet from the inside walls of a pool (outside spa or hot tub) or fountain. A receptacle to supply power for a water-pump for a permanently installed pool or fountain, as permitted in Section 680-7, must be installed between 5 and 10 feet from the inside walls of the pool or fountain. Where so located, the receptacle(s) must be of the single and of the locking and grounding types and must be protected by a GFCI device.

Where a permanently installed pool is installed at a dwelling unit(s), at least one 125-volt receptacle is required to be located a minimum of 10 feet from and not more than 20 feet from the inside wall of the pool. This receptacle must be located not more than 6 foot 6 inches above the floor, platform, or grade level serving the pool. See Section 680-6(a)(2). (Remember that where the word "pool" is used, it also includes outdoor spas and hot tubs.)

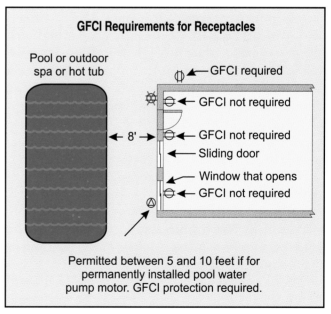

GFCI Requirements for Receptacles

Pool or outdoor spa or hot tub

GFCI required

GFCI not required

8'

GFCI not required

Sliding door

Window that opens

GFCI not required

Permitted between 5 and 10 feet if for permanently installed pool water pump motor. GFCI protection required.

Figure 16-17

All 125-volt receptacles located within 20 feet of the inside walls of a pool or fountain must be protected by a ground-fault circuit-interrupter. See Section 680-6(a)(3).

A fine print note that follows this section provides information on determining the distances from the pool. It reads, "In determining the above dimensions, the distance to be measured is the shortest path the supply cord of an appliance connected to the receptacle would follow without piercing a floor, wall, ceiling, doorway with hinged or sliding door, window opening, or other effective permanent barrier."[N]

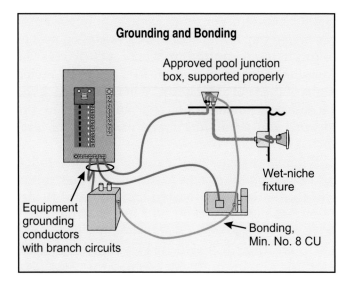

Figure 16-18

Bonding at swimming pools

Swimming pools, spas and hot tubs present special shock hazards to people due to mixing water enriched with chemicals, people and electricity together. Measures as prescribed in Article 680 of the Code, if carefully followed, will reduce the electric shock hazards to an acceptable level. One of these measures is the bonding together of conductive portions of the pool and metal parts of electrical equipment associated with the pool. The goal is to provide a means of equalizing the potential of all equipment and parts so there will be no current flow between parts. This is accomplished by connecting all the parts together by an adequately-sized and properly-connected copper conductor. This concept is often called equipotential bonding or grounding. See Section 680-22 of the NEC®.

These requirements emphasize the purpose of bonding in a swimming pool, spa or hot tub area - that of equalizing the potential (voltage) between various parts of the pool. By keeping the potential difference as low as practicable, the shock hazard is reduced significantly.

Since several pieces of electrical equipment commonly used with pools, hot tubs or spas that are bonded together must also be grounded with an

insulated equipment grounding conductor, an interconnection between the grounded (neutral) conductor and bonding grid exists. This interconnection is often remote from the pool and is not intended to play a part in equipotential bonding.

Section 680-22(a) provides a list of items that are required to be bonded. Included are the following:

(1) All metallic parts of the pool structure, including the reinforcing metal of the pool shell, coping stones and deck. Where reinforcing steel is effectively insulated by a listed encapsulating nonconductive compound at the time of manufacture and installation, it is not required to be bonded.

(2) All metal forming shells and mounting brackets of no-niche fixtures unless a listed low-voltage lighting system is used.

(3) All metal fittings within or attached to the pool structure. (Isolated parts that are not over 4 inches in any dimension and do not penetrate into the pool structure more than 1 inch are not required to be bonded.)

(4) Metal parts of electric equipment associated with the pool water circulating system, including pump motors and metal parts of equipment associated with pool covers, including electric motors. Metal parts of equipment incorporating an approved system of double insulation and providing a means for grounding internal nonaccessible, noncurrent-carrying metal parts shall not be bonded.

5.   Metal-sheathed cables and raceways, metal piping, and all fixed metal parts that are

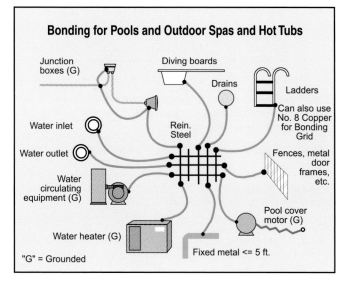

Figure 16-19

within 5 feet horizontally of the inside walls of the pool, and within 12 feet above the maximum water level of the pool, or any observation stands, towers or platforms, or from any diving structures, that are not separated from the pool by a permanent barrier.

Common bonding grid

The parts identified in Section 680-22(a) are required to be connected to a common bonding grid with a solid copper conductor not smaller than No. 8. The conductor can be insulated, covered or bare. Connections to the parts to be bonded and to the common bonding grid must be made by exothermic welding or by pressure connectors or clamps of that are labeled as being suitable for the purpose and are of the following material: stainless steel, brass, copper or copper alloy.

Steel tie wires as being suitable for connecting the reinforcing steel together and welding or special clamping is not required. The steel tie wires must be made up tightly.

Acceptable common bonding grids include as follows:

(1) The structural reinforcing steel of a concrete pool where the reinforcing rods are bonded together properly by the usual steel tie wires or the equivalent.

(2) The wall of a bolted or welded metal pool.

(3) A solid copper conductor not smaller than No. 8. It can be insulated, covered or bare.

(4) Rigid metal conduit or intermediate metal conduit of brass or other identified corrosion-resistant metal conduit.

Bonding of wet-niche light fixtures

Where mounted in a pool or fountain structure, a wet-niche lighting fixture must be installed in a forming shell that is designed to support a wet-niche lighting fixture assembly. The fixture will be completely surrounded by water. [Sections 680-3 and 680-20(b)]

A forming shell must be installed for the mounting of all wet-niche underwater lighting fixtures. The forming shell must also be equipped with provisions for threaded conduit entries.

Conduit permitted to be used to connect the forming shell to a suitable junction box or other permitted enclosure include rigid metal, intermediate metal, liquidtight flexible nonmetallic or rigid nonmetallic. [Section 680-20(b)(1)]

Rigid metal or intermediate metal conduit used to connect the wet-niche fixture housing must be made of brass or other approved corrosion resistant metal, such as stainless steel.

Where rigid nonmetallic or liquidtight flexible nonmetallic conduit is used, a No. 8 insulated copper conductor is required to be installed along with the pool flexible cord assembly so that it can be terminated on a suitable lug in the forming shell, junction box, transformer enclosure or ground-fault circuit-interrupter enclosure.

At the point of this termination, within the forming shell, this No. 8 conductor must be covered with, or encapsulated in, a listed potting compound. [Section 680-20(b)(1)] This compound provides protection from deteriorating effects often caused by the pool water. Where a listed potting compound is not used to encapsulate the No. 8 bonding conductor inside the forming shell, brass or stainless steel conduit must be used.

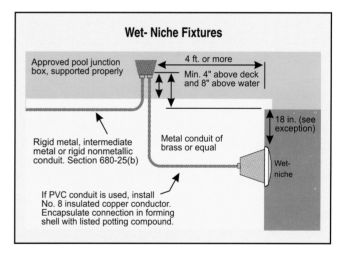

Figure 16-20

Where in contact with the pool water, metal parts of the fixture and forming shell must also be of brass or other approved corrosion-resistant metal.

The end of the flexible cord jacket and the flexible cord conductor terminations located within a wet-niche fixture are required to be covered with, or encapsulated in, a suitable potting compound. [Section 680-20(b)(2)] This will help to prevent the siphoning of water into the fixture through the cord jacket or its contained conductors. This requirement is met by the manufacturer of the wet-niche fixture.

In addition, an equipment grounding conductor connection within a wet-niche fixture must be similarly treated to protect it from the deteriorating effect of pool water in the event of water entry into the fixture.

The fixture must be bonded to and secured to the forming shell by a positive locking device that will ensure a low resistance contact. A special tool is required in order to remove the fixture from the forming shell. Bonding is not required for fixtures listed for the application, having no noncurrent-carrying metal parts.

Underwater wet-niche lighting fixtures include markings that will indicate the proper housing or housings with which they are to be used, and the fixture housings are marked to indicate the fixture or fixtures with which the housings are to be used. These fixtures are provided with a factory-installed permanently-attached flexible cord that extends at least 12 feet outside the fixture enclosure. This will permit removal of the fixture from the forming shell so that it can be lifted onto the pool or spa deck for servicing without lowering the water level or disconnecting the fixture from the branch circuit conductors.

Fixtures with longer cords are available for installation where the junction box or splice enclosure is so located that a 12-foot-long cord will not permit the fixture to be removed from the forming shell and placement on the deck for servicing. To avoid possible cord damage, any cord length in excess of that necessary for servicing should be trimmed from the supply end rather than stored in the forming shell.

Junction boxes

Special requirements are contained in Section 680-21 of the NEC® for junction boxes that are connected to a conduit that extends directly to a forming shell or a mounting bracket for a no-niche swimming pool fixture. Specific requirements include:

(1) The junction box must be listed and labeled for the purpose. This ensures that the box has provisions for the equipment grounding conductor and bonding conductors used for swimming pool equipment.

(2) It must be equipped with threaded entries or hubs or a nonmetallic hib listed for the purpose.

(3) It must be of copper, brass, suitable plastic, or other approved corrosion-resistant material.

(4) It must be provided with electrical continuity between every connected metal conduit and the grounding terminals by means of copper, brass, or other approved corrosion-

resistant metal that is integral with the box.

(5) It must have a number of grounding terminals that shall be at least one more than the number of conduit entries. For example, where there are two conduit entries, the grounding terminal bar must have not less than three terminals.

The junction box or other suitable enclosure must be located not less than 4 inches, measured from the inside of the bottom of the box or other enclosure, above the ground level or pool deck, or not less than 8 inches above the maximum pool water level, whichever provides the greatest elevation. It must also be located so that it is not less than 4 feet from the inside wall of the pool unless it is separated from the pool by a solid fence, wall or other permanent barrier. See Section 680-21(a)(5).

If used on a lighting system operating at 15 volts or less, a flush deck box is permitted provided both of the following conditions are complied with:

(a) An approved (acceptable to the authority having jurisdiction, not listed or labeled) potting compound is used to fill the box to prevent the entrance of moisture.

(b) The flush deck box is located not less than 4 feet from the inside wall of the pool.

Grounding requirements

The following swimming pool and outdoor spa and hot tub equipment is required by NEC® Section 680-24 to be grounded:

(1) Wet-niche and no-niche underwater lighting fixtures other than those low-voltage systems listed for the application without a grounding conductor;

(2) Dry-niche underwater lighting fixtures;

(3) All electric equipment located within 5 feet of the inside wall of the pool;

(4) All electric equipment associated with the recirculating water system of the pool;

(5) Metal junction boxes;

(6) Metal transformer enclosures;

(7) Ground-fault circuit-interrupters;

(8) Panelboards that are not part of the service equipment and that supply any electric equipment associated with the pool.

Methods of grounding

Wet-niche, dry-niche and no-niche lighting fixtures must be connected to an equipment grounding conductor that is sized in accordance with Table 250-122. This equipment grounding conductor must be an insulated copper conductor and can never be smaller than No. 12. [Section 680-25(b)(1)]

The equipment grounding conductor installed between the wiring chamber of the secondary winding of a transformer and a junction box must be sized in accordance with the overcurrent device in this circuit. See Table 250-122.2 The equipment grounding conductor must be an insulated copper conductor and must generally be installed with the circuit conductors in rigid metal conduit, intermediate metal conduit, liquidtight flexible nonmetallic conduit, or rigid nonmetallic conduit. [Section 680-25(b)(2)]

Electrical metallic tubing is permitted to be used for these conductors where installed on or within buildings. Electrical nonmetallic tubing is permitted to be used to protect circuit conductors where installed within buildings as ENT is generally limited to installation inside buildings as it is not suitable for direct-sunlight exposure. [Section 680-25(b)(3)]

Where connecting to transformers for pool lights, liquidtight flexible metal conduit or liquidtight flexible nonmetallic conduit is permitted to be used when installed in accordance with Article 351. Any one length is limited to not more than 6 feet and a total of not more than 10 feet is permitted.

The junction box, transformer enclosure, or other enclosure in the supply circuit that is run to a wet-niche lighting fixture, and the field wiring chamber of a dry-niche lighting fixture, must be grounded to the equipment grounding terminal of a metal panelboard. This terminal must be directly connected to the metal panelboard enclosure, and the equipment grounding conductor must be installed

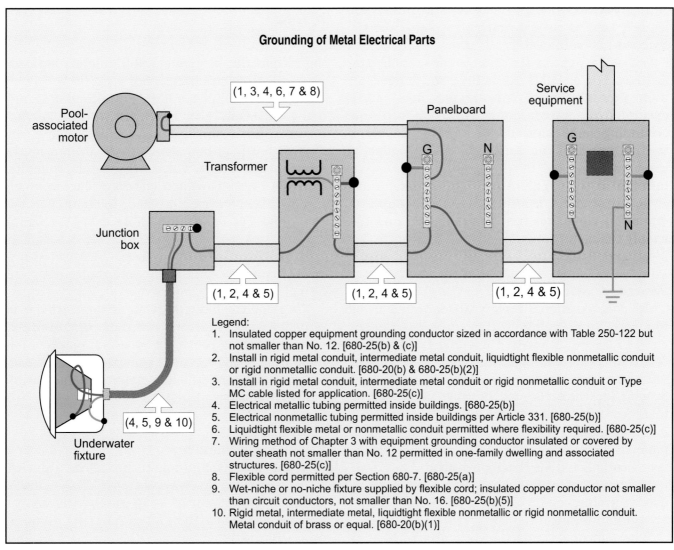

Grounding of Metal Electrical Parts

Legend:
1. Insulated copper equipment grounding conductor sized in accordance with Table 250-122 but not smaller than No. 12. [680-25(b) & (c)]
2. Install in rigid metal conduit, intermediate metal conduit, liquidtight flexible nonmetallic conduit or rigid nonmetallic conduit. [680-20(b) & 680-25(b)(2)]
3. Install in rigid metal conduit, intermediate metal conduit or rigid nonmetallic conduit or Type MC cable listed for application. [680-25(c)]
4. Electrical metallic tubing permitted inside buildings. [680-25(b)]
5. Electrical nonmetallic tubing permitted inside buildings per Article 331. [680-25(b)]
6. Liquidtight flexible metal or nonmetallic conduit permitted where flexibility required. [680-25(c)]
7. Wiring method of Chapter 3 with equipment grounding conductor insulated or covered by outer sheath not smaller than No. 12 permitted in one-family dwelling and associated structures. [680-25(c)]
8. Flexible cord permitted per Section 680-7. [680-25(a)]
9. Wet-niche or no-niche fixture supplied by flexible cord; insulated copper conductor not smaller than circuit conductors, not smaller than No. 16. [680-25(b)(5)]
10. Rigid metal, intermediate metal, liquidtight flexible nonmetallic or rigid nonmetallic conduit. Metal conduit of brass or equal. [680-20(b)(1)]

Figure 16-21

without any joint or splice. [Section 680-25(b)(4)]

Where more than one underwater lighting fixture is supplied by the same branch circuit, the equipment grounding conductor, installed between the junction boxes, transformer enclosures, or other enclosures in the supply circuit to wet-niche fixtures, or between the field-wiring compartment of dry-niche fixtures, is permitted to be terminated on grounding terminals rather than being installed in an unbroken length. [Section 680-25(b)(4)(a)]

Where the underwater lighting fixture is supplied from a transformer, ground-fault circuit interrupter, clock-operated switch, or a manual snap switch that is located between the panelboard and a junction box connected to the conduit that extends directly to the underwater lighting fixture, the equipment grounding conductor is permitted to terminate on equipment grounding terminals on the transformer, ground-fault circuit-interrupter, clock-operated switch enclosure, or an outlet box that is used to enclose a snap switch. [Section 680-25(b)(4)(b)]

Wet-niche lighting fixtures that are supplied by flexible cord or cable assemblies must have all their exposed noncurrent-carrying metal parts grounded by an insulated copper equipment grounding conductor that is an integral part of the cord or cable. This equipment grounding conductor must be connected to an equipment grounding terminal in the supply junction box, transformer enclosure or other enclosure. Also, this equipment grounding conductor cannot be smaller than the supply conductors and never smaller than No. 16. [Section 680-25(b)(5)]

Pool associated motors

All pool associated motors are required to be connected to an equipment grounding conductor sized in accordance with Table 250-122. [Section 680-25(c)] This equipment grounding conductor must be an insulated copper conductor and cannot be smaller than No. 12. It must generally be installed with the circuit conductors in a rigid metal conduit, intermediate metal conduit, rigid nonmetallic conduit or Type MC metal-clad cable specifically listed for the application.

Electrical metallic tubing is permitted to be used to protect conductors where it is installed on or within buildings.

Where flexible connections are necessary at or adjacent to the motor, liquidtight flexible metal or

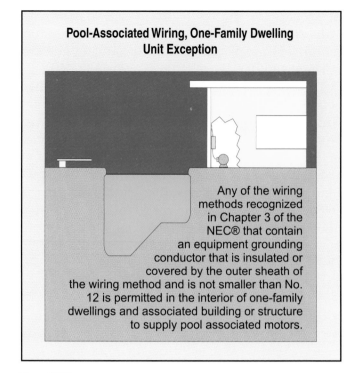

Figure 16-22

liquidtight flexible nonmetallic conduit with approved fittings are permitted.

In the interior of one-family dwellings only, any of the wiring methods recognized in NEC® Chapter 3 containing a copper equipment grounding conductor that is either insulated or covered by the outer sheath of the wiring method, and not smaller than No. 12, are permitted to be used for the wiring method. These wiring methods include Type NM cable and Type UF cable. Note that this wiring method does not apply to the supply of spas and hot tubs that are located outdoors. Flexible cord is permitted to be installed in accordance with Section 680-7.

Permanently installed pools are permitted to be provided with listed cord- and plug-connected pool pumps that are protected by a system of double insulation. They are to be provided with a means for grounding only the internal and nonaccessible noncurrent-carrying metal parts of the pump. (Section 680-28)

Panelboard grounding

A panelboard, and where installed, a disconnecting means, that are not part of the service equipment or source of a separately derived system are required to have an equipment grounding conductor installed between its equipment grounding terminal, and the equipment grounding termi-

Figure caption within image:

Pool-Associated Wiring, One-Family Dwelling Unit Exception

Any of the wiring methods recognized in Chapter 3 of the NEC® that contain an equipment grounding conductor that is insulated or covered by the outer sheath of the wiring method and is not smaller than No. 12 is permitted in the interior of one-family dwellings and associated building or structure to supply pool associated motors.

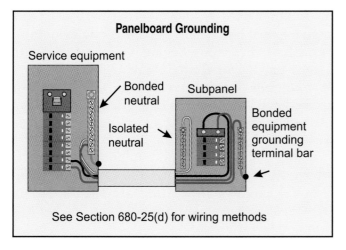

Panelboard Grounding

See Section 680-25(d) for wiring methods

Figure 16-23

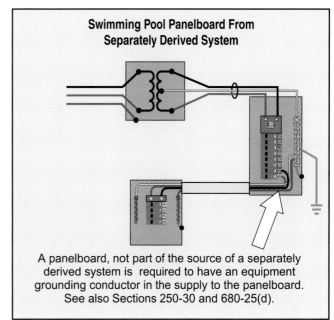

Swimming Pool Panelboard From Separately Derived System

A panelboard, not part of the source of a separately derived system is required to have an equipment grounding conductor in the supply to the panelboard. See also Sections 250-30 and 680-25(d).

Figure 16-24

nal located at the service equipment or source of a separately derived system.

This conductor must be sized in accordance with Table 250-122 and must be an insulated conductor of copper, aluminum or copper-clad aluminum and can never be smaller than No. 12. This conductor must generally be installed with the feeder conductors in a rigid metal conduit, intermediate metal conduit or rigid nonmetallic conduit.

Electrical metallic tubing is permitted to be used to protect conductors where installed on or within buildings in accordance with Article 348. Electrical nonmetallic tubing is permitted to be used to protect conductors where installed within buildings in accordance with Article 331.

The equipment grounding conductor must be connected to an equipment grounding terminal of the panelboard and, where installed, to the enclosure for a disconnecting means.

Panelboards from separately derived systems

There are several ways of grounding swimming pool panelboards that are supplied from a separately derived system. It is important to follow the rules for grounding the separately derived system as given in Section 250-30 and as covered extensively in Chapter 12 of this text.

Section 680-25(d) requires that the equipment grounding conductor from a separately derived system be sized in accordance with Table 250-66 based on the derived phase conductors but not smaller than No. 8. The grounded conductor from the separately derived system is also required to be sized from Table 250-66 based on the derived phase conductors.

Existing panelboards

The equipment grounding conductor installed between an existing remote panelboard and the service equipment is not required to be run in the conduits identified above if the interconnection is by means of a flexible metal conduit, or an approved cable assembly provided with an insulated or covered copper, aluminum or copper-clad equipment grounding conductor. See Section 680-25(d)(1).

The NEC® does not define what is meant by the word "existing." Often, inspection authorities consider an installation to be "existing" after it has had final approval or after a certificate of occupancy has been issued.

Swimming pool panelboards in other buildings

Grounding at the disconnecting means for remote buildings is covered in Section 250-32 and in Chapter Thirteen of this text where the methods for complying with the Code rules are given. Normally, one can choose between grounding the neutral conductor again at the building and not taking the equipment grounding conductor to the building, and taking the equipment grounding conductor to the building, grounding it there and not grounding the neutral but "floating" it there. Specific limitations are given for re-grounding the grounded conductor (may be a neutral) at the remote building. The grounded conductor cannot be re-grounded at the second building where (1) an equipment grounding conductor is run with the

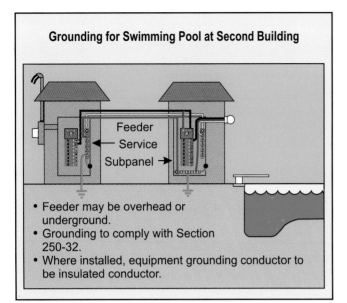

Grounding for Swimming Pool at Second Building

Feeder
Service
Subpanel

- Feeder may be overhead or underground.
- Grounding to comply with Section 250-32.
- Where installed, equipment grounding conductor to be insulated conductor.

Figure 16-25

supply to the building, (2) where there are continuous metallic paths between the buildings that are bonded to the grounding system at each building or (3) where the service has ground-fault protection of equipment.

Where installed between the buildings, the equipment grounding conductor is required to be an insulated conductor.

Cord-connected equipment

Where fixed or stationary equipment is connected with a flexible cord to facilitate removal or disconnection for maintenance, repair, or storage as provided in Section 680-7, the equipment grounding conductor must be connected to a fixed metal part of the assembly. The removable part must be mounted on or bonded to the fixed metal part. [Section 680-25(e)]

All other electrical equipment must be grounded in accordance with Article 250 and must be connected by an approved wiring method covered in NEC® Chapter 3. See Section 680-25(f).

Spa or hot tub installed outdoors

A spa or hot tub installed outdoors must comply with the provisions of Article 680, Part A, General Requirements and Part B covering Permanently Installed Pools. See Section 680-40. This requirement includes those for GFCI protection of receptacles and lighting fixtures as well as required clearances from receptacles and lighting fixtures.

Section 680-40(a) permits listed packaged units (spas or hot tubs) utilizing a factory-installed re-

mote panelboard to be connected with not more than 6 feet of liquidtight flexible conduit (metallic or nonmetallic) or be cord and plug connected with a cord not longer than 15 feet if protected by a ground-fault circuit-interrupter.

Section 680-40(b) permits bonding to be accomplished by metal-to-metal mounting on a common frame or base. This a common construction of a "packaged spa or hot tub equipment assembly" which are commonly referred to as a "skid pack."

Metal bands or hoops that are used to secure wooden staves are not required to be bonded to a common bonding grid.

Spa or hot tub installed indoors

A spa or hot tub installed indoors must conform to the requirements of Article 680, Parts A and B except as modified by Section 680-41. They are required to be connected by wiring methods of NEC® Chapter 3.

Where rated at 20 amperes or less, listed packaged units are permitted to be cord- and plug-connected to facilitate their removal or disconnection for maintenance and repair.

At least one 125-volt, 15- or 20-ampere receptacle on a general purpose branch circuit is required to be located at least 5 feet from and not more than 10 feet from the inside wall of a spa or hot tub.

All receptacles on the property must be located to be at least 5 feet away from the inside walls of the spa or hot tub.

Any 125 volt-rated receptacles that are located within 10 feet of the inside walls of a spa or hot tub must be protected by a ground-fault circuit-interrupter.

The distance measured is the shortest path the supply cord of an appliance connected to the receptacle would follow without piercing a floor, wall or ceiling of a building or other effective permanent barrier.

Receptacles that provide power for a spa or hot tub must also be protected by a ground-fault circuit-interrupter.

Bonding of spa or hot tub installed indoors

The following parts of a spa or hot tub are required to be bonded together:

1. All metal fittings within or attached to the spa or hot tub structure.
2. Metal parts of any electric equipment associated with the spa or hot tub water circulating system, including pump motors.

3. Metal conduit and metal piping that is located within 5 feet of the inside walls of the spa or hot tub and that are not separated from the spa or hot tub by a permanent barrier.

4. All metal surfaces that are located within 5 feet of the inside walls of a spa or hot tub and not separated from the spa or hot tub area by a permanent barrier. Small conductive surfaces not likely to become energized, such as air and water jets and drain fittings, where not connected to metallic piping, towel bars, mirror frames, and similar equipment, are not required to be bonded.

5. Electrical devices and controls not associated with the spas or hot tubs must be located at least 5 feet away from such units or are required to be bonded to the spa or hot tub system.

Methods of bonding of spa or hot tub installed indoors

All metal parts associated with the spa or hot tub are required to be bonded by any of the following methods:

1. The interconnection of threaded metal piping and fittings.

2. metal-to-metal mounting on a common frame or base.

3. By a copper bonding jumper that is insulated, covered or bare and not smaller than No. 8 solid.

GFCI protection of spa or hot tub installed indoors or outdoors

Section 680-42 contains GFCI requirements that apply to both indoor and outdoor spa and hot tub equipment. GFCI protection of the outlet is required if it supplies:

1. A self-contained spa or hot tub.

2. A packaged spa or hot tub equipment assembly.

3. A field-assembled spa or hot tub with a heater load of 50 amperes or less.

GFCI protection of the outlet is not required for a listed self-contained unit or listed packaged equipment assembly that is marked to indicate that integral ground-fault circuit-interrupter protection is provided for all electrical parts within the unit or assembly (pumps, air blowers, heaters, lights, controls, sanitizer generators, wiring, etc.).

The supply for a field-assembled spa or hot tub is not required to be protected by a ground-fault circuit interrupter if the supply is rated greater than 250 volts or rated 3 phase.

A combination pool/hot tub or spa assembly commonly bonded need not be protected by a GFCI.

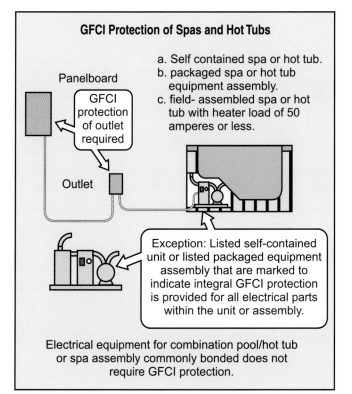

GFCI Protection of Spas and Hot Tubs

Panelboard

GFCI protection of outlet required

a. Self contained spa or hot tub.
b. packaged spa or hot tub equipment assembly.
c. field- assembled spa or hot tub with heater load of 50 amperes or less.

Outlet

Exception: Listed self-contained unit or listed packaged equipment assembly that are marked to indicate integral GFCI protection is provided for all electrical parts within the unit or assembly.

Electrical equipment for combination pool/hot tub or spa assembly commonly bonded does not require GFCI protection.

Figure 16-26

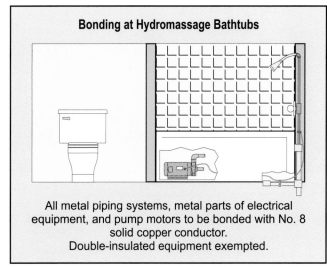

Bonding at Hydromassage Bathtubs

All metal piping systems, metal parts of electrical equipment, and pump motors to be bonded with No. 8 solid copper conductor.
Double-insulated equipment exempted.

Figure 16-27

Bonding hydromassage bathtubs

Section 680-73 contains requirements on bonding of electrical equipment and metal piping systems associated with hydromassage bathtubs. It should be noted that these requirements apply to dwelling units as well as to other than dwelling units.

This section requires that, "All metal piping systems, metal parts of electrical equipment, and pump motors associated with the hydromassage tub shall be bonded together using a copper bonding jumper, insulated, covered, or bare, not smaller than No. 8 solid.

Metal parts of listed equipment incorporating an approved system of double insulation and providing a means for grounding internal nonaccessible, noncurrent-carrying metal parts shall not be bonded."[N]

Electric signs and outline lighting

Section 600-7 contains requirements for grounding of electric signs and metal equipment of outline lighting systems. The rules in Article 600 amend or supplement the general rules for grounding in Article 250 due to the rule of "Code Arrangement" in Section 90-3.

Specific requirements in Section 600-7 are that, "Signs and metal equipment of outline lighting systems shall be grounded."

Section 600-7 continues, "Listed flexible metal conduit or listed liquidtight flexible metal conduit that encloses the secondary wiring of a transformer or power supply for use with electrical discharge tubing shall be permitted as a bonding means in lengths not exceeding 100 feet. Normally, these wiring methods are permitted to serve as an equipment grounding conductor in lengths of not more than 6 feet.

Small metal parts not exceeding 2 inches in any dimension that are not likely to be energized and spaced at least ¾ inches from neon tubing are not required to be bonded.

Where listed nonmetallic conduit is used to enclose the secondary wiring of a transformer or power supply and a bonding conductor is required, the bonding conductor shall be installed separate and remote from the nonmetallic conduit and be spaced at least 1½ inches from the conduit when the circuit is operated at 100 Hz or less. Where the circuit operates at over 100 Hz, the spacing of the bonding conductor must be not less than 1¾ inches.

Bonding conductors are required to be copper and not smaller than No. 14.

Metal parts of the building are not permitted to serve as a grounded or equipment grounding conductor.

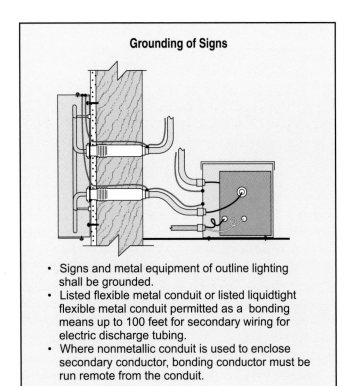

Grounding of Signs

- Signs and metal equipment of outline lighting shall be grounded.
- Listed flexible metal conduit or listed liquidtight flexible metal conduit permitted as a bonding means up to 100 feet for secondary wiring for electric discharge tubing.
- Where nonmetallic conduit is used to enclose secondary conductor, bonding conductor must be run remote from the conduit.

Figure 16-28

Chapter Sixteen: The questions included here were developed using material included in this chapter. The answers can be found by reviewing the text. It is also important that students make use of the 1999 NEC®, where many answers can be found. See pages 279 for answers.

1. Protection from lightning discharges is provided by ____.
   a. building low-rise structures
   b. building structures of wood construction
   c. providing a low-resistance path to earth
   d. providing a high-resistance path to earth

2. It is required that ground terminals for lightning protection systems be ____.
   a. bonded to the electrical system grounding electrode system
   b. bonded to the building water pipe system
   c. isolated from the building electrical system by at least six feet
   d. isolated from the building grounding electrode system

3. Generally, in hazardous (classified) locations, locknuts installed on each side of an enclosure, or a locknut on the outside and a bushing on the inside ____ be used for bonding.
   a. shall
   b. can
   c. cannot
   d. may

4. Bonding locknuts or bonding bushings with bonding jumpers must be used in hazardous (classified) locations to ensure the integrity of the bond and its capability of carrying any ____ that may be imposed.
   a. unidentified currents
   b. neutral currents
   c. objectionable current
   d. fault current

5. Where flexible metal conduit is permitted and used in a hazardous (classified) location, it must have an internal or external bonding jumper installed to supplement the conduit. Where the jumper is installed on the outside, it is limited to ____ feet in length.
   a. 6
   b. 7
   c. 8
   d. 9

6. Where equipment is installed in an agricultural building that requires grounding, a ____ equipment grounding conductor must be installed to ground the equipment.
   a. aluminum
   b. copper
   c. copper-clad
   d. steel

7. Minimizing or elimination of stray voltages in an agricultural building is accomplished by bonding together the reinforcing rods and wire mesh in the concrete and then connecting these to the metal piping systems, stanchions and grounding electrode systems with a copper conductor not smaller than No. ____, .
   a. 8
   b. 6
   c. 4
   d. 2

8. Which of the following statements is true about wiring in an agricultural building?
   a. An 8-ft. ground rod is required at each corner of the building.
   b. GFCI protection is requiured for all 125-volt, 1-phase, 15- and 20-ampere receptacles.
   c. Voltage-gradient ramps at entrances and exist are optional.
   d. Line-to-neutral loads are not permitted.

9. A redundant means of grounding is required in ____ of health care facilities.
   a. limited-care facilities
   b. office areas
   c. all areas
   d. patient care areas

10. Where used in patient care areas, Type MI cable or Type ____ cable is required to have outer metal armor or sheath that is ____ as an acceptable grounding return path.
   a. MV - approved by special permission
   b. SE - permitted
   c. SNM - approved
   d. MC - identified

11. Where metal raceways or Type MC or MI cables are used for feeders supplying critical care areas in health care facilities, grounding of panelboards and switchboards must be assured. Acceptable methods include all of the following EXCEPT ____
   a. grounding bushings with bonding jumpers
   b. threaded bosses
   c. bonding locknuts or bushings
   d. standard locknuts

12. Equipment in the vicinity of swimming pools must be bonded together with a No. 8 or larger copper conductor. This also includes metal raceways and cables, piping and fixed metal parts within ____ feet horizontally or ____ feet vertically of the pool, observation stands, towers, platforms or diving structures.
   a. 5 - 18
   b. 6 - 20
   c. 5 - 12
   d. 9 - 14

13. A common bonding grid at a pool is permitted to be the structural reinforcing steel of a concrete pool, the wall of a bolted or welded pool, or a No. ____ or larger solid copper conductor.
   a. 8
   b. 6
   c. 4
   d. 2

14. A dry-niche lighting fixture is required to be provided with a provision for the drainage of water and a means for accommodating ____ equipment grounding conductor(s) for each conduit entry.
   a. 1
   b. 2
   c. 3
   d. 0 or none

15. All 125-volt receptacles located within ____ feet of the inside walls of a pool are required to be protected by a ground-fault circuit-interrupter.
   a. 30
   b. 25
   c. 22
   d. 20

16. Ground-fault circuit-interrupters used around pools are permitted to be installed in a ____ that contains circuits protected by other than ground-fault circuit-interrupters.
   a. cabinet
   b. metal device box
   c. panelboard
   d. nonmetallic box

17. For other than storable pools, a flexible cord is not permitted to exceed ____ feet in length and must have a copper equipment grounding conductor not smaller than No. 12 with a grounding-type attachment plug.
   a. 3
   b. 4
   c. 5
   d. 6

18. Ground-fault circuit-interrupters used around pools are required to be self-contained units, circuit-breaker types, receptacle-types or other ____ types.
   a. listed
   b. acceptable
   c. approved
   d. labeled

19. A ground-fault circuit-interrupter is required to be installed in the branch circuit supplying underwater lighting fixtures operating at more than ____ volts, so that there is no shock hazard during relamping.
   a. 6
   b. 10
   c. 12
   d. 15

20. The installation of the ground-fault circuit-interrupter installed to protect an underwater lighting fixture must be such that there is no shock hazard with any likely fault-condition combination that involves a person in a conductive path from any ungrounded part of the ____ or the fixture to ground.
   a. service
   b. system
   c. feeder
   d. branch circuit

21. Where metal conduit is used to supply a ____ niche fixture, it is required to be of brass or other approved corrosion-resistant metal.
   a. dry-
   b. wet-
   c. no-
   d. metal-

22. A wet-niche fixture is required to be bonded to and secured to the forming shell by a positive locking device that assures a low-resistance contact and requires a ____ to remove the fixture from the forming shell.
   a. tool
   b. permit
   c. fitting
   d. helper

23. Transformers used for the supply of underwater fixtures, together with the transformer enclosure, are required to be ____ for the purpose.
   a. suitable
   b. identified
   c. acceptable
   d. approved

24. The transformer used to supply underwater lighting fixtures is required to be a ____ winding type having a grounded metal barrier between the primary and secondary windings.
    a. three
    b. isolated
    c. one
    d. four

25. Signs and ___ must be grounded.
    a. all line-side equipment
    b. electrode receptacle housings
    c. metal equipment of outline lighting
    d. electrical-discharge tubing

26. Where flexible nonmetallic conduit is used on the secondary of a sign transformer, the bonding conductor must be:
    a. installed inside the conduit
    b. not more than 6-feet long
    c. installed outside the conduit
    d. listed GTO cable

27. Listed flexible metal conduit that is used to enclose the secondary wiring of a sign transformer is permitted as the bonding means for lengths not exceeding:
    a. 6 feet.
    b. 25 feet
    c. 50 feet
    d. 100 feet

Notes:

Over 600 Volt Systems

Objectives

After studying this chapter, the reader will be able to understand:

- Grounding rules for high voltage (1kV and over).
- Use of surge arresters.
- Grounding of outdoor industrial substations.

High voltage (1kV and over)

For grounding of systems or circuits of 1kV and over, the NEC® indicates that where such high-voltage systems are grounded, they shall comply with Article 250 plus other rules in Part K of Article 250 which supplement or modify other rules in Article 250.

Most medium-voltage systems (2.4 to 15kV) are resistance grounded while others are high resistance grounded. The only difference in the various methods of resistance grounding is the amount of ground-fault current that will flow due to the varying resistance.

Most high-voltage systems (above 15kV) are effectively grounded utilizing surge arresters because the transmission lines are open and commonly subjected to lightning surges and transient overvoltages.

Systems rated 2,400 volts to 13,800 volts

Grounding may be solid or low resistance depending on the available ground-fault current and the size of the system. Reactance grounding in this voltage range is preferred if the circuits are overhead and thus subject to lightning exposure, if they serve rotating machinery and if excessive ground-fault current will not develop. It is common in industrial systems to ground the neutral of systems rated 2,400 volts and above through a resistor.

Owing to the higher voltage of these systems, compared to systems of 600 volts and less, the ground-fault current levels that will flow in the grounding electrode conductor and grounding electrode is sufficiently high that a current transformer may be placed in the grounding electrode conductor to give positive indication that a ground fault exists or may be used to clear the circuit in the event the ground fault persists over a predetermined time. Compare this with a system of 600 volts or less. There, the current transformer cannot be placed satisfactorily in the grounding electrode conductor, for little ground-fault current will flow in that conductor.

Systems rated 15,000 volts or more

Such systems are nearly always outdoors and have no rotating equipment served by them. Usually they are solidly grounded. Such solid grounding permits the use of grounded-neutral type surge arresters, which cost less as well as provide better protection from overvoltage.

In systems over 600 volts, there will be different voltage levels for various utilization purposes. Grounding must be resorted to at each voltage level as previously described for systems of 600 volts or less.

Grounding outdoor industrial substations

The grounding of outdoor industrial substations involves grounding the fence and other supporting structures, all equipment and conductor enclosures and the grounding of the neutral.

A ground bus or grid should be established extending about 3 feet outside the periphery of the fence. The bus should be connected to many ground rods around its periphery and, in addition, should be connected to a metallic underground water piping system where available. The ground bus should have a minimum size of No. 4/0 and be approximately 25 percent of the capacity of the system. Cable may be sized on the basis of the capacity of bare conductor in free air and copper bus on the basis of 1,200 amperes per square inch. Connections from the fence and from all equipment within the fence, that is, transformer cases, steel structures, switchgear including operating mechanisms for gang-operated disconnects, etc., should be not less than No. 4/0 nor less than 25 percent of the capacity of the secondary conductors. To be sure of having a good, permanent neutral ground, it is best to connect the neutral to two points on the ground bus.

A grounding bus and connections are effective only if the mechanical construction is sound and as permanent as possible. No connection should be soldered. It is preferable to properly braze or weld all connections and to protect all cable from mechanical injury. If a metallic enclosure is used for the mechanical protection of the conductor, the enclosure should be connected in parallel with the cable being protected from end to end of the enclosure.

There is some difference of opinion as to whether the fence ground should be separated from the ground bus used for the station. If the fence is connected to the station bus and a fault occurs, the fence will be elevated above earth potential by the IZ drop. The potential gradient in the earth near the fence drops off very rapidly within the first few feet. Anyone contacting the fence may thus be subjected to a hazardous step-voltage under fault conditions.

If the fence ground was isolated from the station

ground, then the fence will be elevated from the station ground, and under fault conditions anyone contacting the fence and the equipment will be subjected to a hazard.

In most industrial substations, the fence and equipment are close. In such cases, it is best to connect the fence ground and the station ground bus. The shock hazard when using a common ground for fence and equipment can be kept to a minimum by having the ground connection resistance as low as possible.

Derived neutral systems

NEC® Section 250-182 permits a system neutral to be derived from an ungrounded system by a grounding transformer. These grounding transformers include zigzag, wye-delta, or T-connected types. Ground-fault protection may be provided, and positive tripping can be accomplished with low magnitudes of ground-fault current.

Solidly grounded neutral systems

Section 250-184 of the NEC® also permits solidly grounded neutral systems. Multiple grounding is permitted at more than one point for services, direct buried portions of feeders employing a bare copper neutral and for overhead portions of a system. The neutral conductor of solidly grounded systems is generally required to have an insulation level of not less than 600 volts. Bare copper conductors are permitted

Impedance grounded neutral systems

NEC® Section 250-186 permits impedance grounded neutral systems. Such impedance grounding can be accomplished by reactance grounding, resistance grounding or high-resistance grounding.

Grounding of systems supplying portable or mobile equipment

Special requirements apply to electrical systems that supply portable or mobile high-voltage equipment, other than substations installed on a temporary basis. The requirements are found in Section 250-188 of the NEC®.

Portable or mobile equipment is required to be supplied from a system having its neutral grounded through an impedance. Where a delta-connected high-voltage system is used to supply portable or mobile equipment, a system neutral is required to be supplied.

Exposed noncurrent-carrying metal parts of portable or mobile equipment is required to be grounded by an equipment grounding conductor that is connected to the point at which the system neutral impedance is grounded.

The voltage developed between the portable or mobile equipment frame and ground by the flow of maximum ground-fault current must not exceed 100 volts. This somewhat limits the shock hazard.

Ground-fault detection and relaying must be provided to automatically de-energize any high-voltage system component that has developed a ground fault. The continuity of the equipment grounding conductor must be continuously monitored so the high voltage circuit to the portable or mobile equipment will be automatically de-energized upon loss of continuity of the equipment grounding conductor.

The grounding electrode to which the system neutral impedance is connected must be isolated from and separated in the ground at least 20 feet from any other system or equipment grounding electrode. No direct electrical connection between this and other grounding electrodes or fences is permitted.

High-voltage trailing cable and couplers for interconnection of portable or mobile equipment must meet the requirements of Part C of Article 400 for cables and Section 490-55 for couplers.

Grounding of high-voltage equipment

Section 250-190 generally requires grounding of all noncurrent-carrying metal parts of fixed, portable or mobile equipment of high voltage equipment including fences, housings, enclosures and supporting structures.

The exception to this section excludes from the grounding requirement equipment that is isolated from ground and located so as to prevent any person who can make contact with the ground from contacting such metal parts when the equipment is energized.

Surge arresters

Where surge arresters are used on secondary services of less than 1,000 volts, it is required that the connections to the service conductor and to the grounding conductor be as short as practicable. The Code specifies that the grounding conductor may be (1) the grounded service conductor; (2) the common grounding electrode conductor; (3) the service grounding electrode or (4) the equipment grounding terminal in the service equipment. The use of method (3) is practical only on an ungrounded system. A minimum size of No. 14 copper or No. 12 aluminum is specified. Note that only the minimum size is specified.

If a surge arrester protects the primary of a transformer which supplies a secondary distribution system, the grounding conductor may be interconnected to the secondary neutral, provided that in addition to the direct grounding connection at the arrester, the grounded conductor of the secondary has elsewhere a grounding connection to a continuous metallic underground water piping system. However, where there are not less than four secondary connections to any underground metallic water piping system per mile, the direct grounding connection at the arrester may be omitted.

The grounding electrode conductor of a lightning arrester also may be connected to the secondary neutral. This is true if, in addition to the direct grounding connection at the arrester, the grounding conductor of the secondary system is part of a multi-grounded neutral system of which the primary neutral has at least four grounding connections in each mile of line, as well as a ground at each service.

Where the secondary is not grounded to a metallic water system but uses other available electrodes, an interconnection between surge arrester grounding conductor and the secondary neutral shall be made through a spark gap. The spark gap must have a breakdown voltage of at least twice the primary circuit voltage but not necessarily more than 10kV. There shall be at least one other ground on the grounded conductor of the secondary not less than 20 feet away from the surge arrester grounding electrode.

No other connection between a surge arrester ground and a secondary neutral is allowed by the Code except by special permission.

Chapter Seventeen: The questions included here were developed using material included in this chapter. The answers can be found by reviewing the text. It is also important that students make use of the 1999 NEC®, where many answers can be found. See page 279 for answers.

1. Most medium-voltage systems ____ to ____ are resistance grounded while others are high resistance grounded.
   a. 5.4 - 20 kV
   b. 2.4 - 15 kV
   c. 3.3 - 17 kV
   d. 4.6 - 18 kV

2. Most high-voltage systems ____ are effectively grounded utilizing surge arresters because the transmission lines are open and commonly subjected to lightning surges and transient overvoltages.
   a. above 15 kV
   b. above 12 kV
   c. above 10 kV
   d. above 11 kV

3. It is common in industrial systems to ground the neutral of systems rated ____ volts and above through a resistor.
   a. 2,000
   b. 2,400
   c. 1,800
   d. 1,380

4. Systems rated ____ volts or more are nearly always outdoors and have no rotating equipment served by them.
   a. 10,000
   b. 12,000
   c. 15,000
   d. 13,000

5. In systems over ____ volts, there will be different voltage levels for various utilization purposes.
   a. 1,000
   b. 2,000
   c. 600
   d. 1,500

6. The grounding of outdoor industrial substations is concerned with grounding the fence and other supporting structures, all equipment and conductor enclosures and the grounding of the neutral. A ground bus or grid should be established extending about ____ feet outside the periphery of the fence.
   a. 3
   b. 4
   c. 6
   d. 8

7. Cable can be sized on the basis of the capacity of bare conductor in free air and copper bus on the basis of ____ amperes per square inch.
   a. 1,250
   b. 1,000
   c. 1,200
   d. 1,300

8. Connections from the fence and from all equipment within the fence, that is, transformer cases, steel structures, switchgear including operating mechanisms for gang-operated disconnects, etc., should be not less than No. ____ nor less than ____ percent of the capacity of the secondary conductors.
   a. 3/0 - 50
   b. 4/0 - 25
   c. 2/0 - 40
   d. 1/0 - 30

9. To be sure of having a good, permanent neutral ground, it is best to connect the neutral to ____ points on the ground bus.
   a. four
   b. three
   c. two
   d. five

10. Impedance grounded neutral systems are permitted and can be accomplished by all but one of the following methods:
   a. reactance grounding
   b. resistance grounding
   c. high-resistance grounding
   d. low-resistance grounding

11. Without exception, where surge arresters are installed either indoors or outdoors they are required to be ____ to unqualified persons.
   a. inaccessible
   b. accessible
   c. readily accessible
   d. available

12. Where there are not less than four secondary connections to any underground metallic water piping system per each ____, the direct grounding connection at the arrester may be omitted.
   a. ¼ mile
   b. mile
   c. ½ mile
   d. ¾ mile

Notes:

Tables

Objectives

After studying this chapter, the reader will be able to understand

- Application of tables to determine proper types and sizing of grounding electrode conductor.
- How to determine where voltage drop will affect sizes.
- Determine minimum size of equipment grounding conductors.
- Dimensions and lengths of conduits and other race ways.

Included in this chapter are several tables that are vital or useful in studying the subject of grounding and correctly applying the NEC® rules in that regard.

In addition, several tables are included to assist the installer or inspector of electrical systems in properly designing, installing and maintaining these systems.

By carefully utilizing the information in these tables, the electrical system will benefit from additional safety and should serve the owner for years to come.

The following tables are included:

Table 1.   Table 250-66, Grounding Electrode Conductor for AC systems from the NEC®.

Table 2.   Analysis of NEC® Table 250-66 grounding electrode conductor and service conductor compared as to relative size.

Table 3.   Comparison of copper grounding electrode conductor per NEC® Table 250-66 when used with and without steel conduit for physical protection.

Table 4.   Comparison of aluminum grounding electrode conductors per NEC® Table 250-66 when used with and without aluminum conduit for protection.

Table 5.   Rating of grounding electrode conductors specified in NEC® Table 250-66 and voltage drop under maximum short-time rating.

Table 6.   Table 250-122 from the National Electrical Code® "Minimum size Equipment Grounding Conductors for Grounding Raceways and Equipment."

Table 7.   Analysis of Table 250-122 of the National Electrical Code®.

Table 8.   Table 8 from Chapter Nine of the National Electrical Code® "Conductor Properties."

Table 9.   Not used.

Table 10. Not used.

Table 11. Maximum length of electrical metallic tubing, intermediate metal conduit and rigid steel conduit that may safely be used as an equipment grounding circuit conductor.

Table 12. Not used.

Table 13. Not used.

Table 14. Aluminum conduit used as an equipment grounding conductor. Compared with equipment grounding conductors as specified in Table 250-122 of the National Electrical Code®.

Table 15. Sizes of conductor (copper bar) for use in the equipment grounding strap for overcurrent devices from 1,600 amperes through 5,000 amperes.

Table One

Table 250-66 — Grounding Electrode Conductor for AC Systems			
Size of Largest Service-Entrance Conductor or Equivalent Area for Parallel Conductors[1]		Size of Grounding Electrode Conductor	
Copper	Aluminum or Copper-Clad Aluminum	Copper	Aluminum or Copper-Clad Aluminum[2]
2 or smaller	1/0 or smaller	8	6
1 or 1/0	2/0 or 3/0	6	4
2/0 or 3/0	4/0 or 250 kcmil	4	2
Over 3/0 thru 350 kcmil	Over 250 kcmil thru 500 kcmil	2	1/0
Over 350 kcmil thru 600 kcmil	Over 500 kcmil thru 900 kcmil	1/0	3/0
Over 600 kcmil thru 1100 kcmil	Over 900 kcmil thru 1750 kcmil	2/0	4/0
Over 1100 kcmil	Over 1750 kcmil	3/0	250 kcmil

Notes:

1. Where multiple sets of service-entrance conductors are used as permitted in Section 230-40, Exception No. 2, the equivalent size of the largest service-entrance conductor shall be determined by the largest sum of the areas of the corresponding conductors of each set.

2. Where there are no service-entrance conductors, the grounding electrode conductor size shall be determined by the equivalent size of the largest service-entrance conductor required for the load to be served.

[1] This table also applies to the derived conductors of separately derived ac systems.

[2] See installation restrictions in Section 250-64(a).

FPN: See Section 250-24(b) for size of ac system conductor brought to service equipment.

Table Two

Analysis of Table 250-66
*Grounding Electrode Conductor and Service Conductor Compared As To Relative Size

Size of Copper Service Conductor AWG No.	Continuous Rating of Service Conductor 75° C Wire Ampacity	Size of Copper Grounding Electrode Conductor AWG No.	Continuous Rating Grounding Electrode Conductor 75° C Wire Ampacity	Grounding Electrode Conductor Expressed as a percent of Service Conductor
2 or smaller	115	8	50	43%
0	150	6	65	43%
3/0	200	4	85	43%
350 kcmil	310	2	115	37%
600 kcmil	420	1/0	150	36%
1000 kcmil	545	2/0	175	32%
1250 kcmil (over 1100 kcmil)	590	3/0	200	34%

* This grounding Electrode Conductor is that part of the Grounding System which is the sole connection between the grounding electrode and the grounded system conductor.

Table Three

Comparison of Copper Grounding Electrode Conductors per Table 250-66 When Used With and Without Steel Conduit for Physical Protection.					
Copper Wire Size. AWG No.	DC Resistance of Bare Wire. Ohms/100'	Size of conduit to enclose conductor. Inch	*Approximate DC Resistance of conduit. Ohms/100'	*Approximate Impedance of Conductor in Conduit. Ohms/100' at 500 Amps Sq. In. density	
				Conduit bonded at both ends.	Conduit NOT bonded.
8	0.07780	½	0.0321	0.105	0.210
6	0.04910	¾	0.0242	0.078	0.156
4	0.03080	¾	0.0242	0.078	0.156
2	0.01940	1	0.0154	0.056	0.112
0	0.01220	1	0.0154	0.056	0.112
00	0.00967	1	0.0154	0.056	0.112
000	0.00766	1¼	0.0120	0.044	0.088

* These values are approximate but sufficiently accurate for all practical purposes. Manufacturing tolerances for conduit, although quite satisfactory for its physical requirements, can vary the resistance up to about 15 percent compared to standard dimensions.

Table Four

Comparison of Aluminum Grounding Electrode Conductors per Table 250-66 When Used With and Without Aluminum Conduit for Physical Protection.			
Aluminum Wire Size	DC Resistance Ohms/100'	Size of Conduit to Enclose Conductor In.	Approximate DC Resistance of Aluminum Conduit Ohms/100'
6	0.08080	½	0.00800
4	0.05080	½	0.00800
2	0.03190	¾	0.00605
1/0	0.02010	1	0.00385
3/0	0.01260	1	0.00385
4/0	0.00100	1¼	0.00300
250 kcmil	0.00847	1¼	0.00300

*Must be installed to comply with 250-64, that is, made electrically continuous from the point of attachment to cabinets or equipment to the grounding electrode and shall be securely fastened to the ground clamp.

Table Five

Rating of Grounding Electrode Conductors Specified in Table 250-66 and Voltage Drop under Maximum Short Time Rating.

Grounding Conductor Size AWG	DC Resistance per 100 ft. Copper Conductor	Ampacity 75° C Wire Continuous Rating	Circular Mils	*Short-Time Rating Amps	Ratio Short time Rating to Continuous Rating	Voltage Drop per 100 ft. at Short time Rating
8	0.077800	50	16510	391	7.8	30.4
6	0.049100	65	26240	621	9.6	30.5
4	0.030800	85	41740	988	11.6	30.4
3**	0.024500	100	52620	1,245	12.5	30.5
2	0.019400	115	66360	1,571	13.7	30.5
1**	0.015400	130	83690	1,981	15.2	30.5
1/0	0.012200	150	105600	2,499	16.7	30.5
2/0	0.009670	175	133100	3,150	18.0	30.5
3/0	0.007660	200	167800	3,972	19.9	30.4
4/0**	0.006080	230	211600	5,008	21.8	30.5
kcmil						
250**	0.005150	255	250000	5,917	23.2	30.5
300**	0.004290	285	300000	7,101	24.9	30.5
350**	0.003670	310	350000	8,284	26.7	30.4
400**	0.003210	335	400000	9,467	28.3	30.4
500**	0.002580	380	500000	11,834	31.1	30.5

* Based on 1 ampere per 42.25 of circular mil area for five seconds.

** Not shown in Table 250-66 but listed here to assist in sizing grounding conductor lengths over 100 feet long.

If length of grounding conductor exceeds 100 feet, the conductor should be increased in size so that voltage drop based on short-time current rating of the conductor specified should not exceed 40 volts. In other words, the resistance of the grounding conductor used when over 100 feet long should not exceed the resistance of 100 feet of the conductor specified in the table.

Table Six

Table 250-122. Minimum Size Equipment Grounding Conductors for Grounding Raceway and Equipment.

Rating or Setting of Automatic Overcurrent Device in Circuit Ahead of Equipment, Conduit, etc., Not Exceeding (Amperes)	Size (AWG or kcmil)	
	Copper	Aluminum or Copper-Clad Aluminum*
15	14	12
20	12	10
30	10	8
40	10	8
60	10	8
100	8	6
200	6	4
300	4	2
400	3	1
500	2	1/0
600	1	2/0
800	1/0	3/0
1000	2/0	4/0
1200	3/0	250 kcmil
1600	4/0	350 kcmil
2000	250 kcmil	400 kcmil
2500	350 kcmil	600 kcmil
3000	400 kcmil	600 kcmil
4000	500 kcmil	800 kcmil
5000	700 kcmil	1200 kcmil
6000	800 kcmil	1200 kcmil

Note: Where necessary to comply with Section 250-2(d), the equipment grounding conductor shall be sized larger than this table.

* See installation restrictions in Section 250-120.

Table Seven

Analysis of Table 250-122

Rating of O. C. Device in Amperes	EGC Copper Conductor		*Short-Time Rating of EGC Conductor in Amps	** K-factor	Percent Rating of EGC Conductor to OC Device	*** Ampacity per Table 310-16
	AWG No.	Cir. Mils.				
20	12	6530	155	7.7	125%	25
30	10	10380	246	8.2	117%	35
40	10	10380	246	6.1	88%	35
60	10	10380	246	4.1	58%	35
100	8	16510	391	3.9	50%	50
200	6	26240	621	3.1	33%	65
400	3	52620	1,245	3.1	25%	100
600	1	83690	1,981	3.3	22%	130
800	1/0	105600	2,499	3.1	19%	150
1000	2/0	133100	3,150	3.2	18%	175
1200	3/0	167800	3,972	3.3	17%	200
1600	4/0	211600	5,008	3.1	14%	230
2000	250 kcmil	250000	5,914	3.0	13%	255
2500	350 kcmil	350000	8,285	3.3	12%	310
3000	400 kcmil	400000	9,467	3.2	11%	335
4000	500 kcmil	500000	11,834	3.0	10%	380
5000	700 kcmil	700000	16,568	3.3	9%	460
6000	800 kcmil	800000	18,935	3.2	8%	490

*   One ampere per 42.25 circular mils for five seconds.

**   K factor. Ampere rating of overcurrent device times K equals short-time rating of E.G.C. conductor in amperes.

*** Based on 75° C copper wire, Table 310-16.

Table Eight

Chapter 9, TABLE 8. Conductor Properties								
		Conductors				DC Resistance at 75°C (167°F)		
Size	Area	Stranding		Overall		Copper		Aluminum
AWG/	Cir.	Quan-	Diam.	Diam.	Area	Uncoated	Coated	ohm/
kcmil	Mils	tity	In.	In.	In.²	ohm/kFT	ohm/kFT	kFT
18	1620	1	-	0.040	0.001	7.77	8.08	12.8
18	1620	7	0.015	0.046	0.002	7.95	8.45	13.1
16	2580	1	-	0.051	0.002	4.89	5.08	8.05
16	2580	7	0.019	0.058	0.003	4.99	5.29	8.21
14	4110	1	-	0.064	0.003	3.07	3.19	5.06
14	4110	7	0.024	0.073	0.004	3.14	3.26	5.17
12	6530	1	-	0.081	0.005	1.93	2.01	3.18
12	6530	7	0.030	0.092	0.006	1.98	2.05	3.25
10	10380	1	-	0.102	0.008	1.21	1.26	2.00
10	10380	7	0.038	0.116	0.011	1.24	1.29	2.04
8	16510	1	-	0.128	0.013	0.764	0.786	1.26
8	16510	7	0.049	0.146	0.017	0.778	0.809	1.28
6	26240	7	0.061	0.184	0.027	0.491	0.510	0.808
4	41740	7	0.077	0.232	0.042	0.308	0.321	0.508
3	52620	7	0.087	0.260	0.053	0.245	0.254	0.403
2	66360	7	0.097	0.292	0.067	0.194	0.201	0.319
1	83690	19	0.066	0.332	0.087	0.154	0.160	0.253
1/0	105600	19	0.074	0.373	0.109	0.122	0.127	0.201
2/0	133100	19	0.084	0.419	0.138	0.0967	0.101	0.159
3/0	167800	19	0.094	0.470	0.173	0.0766	0.0797	0.126
4/0	211600	19	0.106	0.528	0.219	0.0608	0.0626	0.100
250	-	37	0.082	0.575	0.260	0.0515	0.0535	0.0847
300	-	37	0.090	0.630	0.312	0.0429	0.0446	0.0707
350	-	37	0.097	0.681	0.364	0.0367	0.0382	0.0605
400	-	37	0.104	0.728	0.416	0.0321	0.0331	0.0529
500	-	37	0.116	0.813	0.519	0.0258	0.0265	0.0424
600	-	61	0.099	0.893	0.626	0.0214	0.0223	0.0353
700	-	61	0.107	0.964	0.730	0.0184	0.0189	0.0303
750	-	61	0.111	0.998	0.782	0.0171	0.0176	0.0282
800	-	61	0.114	1.030	0.834	0.0161	0.0166	0.0265
900	-	61	0.122	1.090	0.940	0.0143	0.0147	0.0235
1000	-	61	0.128	1.150	1.040	0.0129	0.0132	0.0212
1250	-	91	0.117	1.290	1.300	0.0103	0.0106	0.0169
1500	-	91	0.128	1.410	1.570	0.00858	0.00883	0.0141
1750	-	127	0.117	1.520	1.830	0.00735	0.00756	0.0121
2000	-	127	0.126	1.630	2.090	0.00643	0.00662	0.0106

1. These resistance values are valid ONLY for the parameters as given. Using conductors having coated strands, different stranding type, and especially, other temperatures, change the resistance.

2. Formula for temperature change: $R_2 = R_1 [1+a(T_2-75)]$ where: $a_{cu} = 0.00323, a_{AL} = 0.00330$.

3. Conductors with compact and compressed stranding have about 9 percent and 3 percent, respectively, smaller bare conductor diameters than those shown. See Table 5A for actual compact cable dimensions.

4. The IACS conductivities used: bare copper = 100% aluminum = 61%.

5. Class B stranding is listed as well as solid for some sizes. Its overall diameter and area is that of its circumscribing circle. (FPN): The construction information is per NEMA WC8-1992. The resistance is calculated per National Bureau of Standards Handbook 100, dated 1966, and Handbook 109, dated 1972.

Table Eleven

Conduit Size In.	AWG/ kcmil	OC 75°C	Fault 500%	EMT Length	IMC Length	RGC Length
½	14	15	75	231	232	227
	12	20	100	246	255	247
	10	30	150	226	250	240
¾	12	20	100	266	262	257
	10	30	150	255	259	250
	8	40	200	261	275	264
	8	50	250	211	229	222
1	8	40	200	288	284	275
	8	50	250	236	237	229
	6	60	300	263	270	257
	6	70	350	228	240	228
	4	80	400	254	274	259
	4	90	450	228	251	238
	3	100	500	227	256	244
1¼	3	100	500	268	274	258
	2	110	550	273	283	265
	2	125	625	244	257	243
	1	125	625	266	281	263
	1	150	750	226	245	234
1½	1	125	625	285	292	273
	1	150	750	243	256	241
	1/0	150	750	264	279	260
2	2/0	175	875	280	287	268
	3/0	200	1000	268	281	261
	4/0	225	1125	257	277	257
	4/0	250	1250	234	257	240
2½	250	250	1250	289	269	261
	250	300	1500	249	238	229
	300	250	1250	302	280	271
	300	300	1500	261	248	239

Maximum length of steel conduit or tubing that may safely be used as an equipment grounding circuit conductor.

Based on a clearing ground-fault current of 500 percent of overcurrent device rating; circuit 120 volts to ground; 50 volts drop at the arc; 30° C ambient temperature; 75° C conductor temperature.
Calculated with "Steel Conduit Analysis Vs 1.2" by Georgia Institute of Technology.

Table Eleven (continued)

Maximum length of steel conduit or tubing that may safely be used as an equipment grounding circuit conductor.						
Conduit Size In.	AWG/ kcmil	OC 75°C	Fault 500%	EMT Length	IMC Length	RGC Length
2½	400	300	1500	278	263	253
	400	350	1750	244	239	227
	500	350	1750	255	249	236
	500	400	2000	227	230	216
3	500	350	1750	280	262	253
	500	400	2000	251	240	231
	600	400	2000	260	248	238
	600	450	2250	235	231	220
	700	450	2250	242	237	225
	700	500	2500	221	222	209
3½	600	400	2000	280	256	248
	600	450	2250	257	237	229
	700	450	2250	264	244	235
	700	500	2500	243	228	218
	750	450	2250	267	246	237
	750	500	2500	246	230	221
	900	500	2500	254	237	227
	900	600	3000	220	212	200
4	750	450	2250	277	253	245
	750	500	2500	255	236	228
	900	500	2500	264	243	234
	900	600	3000	229	216	206
	1000	500	2500	269	247	238
	1000	600	3000	233	219	210

Based on a clearing ground-fault current of 500 percent of overcurrent device rating; circuit 120 volts to ground; 50 volts drop at the arc; 30° C ambient temperature; 75° C conductor temperature.
Calculated with "Steel Conduit Analysis Vs 1.2" by Georgia Institute of Technology.

Table Fourteen

Aluminum Conduit used as an equipment grounding circuit conductor (EGC) compared with equipment grounding circuit (EGC) conductors as specified in Table 250-122 of the National Electrical Code®.						
Size of Conduit In.	Area of Conduit Wall Sq. In.	Area of Conduit Wall Based on NEMA Standards Sq. In.	DC Resistance Ohms/M ft.	Largest EGC Conductor as specified in Table 250-95 Aluminum Wire No.	Nearest Wire No. Equivalent to Conduit	
					Aluminum Wire No.	Copper Wire No.
½	0.254	0.254	0.08000	12	4/0	2/0
¾	0.337	0.337	0.06050	8	300 kcmil	4/0
1	*0.530	0.500	0.03850	6	500 kcmil	300 kcmil
1¼	*0.680	0.670	0.03000	4	600 kcmil	350 kcmil
1½	*0.790	0.790	0.02580	4	700 kcmil	500 kcmil
2	*1.030	1.080	0.01980	1	900 kcmil	600 kcmil
2½	1.710	1.710	0.01190	1	1500 kcmil	1000 kcmil
3	*2.310	2.250	0.00880	2/0	2000 kcmil	1250 kcmil
3½	2.700	2.700	0.00752	2/0	over	1500 kcmil
4	*3.400	3.180	0.00598	2/0	2000 kcmil	2000 kcmil
5	4.300	4.300	0.00473	3/0	over 2000 kcmil	over 2000 kcmil

*Values as measured from a sample.

Column 3 gives values as calculated on basis of NEMA standards.

Table Fifteen

	Sizes of conductor (Copper Bar) for use in the equipment grounding circuit and for the Equipment Grounding Strap for overcurrent devices from 1600 amperes through 5000 amperes.					

Rating of OC Device in Amps.	Copper EG Conductor		* Conductor Rating Amps	** Short-Time Rating of EGC Conductor in Amps	*** K factor	**** Percent Rating of EGC Conductor to OC Device
	Copper Bar - Inch	Cir. Mils.				
1600	¼ x 1	318300	366	10,610	5.3 (6.6)	22.9
2000	¼ x 1¼	397900	457	13,260	6.0 (6.6)	22.9
2500	¼ x 1½	477450	549	15,920	6.0 (6.4)	21.9
3000	¼ x 2	636600	647	21,220	6.0 (7.1)	21.6
4000	¼ x 2½	795750	809	26,525	6.3 (6.6)	20.2
5000	¼ x 4	1273000	1,220	42,430	6.8 (8.5)	24.4

* Based on 30° C rise, 40° C ambient.

** One ampere per 30 circular mils for 5 seconds. Use this value ONLY where insulated, current-carrying conductors will NOT come in contact with the bare equipment grounding bus bars. Where there is likelihood of contact, then recalculate the 5 second withstand rating on the basis of one ampere...for five seconds...for every 42.25 circular mils of copper conductor. This is discussed in Chapter 11 of this text.

*** K-factor. Ampere rating of O. C. device times K-factor equals short-time rating of EGC conductor in amperes. Figures in brackets are based on the actual EGC conductor used.

**** Based on 30° C rise, 40° C ambient for EGC conductor.

Appendix

The History and Mystery of Grounding

In 1941, Howard S. Warren completed work on the manuscript "How The Code Came." This material was turned over to then president, Alvah Small, Underwriters Laboratories, to be made available to anyone should it serve a useful purpose. Alvah Small was chairman of the NFPA Electrical Committee from 1925 to 1950. This tome on the history of the Code was dedicated to Alvah for his work as chairman in keeping the Code abreast of developments in the electrical industry and always in the public interest. Warren represented the telephone group on the Electrical Committee from 1907 until 1939. Warren was preceded by C.J.H. Woodbury and followed by A.H. Schirmer, L.S. Inskip and L.H. Sessler.

The subject of grounding has always been confusing to many, and as such, precipitates much interest wherever sessions are held on Code discussions. Grounding has always had its controversial aspects. Part of this stems from the fact that the early sponsors of the Code, the Underwriters' National Electric Association, represented the insurance interests and the grounding of electrical circuits had much to offer for personnel safety, but little fire safety for the building could be obtained.

Many of the associations' names have changed today. The Underwriters National Electric Association was later to become the National Board of Fire Underwriters. The National Conference on Standard Electrical Rules was later to become the National Electrical Code Committee. AIEE was later to become the Institute of Electrical and Electronics Engineers (IEEE). The National Electric Light Association was later to become the Electric Light and Power Group.

The applications of grounding and the details of accomplishing it are many. The grounding section of the 1940 Code (Article 250) occupies over 18 pages. Curiously enough, grounding has always had a controversial aspect.

In the first (1897) National Electrical Code®, as in the previous "Rules and Requirements" issued by the Underwriters' National Electric Association, the only mandatory grounding requirement was that for lightning arresters or protectors. There couldn't be much argument about that, since from its very nature a protector or lightning arrester must be provided with a path to earth.

In the July 1, 1894, issue of "Rules for Safe Wiring," there were included directions for installing communication circuit protectors, which covered the following: (1) The location of protector; (2) Type of wiring to be used from the last outside support to the protector; (3) Conditions which grounding conductor must comply with and (4) Means for connecting grounding conductor to earth.

These provisions were carried over into the first (1897) edition of the National Electrical Code®. In addition, the frames of generators and motors were required to be grounded when, and only when, it was impracticable to insulate them.

The fireworks began about three years later, when the question of grounding secondary distribution circuits was precipitated.

Circuit grounding

Consider for a moment what the grounding of a secondary circuit involves. Take the simple case of a two-wire, 120-volt circuit fed by a transformer. The transformer keeps the difference of potential between the two wires of the circuit at 120 volts, but, in the absence of grounding, the potential of the circuit with respect to earth is indefinite. If there is no foreign electrical influence, the average potential of the two conductors will hover around zero or earth potential, but if one of the conductors becomes crossed with a foreign wire of high potential to earth, it will immediately become charged to the same high potential (assuming that no breakdown of insulation occurs). This brings about a hazardous state of affairs, for if a person in contact with earth or with any grounded metal should touch the charged conductor, he would be liable to receive a dangerous shock. Such hazardous condition might be produced by an accidental cross with a foreign wire or by breakdown of insulation of the transformer in such way as to impress potential from the primary circuit upon a secondary circuit conductor.

As a safeguard against such hazard, it was proposed that secondary circuits be grounded, i.e., that one conductor, in case of a two-wire circuit, or the neutral conductor, in case of a three-wire circuit, be connected to earth, thus assuring that no conductor of such grounded secondary circuit could have a potential to earth greater than the normal circuit voltage.

Such grounding seemed all right for cases where the normal secondary voltage was not in itself hazardous, but as grounding increases the liability of shocks at normal circuit voltage, it did not look so good for circuits of more than about 150 volts. For example, a 220/440-volt secondary, if grounded, would increase the chance of a 220-volt shock, which of itself would involve a serious hazard. On the other hand, the grounding would protect against the probably much higher voltages due to accidental crosses with foreign wires or to transformer breakdowns. So, in this case, it was a question of balancing the increase in hazard from 220-volt shocks against the decrease in hazard from shocks at higher voltages. The 220/440-volt secondary circuit was a borderline case, some experts favored grounding while others opposed it.

As this was a technical electrical question, the Underwriters, who at this period were in control of Code revisions, wisely declined to put a mandatory circuit-grounding rule into the Code until the electrical people could agree on what the rule should be, all the more so as such a rule was considered to be more concerned with life hazard than with fire hazard. Differences of opinion on the part of electrical experts as to the circuit voltage below which grounding ought to be made mandatory produced a stalemate which persisted for several years.

Discussions of electrical circuit grounding go back to the beginnings of the ac distribution system. In England, as well as in this country, such grounding (our British cousins call it "earthing") was proposed. Thus Killingsworth Hedges said, in 1889:[7]

"One precaution is to earth the secondary circuit. Another is to connect one or both leads to a safety appliance which would automatically divert any excess current to earth and at the same time shut off the supply." Prof. Elihu Thomson, in 1890, suggested grounding the transformer secondary winding or surrounding it with a grounded sheath. As alternatives, he suggested cutting off the secondary by automatic means or grounding it automatically by film cutouts when excessive voltage appeared. The placing of a grounded metal sheath between the primary and secondary windings of a transformer was called by the English, "Kent's sheath," although the idea had its origin in America. Thomson regarded this device as very effective for securing a safety grounding connection for the secondary circuit when there was a leak from the primary. Discussing this matter at an NELA convention, he said:[1]

"In the ac system with low voltage secondary, good workmanship is all that is necessary to abolish risks if we leave out the high voltage primary which must be thoroughly insulated. Even some static capacity which may give severe if not fatal shock is to be looked out for. The writer early recognized this danger and felt it must be provided against. Hence came the expedients of grounding the secondary or surrounding it with a grounded sheath or cutting off the secondary absolutely by automatic means."

But the grounding of secondary circuits had plenty of opposition in the early days. Woodbury, in giving his suggestions about methods for reducing the electrical fire hazard, considered "it undesirable to permanently ground one wire of the secondary circuit."[2]

On March 10, 1892, the New York Board of Fire Underwriters issued a pamphlet report on "Grounding of Electric Wires," prepared by its Committee on Police and Origin of Fires. At a special meeting of the board, held on that date, it was voted that this report be printed and circulated to the members, together with certain parts of a report by Professor Henry Morton, and a reply thereto by the Edison Electric Illuminating Company of New York.

The electrical world announced:

"The New York Board of Fire Underwriters has condemned the practice of grounding the neutral as dangerous and orders it to be stopped.

"There can be no doubt that the practice of grounding the neutral is not as safe as a completely insulated system. To change it now, however, may require much work and, therefore, may interfere somewhat with the lighting. It is a case in which prevention would have been better than cure."

The following resolutions were also adopted by the board and circulated as aforesaid:

"RESOLVED, that the Committee on Police and Origin of Fires be and they are hereby directed to require all electric companies furnishing current for power and lights (to parties that have received certificates of approval of electric equipments from this Board), to make regular weekly reports of the tests of their currents, as called for in the requirements of this Board, and in the event of failure on the part of the Electric Companies so to do, then the Superintendent of the Survey Department is hereby directed to decline to grant certificates, or make inspections of equipments supplied with current by said delinquent companies, and further.

"RESOLVED, that said Committee give notice to all Electric Companies receiving certificates from this Board that the intentional grounding of any portion of the equipment is a violation of the rules of this Board and such a practice must be discontinued.

"This Board will refuse certificates of approval to all companies who do not comply with this rule, and further

"RESOLVED that the committee require of Electric Companies that all grounds now existing on electric circuits shall be removed on or before October 1st, 1892."

These resolutions were promulgated over the name of William Del. Boughton, chairman of the board.

The extracts from Professor Morton's report on fire hazards from grounding of electric wires, particularly the neutral wire in the Edison system, included the following: "The grounding of the middle wire, in my opinion, decidedly increases the fire risk, for the following reasons:

"First. If all the wires are insulated, then two ground contacts must occur in order that fire should be produced by the contact of a conductor with the gas, water or steam pipe, or other conducting substance connected with the ground. If, on the other hand, the middle wire is "grounded," then every gas, water or steam pipe becomes in fact a 'live wire,' contact with which results in a current, only limited in amount by the capacity of the conductors (including fusible catches and the like) between the point of contact and the general network of supply wires.

"Second. It renders entirely impossible any testing by which from time to time the insulation of the systems can be watched and measured.

"As a result when a ground occurs, unless it attracts attention in its own neighborhood or is enormous in amount, it can go on indefinitely without being discovered, while it may be doing mischief all the time, whereas with an insulated system, a ground will instantly show itself by the proper instruments at the central station or elsewhere, and at the same time will be harmless until another ground, on the wire of opposite polarity, is developed.

"If the middle wire of the Edison system is grounded the entire system is then in such connection with the earth that the contract of either positive or negative conductors with water, gas or steam pipe, or iron parts of a building or other conductor, will establish a connection through which a large current will flow."

Commenting on the foregoing, R. R. Bowker, first vice president of the Edison Electric Illuminating Company of New York, said:[13]

"In view of the fact that many of the questions at issue are comparatively new to members of your board, we have thought it best to make a general explanation of the nature of the three-wire system, and of the practice involved in 'grounding the neutral,' as is done in the accompanying statement, which while it does not take up the leading questions of the report in regular order, does, it will be found from the references, present a direct and sufficient answer to each of these questions. The careful and specific attention of each member of your Board is requested for this statement as the question involved is one of such large and increasing importance.

"We would emphasize to you that our street system is in no sense dependent on the practice of grounding the neutral and that contrary to the inference of Professor Morton's report, no copper or other element of cost is saved with that practice in view; the motive of the practice is not commercial but precautionary. We remind you that the practice is in vogue in other cities with the approval of Underwriters. We would also beg to correct the misapprehension that there is any grounding of the neutral wire in houses, or any decrease of insulating precautions because of that practice.

"We ask your Board for such modifications of your rules as will recognize the features of the three-wire system and provide for the practice thus adopted. We beg herewith, however, to say that should the Board, in full view of the protest of this company, decide to take the responsibility, with the insuring community and before the public, of directing this company to discontinue the practice, the superintendents of the company will have instructions to prepare to disconnect the neutral grounds at the junction boxes throughout this city, which would require, however, considerable time, and would not, we believe, produce the result desired of giving opportunities for insulating tests on lines of laboratory practice and of special circuit systems. We would, moreover, enter definite protest against the overruling of this company by your board in a question of practice which we believe can best be determined by working electrical engineers familiar with the needs and conditions of actual station practice."

A statement submitted to the New York Board of Fire Underwriters by the Edison Electric Illuminating Company contained the following:

"In the Edison 3-wire system, two dynamos are coupled together both in series by a short copper conductor connecting the negative side of one with the positive side of the other. If each is running at 100 volts the product of the two thus in series is at 200 volts. The positive conductor runs from the outer side of one dynamo; the difference of potential between the positive and negative wire is therefore 200 volts. The middle or neutral wire runs out from the short connecting link between the two dynamos and like the positive and negative wire, is absolutely continuous from the dynamo in the station to the extreme point of every house installation. Incandescent lamps are placed 'in multiple' between the neutral wire and the positive conductor and between the neutral wire and the negative conductor. In each house and throughout the system lamps are 'balanced' as evenly as possible between the two sides of the system - that is, when a hundred lamps are placed in a house, it is the endeavor to put fifty of these on the positive side and fifty on the negative side of the system, so placed that the actual burning is about the same on either side. When one thousand lamps are burning on the positive side and one thousand lamps on the negative side of the system, the neutral wire brings back no current whatever to the dynamo at the station. The current at 200 volts flows through the lamp between the positive conductor and the neutral wire, giving this 100 volts, and thence through the lamp between the middle wire and the negative conductor giving that the other 100 volts, and the return is all on the negative side. If five of these lamps are taken out or turned off on the positive side, the neutral wire would become a positive conductor to the extent of the current for these five lamps, or if five lamps were taken out or turned off on the negative side, the neutral wire would become a negative conductor to the extent of the current for those five lamps.

"On the New York system the neutral is grounded at the junction boxes in the streets, and not at the stations; on the Brooklyn system the grounding is at the station. Authorities differ as to which is the better course to obtain the advantages of grounding the neutral. In New York the neutral is grounded at about 250 junction boxes, something less than 1/3 of the number of junction boxes on the whole system, at a distance from the station varying from 1/4 to 1-1/2 miles. The neutral is grounded, as it were, on the other side of the junction box from the station, so that if in any possible event the current should flow through the ground it would seek the shortest course to the nearest junction box grounded and thence through the regular neutral wire to the station. That the ground is practically not used as the neutral conductor is sufficiently evidenced by the fact that no electrolysis has yet been discovered by our electricians at the points where the neutral wire is grounded, as would certainly be the case were the ground connection used to any extent.

"The neutral wire is not in any case grounded within house installations or in any way in connection with houses.

"It should be clearly recognized moreover that it is not practicable on any large system of electrical conductors, whether overhead or underground, especially so large as is required for the city of New York, that absolute insulation, such as is possible in laboratory tests, can be obtained or maintained. There is no known material which is an absolute insulator and even electrical appliances and adjoining woodwork are themselves conductors to an infinitesimal degree. In any known system of considerable aggregate length, while no single part can be definitely criticized, the mean level of insulation of the system is necessarily brought into comparatively close association with the ground by numerous small leaks, each of individual importance but all together making a considerable aggregate.

"I desire to make perfectly clear that if any given system could be made and kept absolutely free from leakage and best electrical authorities, including Edison himself, would agree that the balance of advantage would be found in keeping the system in that condition and not in grounding the neutral. But the difference between laboratory tests and actual working conditions is such as to make their practical conclusions almost contradictory and we have the authority of Edison's chief electrician for the assertion that 'in endeavoring to attain the perfection of safety' the report 'advocates a course of procedure which would introduce greater risk and defeat its own purpose.'

"It would seem scarcely necessary after these remarks to reply more specifically to Professor Morton's position that there should be no difference between the rules made to apply to the two-wire and three-wire systems and no reason to alter previous rules of the Board to recognize or apply to the present situation. If Rule 40, which has often been quoted in this connection, is understood by your board to apply to the grounded neutral of the Edison three-wire system, it should certainly be modified. As a matter of fact, having been adopted before the three-wire system came into operation, this company asserts that it does not so apply and in this respect no alteration would then be needed.

"It is the general opinion of the Edison interest that, while absolute insulation, if it can be had and where it can be had, is preferable, the advantages of grounding the neutral under certain conditions and particularly on large systems, are such as to make that practice in those cases the best working method particularly as a precaution against fire risk."

Discussion at national conference.

At the meeting of the National Conference on Standard Electrical Rules in March 1896, Professor Kennelly suggested a rule requiring "that the secondary coil of a converter should be grounded at its center." This led to the following discussion:

James I. Ayer:

"Relative to the question of Mr. Kennelly as to the grounding of the secondary coil of the converter at its center, it seems to me that is putting a severe strain on the insulation. I assume a low resistance ground is meant, so that in event of a ground occurring on the line and a breaking down of the insulation of the transformer the current would pass through so as to destroy the converter and cut out the primary current from the secondary. There is no doubt about the advisability of getting some sort of transformer protection, especially on the primary lines where 2000 volts and upwards are transformed directly to the house service low potential.

"In a discussion of this question some time last spring, Professor Puffer recommended the same thing. I suggested carrying out a practice which I adopted years ago in a good many cases of using a magnetic cutout. This magnetic cutout was made with two solenoids, each of which have one connection to one of the secondary lines, one on one side and one on the other, the other terminal being connected to ground. These solenoids are of very high resistance so that the leakage of a 50-volt current amounts to 8 watts and for 110 volts to about 11 watts. This gives absolute protection. Any rise of pressure on the secondary mains above 500 volts would operate the circuit breaker. It is very positive. It rather straddles the question of grounding secondaries or grounding house mains by grounding them through a fixed high resistance ground connection. It gives an absolute, positive-acting device, not necessarily one that burns up transformers. In fact, it leaves the transformers to take care of itself but cuts out the primary from the secondary main.

"This is a matter to which I have given a great deal of thought. I found in the early days of the use of transformers the necessity of some such protection where there are many of them connected with the line. Thunderstorms would readily break them down and create punctures which would give a connection from one side of the transformer to one side of the secondary circuit. The lights would burn and things continue on apparently all right. Transformers do break down with the high potential being used and I believe the right thing to recommend, and in time to demand, is the proper protection for isolated transformers, to give you protection between the secondary and primary, but the question of grounding secondary mains and low resistances IS a very serious one and should receive a very full discussion before action is taken."

Prof. Kennelly:

"I think the matter is so important that it is worth while to give it a few moments. I should be sorry to advocate anything which would unduly harass the manufacturers, but I believe that I make this suggested recommendation in the interests of the manufacturers just as much as in the interests of the parties who receive lighting from electric systems.

"If we exclude lightning, there can be no objection to grounding the secondary because it is obvious that the system should be capable of standing the working pressure. Then no one can fairly object to grounding the center of the secondary on the score of danger of puncture because, if the transformers are going to be so weakly insulated that grounding the secondary is going to injure them, those transformers had better be replaced. It may not be necessary, however, to actually ground the secondaries. A simple protective device that will ground the secondary if a cross develops would be preferable, and if Mr. Ayer's device is simple and practical, its adoption would seem worthy of consideration. A cross between the primary and coils of a transformer is, we know, of very rare occurrence, yet we all know that no one would be justified in permitting a person standing, say on a cellar floor, to handle the socket of a lamp if the operation involved making contact with one of the secondary conductors, unless the transformer had just been tested free from leakage or cross. If, however, the secondary coil of the transformer were grounded permanently, or protected in an equivalent degree there would be no such danger from accidental contact."

Mr. Ayer:

"In considering this question of the protection of the transformers, the practical conditions to be met with have got to be well understood. You take the case of distribution with the overhead system, where there are any material number of transformers in a network covering a large area, there is no such thing as getting an insulated circuit. You have got a grounded circuit, always.

"There has been a great deal of discussion especially among the Edison Companies and Underwriters, as to the wisdom of grounding the third wire of the Edison system. That has been a very fruitful topic of discussion. I think it is useless to try to enumerate the details or discuss the pros and cons of this here.

"In a resume of the English rules, you see that they contemplate at all times an automatic device in the secondary circuit, or in the primary wire, if not in the secondary, which will automatically cut off the primary current from the transformer or the secondary from the house. That is the English practice. They call for some intervening device which shall open the secondary circuit in the event of the primary forming a connection with high tension. It is a simple thing to make. I have one in my residence today and have had for five years, which has been in service all the time. The thing is operated with a leakage of 500 volts from the primary line to the secondary. It introduces a high resistance ground on the secondary and a leakage on the secondary of about 8 watts. I believe the solution of the problem of the protection of the secondary coil is the introduction of this intermediate automatic device. Devices have been made in England operated by clock work, fuses, etc., but not similar to this to my knowledge."

The protective device for secondary circuits, which Ayer thus commended, did not meet with much favor and the solution of the problem was long delayed; indeed, it has not been solved even yet in all its aspects.

The Underwriters were slow to reverse their stand against allowing any permanent grounds on electric circuits.

The electric companies, on the other hand, were generally in favor of grounding the neutral wire. W.L.R. Emmet of the General Electric Company said, in 1899:

"The permanent grounding of transformer secondaries is prohibited by the fire underwriters' rules, which pro prohibition is generally respected throughout the country, so that most secondary circuits are entirely unprotected."[15]

AIEE talks it over.

In 1899, Dr. Cary T. Hutchinson presented an AIEE paper on the protection of secondary circuits, in the course of which he heartily endorsed the practice of grounding and reproved the Underwriters for their stand against it saying,[8]

"The need of some effective method for protecting secondary circuits is well recognized yet there is at present probably no device for the purpose that is thoroughly trustworthy and efficient in all circumstances. The means that have been employed for this object may be classified under three general heads: (1) Devices intended to ground, short-circuit or open-circuit the secondary circuit when subjected to abnormal difference of potential; (2) grounded metallic shields imposed between the primary and secondary coils of the transformer; (3) permanent grounding of the secondary system.

"The third method, grounding the secondary permanently, is the only sure way to prevent the potential above earth of the secondary system rising above the voltage of the circuit.

"The grounding of the secondary circuit thus absolutely insures the safety of the circuit as regards abnormal pressure.

"It would seem that a remedy as simple as this, one generally applicable, would have been applied in all cases but the fact is that it has attained a limited use only. The reason for this state of affairs is the refusal of the Board of Fire Underwriters to authorize the practice of grounding any part of a circuit carrying current. The Underwriters take the position that to ground one side of a current

brings about an increased liability to fire, because the full voltage of the circuit continually acts upon the insulation of the circuit, instead of one-half of this voltage as in an insulated system, and because one accidental ground on the insulated side may cause a fire.

"The Association of Edison Illuminating Companies has had a committee on grounding the neutral since 1890. The recommendations of this committee being based on extensive experience are entitled to great weight. Their recommendation of uniformity being the greatest safety was assured by the practice of grounding the neutral. This is the practice of nearly all the large Edison Companies as is well known. In some cases the companies have been forced to this position because they were not able to free the neutral from ground. In other cases they have deliberately adopted this as the best remedy for many troubles due to operation dangers from fire and to abnormal pressures.

"It is notorious fact that the systems in nearly all of the large cities, particularly the Edison system, are grounded. The neutrals in most cases are permanently grounded at the junction boxes. It is equally well known that in several of the large ac distributing systems the neutral wires are grounded. The authorities of the Board of Underwriters certainly should, and probably do know, these facts. To keep rules in print prohibiting such practices under the circumstances must weaken their authority and lessen the respect for all their rules.

"I therefore offer the following resolutions:

"RESOLVED that the AIEE favors a rule permitting the grounding of one wire of every low potential consumption system, and

"RESOLVED that the committee of the AIEE on the National Electrical Code is hereby requested to confer with the Electrical Committee of the Underwriters National Electric Association in relation to recommending to the National Board of Fire Underwriters the passage of a rule permitting the permanent grounding of one wire of such systems under suitable restrictions."

Hutchinson's resolution was passed, after discussion, some of which is noted below: Professor Elihu Thomson: "I endorse everything that is said and I hope some action will be taken by a body so influential as this, toward getting some such thing adopted."

C. M. Goddard:

"I should like to second the resolution and for several reasons; not, however, because the Underwriters are fully convinced as yet that one of their rules, almost as unchangeable as the laws of the

Medes and Persians, that no circuit should be permanently grounded, has been wrong all the time. I am not convinced of that fact but because we are delighted to have the institute take up any subject in relation to our rules and try to help us arrive at a right solution.

"I think it very probable that this grounding of the secondary may be the best solution but the Underwriters are always conservative in such matters.

"We would be delighted to have you help us in making our rules. We modified this rule last year at the request of the Institute and we will certainly modify it further if you will only come to us and show us that you are right, for we have a great deal of respect for the opinions of the Institute."

Professor W. E. Goldsborough:

"I hardly feel like saying anything in opposition but I am very much interested in the subject and all my experience has tended to lead me to feel it is not best to ground the secondary."

Professor W. L. Puffer:

"I would carry grounding to such an extent that I would compel the absolute grounding immediately at the point of entrance in buildings of every system (except series arcs which ought not to be allowed in buildings) and including such things as water pipes and gas pipes."

While the consensus of this discussion was decidedly in favor of grounding the secondaries, there was some adverse opinion and the Underwriters seem to have continued other opposition to the practice, although the Code, i.e., the 1897 and 1899 editions, did not specifically prohibit it.

Father defers to mother.

The Electrical World and Engineer, reporting the December, 1900, meeting of the Electrical Committee, said:[11]

"The motion to change the limit of low tension circuits from 300 to 550 volts was carried at the first session. A long discussion took place on the advisability of grounding one of the wires in low tension circuits in which numerous authorities were quoted and prominent electrical engineers spoke in its favor. The resolution was finally adopted to allow the grounding of one wire in every low potential circuit but making it optional with the installing engineer."

In accordance with this decision, the 1901 edition of the Code contained the following section:

"13A. Grounding Low Potential Circuits.

The grounding of low potential circuits under the following regulations is only allowed when such circuits are so arranged that under normal conditions of service there will be no passage of current over the ground wire.

"Direct-Current 3-Wire Systems:

a. Neutral wire may be grounded and when grounded the following rules must be complied with:

   1. Must be grounded at the central station on a metal plate buried in coke beneath permanent moisture level and also through all available underground water and gas-pipe systems.

   2. In underground systems the neutral wire must also be grounded at each distributing box through the box.

   3. In overhead systems the neutral wire must be grounded every 500 feet, as provided in Sections c, e, f, and g.

Inspection departments having jurisdiction may require grounding if they deem it necessary.

Two-wire direct-current systems having no accessible neutral point are not to be grounded.

"Alternating-Current Secondary Systems

b. The neutral point of transformers or the neutral wire of distributing systems may be grounded and when grounded, the following rules must be complied with:

   "1. Transformers feeding 2-wire systems must be grounded at the center of the secondary coils, as provided in sections d, e, f and g.

   "2. Transformers feeding systems with a neutral wire must have the neutral wire grounded as provided in sections d, e, f and g, at the transformer and at least every 250 feet for overhead systems and every 500 feet for underground systems.

"Inspection departments having jurisdiction may *require* grounding if they deem it necessary."

Thus, by making the practice permissive, the Underwriters began a strategic retreat from their early position against the grounding of electrical circuits.

At the Electrical Committee meeting in December, 1901, the following proposal was considered:[3]

"A letter was also referred for conference with the AIEE Committee, asking for modification of the rules with regard to grounding low potential circuits so that the ground might be on one side of the circuit instead of at the neutral point of the transformer."

The Underwriters held to their position that the grounding of secondaries was a question for the electrical people to decide. A special subcommittee on the subject reported to the Electrical Committee meeting of December 9, 1902:[9]

"It was also agreed that Rule 13A-b should be amended to read as follows:

"(b) Transformer secondaries of distributing systems should preferably be grounded and when grounded the following rules must be applied:

"1. The grounding must be made at the neutral point or wire, whenever a neutral point or wire is accessible.

"2. When no neutral point or wire is accessible, the maximum difference of potential from the grounded point of the secondary circuit to any other point there of must not exceed 250 volts.

"3. To be identical with the present clause marked (2)." Here was a proposal to go further and take a position favoring circuit grounding. The meeting declined to either approve or reject this recommendation but voted to submit it to the Board of Directors of the AIEE and to the Underwriters' National Electric Association.

This recommendation to change the permissive "may be grounded" to "should preferably be grounded," was evidently accepted by the AIEE, for the rule in the 1903 Code was substantially the same as the subcommittee had recommended, namely:

"b. Transformer secondaries of distributing systems should preferably be grounded, and when grounded, the following rules must be complied with:

1. The grounding must be made at the neutral point or wire, whenever a neutral point or wire is accessible.

2. When no neutral point or wire is accessible, one side of the secondary circuit may be grounded, provided the maximum difference of potential between the grounded point and any other point in the circuit does not exceed 250 volts.

3. The ground connection must be at the transformer as provided in sections d, e, f, g and when transformers feed systems with a neutral wire, the neutral wire must also be grounded at least every 250 feet for overhead systems, and every 500 feet for underground systems. "Inspection Departments having jurisdiction may require grounding if they deem it necessary."

This rule in the 1903 Code represented quite a step away from the old position prohibiting circuit grounding, but still it was unsatisfactory in that it left the question open as to whether secondaries should be grounded or not. There was a substantial body of opinion in favor of making the grounding mandatory under certain conditions, but there were differences of opinion as to what the conditions were which warranted making grounding mandatory. It was generally agreed that where the voltage between conductor and ground did not exceed 150 volts, safety was promoted by grounding, but where the voltage was from 150 volts to 250 volts, the advantage of grounding was in a doubtful region. For voltages to ground over 250, the consensus of expert opinion was against grounding.

At the Electrical Committee meeting on December 2-3, 1903, a suggestion that the grounding of secondary neutrals be made compulsory was referred to a committee of three to be appointed by the chair, to report back.

More discussion.

At an AIEE meeting on December 19, 1903, H. G. Stott took up the discussion, saying:

"Some years ago a committee of the Institute met with the Board of Fire Underwriters and a rule was passed making it permissible to ground the neutral at the central point of the transformer or at one side of the transformer. The grounding of any part of the secondary network, especially the neutral point, is the best form of protection obtainable because in case of a cross with an electric light wire carrying perhaps 8,000 volts, the pressure of the wire is immediately reduced to the pressure of the circuit as the carrying capacity of the secondary network is greatly in excess of the amount of current passing through the average arc light circuit."

Discussion of this question continued. It was one of the subjects considered at the meeting of the National Conference on April 21, 1905. On that occasion, Dr. F. A. C. Perrine (formerly Professor of Electrical Engineering at Stanford University) stressed—perhaps overstressed—the opposing effects of grounding on the fire and life hazards, saying:

"For example, we have heard a good deal of the rules for grounding secondaries. We know the Underwriters have admitted that, at the request of the electrical interests, in spite of the fact that if anything it increases the fire hazard, decreases the life hazard. Now there is a condition in which the Underwriters have acceded to the request of the electrical interests but it is a condition in which the fire hazard risk is distinctly opposed to the life hazard risk."

Benallack confirmed the influence exerted on the Underwriters by the electrical people:[4]

"Other speakers had called attention to the importance of grounding secondary circuits as a good way, and in fact the only way, of making these circuits safe from the leakage of currents from the primary circuits. The Underwriters had been slow to permit this at first and it was largely due to the central station men who appreciated the danger that the practice was permitted and finally positively recommended by the Underwriters."

At the National meeting on December 4, 1905, the subject of grounding secondaries was discussed further. H.C. Wirt remarked:

"The most important rule here is the rule on grounding of the secondary to make it compulsory." After discussion, the following resolutions were adopted:

"RESOLVED, that it is the sense of this Conference that the grounding of low tension secondary ac systems be strongly recommended in all cases wherever reliable grounding connections can be secured. "RESOLVED, that in cases of three-wire systems with grounded neutral, solid connections without fuse be permitted on the neutral wire.

"RESOLVED, that it is the sense of this Conference that the grounding of secondary ac systems inside of buildings to water pipes, provided such connection is made at the nearest point to cellar wall on water pipes and outside of meter, is not only safe but places no additional burden or menace on such water pipes."

Woodbury referred to this meeting in his Annual Report for 1905, to H. V. Hayes:

"At the second meeting of the Conference on December 4, results of importance to the electrical interests were adopted in regard to grounding of secondaries on ac systems and also the resolution that grounding did not furnish any hazard or menace to water pipes to which they were attached. Although these results were filed by others yet I was concerned in their operation, particularly in relation to the last clause, on account of difficulties which some of the operating companies had experienced through claims of town officials for reimbursement for damage to water mains by lightning, alleging that the grounding of telephone lines to water pipes was the cause."

However, the Electrical Committee at its December 1905, meeting did not adopt the National Conference resolutions, as no formal endorsement of them by the AIEE or NELA had been received. The grounding of low-potential circuits was discussed but no changes in the rule were made and the matter was referred back to the subcommittee for further consideration. Nor was any such endorsement, or any further recommendation as to the grounding rule, received from the electrical people in 1906. Accordingly, in the 1907 edition of the Code, rule 13A remained practically unchanged from the previous edition. The year 1908, however, saw some progress toward a mandatory grounding rule for the Code. At the Electrical Committee meeting on March 25, the Committee on Rule 13A reported:

"The sense of the meeting was that, when suitable rules are prepared, the grounding of secondary systems shall be made mandatory."

At the meeting of the National Conference on March 27, 1908, C.M. Goddard made these remarks: "The Underwriters have all along felt that it (grounding) was a rule which did not very seriously affect the fire hazard but I think the more it has been looked into, that the Underwriters have perhaps decided that there are certain fire hazards guarded against by the grounding although, of course, we all know that it is principally for guarding against life hazard."

"It is apparent that when the rule has been whipped into shape so that it will be satisfactory to all, it shall be made mandatory by the Underwriters instead of being simply permissive.

"Now the Underwriters want to be very sure that it is the desire of the electrical interests as a whole, both central station men and the engineering branch, also the municipal departments. We want to be very sure that we are not making the rule mandatory against their wishes, I should think the Underwriters would be perfectly satisfied to leave the rule as it stands."

It was voted to be the sense of the meeting that the grounding rule be made mandatory; also that instructions be given for making the grounds.

At the 17th annual meeting of the Electrical Committee, held in New York on March 24 and 25, 1909, the Committee on Rule 13A reported:[6]

"Your committee appointed to consider changes in grounding rule, and question as to whether it should be made mandatory, begs to report as follows:

"First—As we are asked only to make this rule mandatory up to and including 150 volts for alternating current secondary circuits, and to prohibit grounding of such circuits carrying in excess of 150 volts, it is evident that this suggestion cannot be based on the question of fire hazard because the fire hazard must certainly be as great from an ungrounded circuit in excess of 150 volts as from such circuit carrying not over 150 volts.

"Second—Your committee believes that the Underwriters have no right to enforce a rule which they cannot defend on the ground that it is adopted for the purpose of lessening the fire hazard.

"Third—Your committee believes that the Underwriters should use all of their moral influence in supporting any rule which may seem advisable in order to lessen the life hazard.

"Your committee would, therefore, recommend no change in the present rule but suggest that the rule, if made mandatory, might be followed in the rules of the National Board of Fire Underwriters for the government of insurance inspectors by a note saying, that in order to lessen the life hazard, compliance with the rule is recommended, but as the rule is not adopted for the purpose of lessening the fire hazard, no penalty could be enforced by the Underwriters for failure to observe its provisions.

"Your committee regrets that they have received no information as to the position of the American Institute of Electrical Engineers or the Associated Edison Illuminating Companies in regard to the subject."

This report was adopted by the meeting and the grounding section was again left unchanged in the 1909 Code.

In its annual report for 1909, the Board of Directors of AIEE included this passage:[14]

"The Code Committee has held several meetings, the most important matter considered being in connection with Rule 13A of the Board of Fire Underwriters. This rule deals with the grounding of secondary circuits. The committee agreed that the rules should be made mandatory up to and including 150 volts between any wire and ground, and optional up to, and including, 250 volts between any wire and ground. This resolution was handed to the president of the National Conference on Standard Electrical Rules. Final action on the resolution was deferred pending an arrangement for a meeting of the Code Committee with a committee of the National Board of Fire Underwriters."

So, in the 1911 Code the grounding rule, which in revision became Rule 15, was kept about as it had been, to wit:

"15. Grounding Low Potential Circuits

"Alternating-Current Secondary Systems

b. Transformer secondaries of distributing systems should preferably be grounded, and when grounded the following rules must be complied with:

1. The grounding must be made at the neutral point or wire, whenever a neutral point or wire is accessible.

2. When no neutral point or wire is accessible, one side of the secondary circuit may be grounded, provided the maximum difference of potential between the grounded point and any other point in the circuit does not exceed 250 volts.

3. The ground connection must be at the transformers or on the individual service as provided in sections c to g, and when

transformers feed systems with a neutral wire, the neutral wire must also be grounded at least every 500 feet."

"Inspection Departments having jurisdiction may require grounding if they deem it necessary."

But the time was now approaching when the grounding of low-tension secondaries was to be made mandatory.

In its report for 1911, the Board of Directors of the AIEE included the following passage:[10]

"The Code Committee through its chairman, represented the institute at the annual meeting of the National Board of Fire Underwriters held at New York on March 20-21, 1911. The only matter of interest to the institute taken up at this meeting was the grounding of secondaries and the work of the institute's representative resulted in the passing of a resolution by the underwriters conference enforcing the practice of the grounding of secondaries and recommending that municipalities and lighting companies make such a rule mandatory with the further resolution that the institute use its efforts to bring about an agreement with the NELA in the matter of grounding secondaries up to 250 volts instead of at 150 volts, the present adopted standard of the Association."

In March 1911, the Electrical Committee held its first meeting as the adopted child of the National Fire Protection Association. The following is an extract from a report of that meeting:[12]

"The following resolution, presented by the Committee on Rule 13A, was adopted:

"WHEREAS—The grounding of secondary alternating-current circuits having a normal difference of potential of not over 150 volts does effectually eliminate the life hazard of such circuits due to their accidental contact with circuits of a dangerous potential without introducing any hazard due to grounding, and

"WHEREAS—There is a difference of opinion as to whether the grounding of such circuits having a normal difference of potential of over 150 volts introduces hazards due to such grounding, and

"WHEREAS—Such grounding does not increase but rather tends to decrease, the fire hazard, therefore be it

"RESOLVED—That municipal departments are urged to make the grounding of secondary circuits up to 150 volts mandatory as a necessary safeguard to life; and

"RESOLVED—That Underwriters Inspection Departments be urged to recommend at all times such grounding as a proper and desirable precaution which introduces no fire hazard, and

"RESOLVED—That the National Electric Light Association be urged to see that all of its member companies be brought to realize the necessity of such grounding for the protection of their customers, and

"RESOLVED—That departments in charge of water works be urged to allow the attaching of such ground wires to their piping system in the full confidence that the integrity of such piping systems will in no way be affected, whatever may be the normal voltage, and

"RESOLVED—That all concerned give careful thought to other methods of obtaining satisfactory grounds wherever metallic water pipes are not available, to the end that this most necessary precaution of grounded secondary circuits may be available in all localities and under all conditions, and

"RESOLVED—That the American Institute of Electrical Engineers be urged to use its best endeavors to harmonize the present difference of opinion as to the limit of voltage at which grounding ceases to be desirable by determining a limit which shall meet with general approval."

During the year following, the American Institute of Electrical Engineers joined with other prominent national electrical societies in a final effort to secure a mandatory requirement for the grounding of low-potential secondary circuits. The AIEE Board of Directors' report for 1912, contained this passage:

"The Code Committee held a meeting on March 12, 1912, with representatives of the National Electric Light Association, Association of Edison Illuminating Companies, the National Inspectors Association, and concurred in a joint recommendation to the NFPA in regard to the grounding of secondaries."

With this substantial backing, the Electrical Committee was encouraged to approve, at its meeting on March 26 and 27, 1913, the report of its Committee on Rule 15, which made "the grounding of transformer secondary circuits obligatory."

Circuit Grounding Made Compulsory

So, at last, in the 1913 Code, a mandatory grounding rule was included, reading as follows:

"15. Grounding Low-Potential Circuits

"Alternating-Current Secondary Systems b. Transformer secondaries of distributing systems (except when supplied from private industrial power or lighting plants when the primary voltage does not exceed 550 volts) must be grounded, provided the maximum difference of potential between the grounded point and any other point in the circuit does not exceed 150 volts and may be grounded when the maximum difference of potential between the grounded point and any other

point in the circuit exceeds 150 volts. In either case the following rules must be complied with:

"1. The grounding must be made at the neutral point or wire, whenever a neutral point or wire is accessible.

2. When no neutral point or wire is accessible, one side of the secondary circuit must be grounded.

3. The ground connection must be at the transformers or on the individual service as provided in sections c to g, and when transformers feed systems with a neutral wire, the neutral wire must also be grounded at least every 500 feet."

This rule has not been changed much, in substance, since that time. The main difference is that in the 1940 Code, grounding for voltages from 150 to 300 is recommended, instead of being merely permitted. This change first appeared in the 1923 edition.

References

1. *"Safety and Safety Devices in Electrical Installations" by Professor Elihu Thomson, Electrical World, Feb. 22, 1890.*

2. *Address at Cornell University by C. J. H. Woodbury, Electrical Engineer, November 18, 1891.*

3. *Electrical World and Engineer, December 14, 1901.*

4. *"The Electrical Fire Hazard" by William T. Bennallack, Journal Western Society of Engineers, Chicago IL., March 10, 1905.*

5. *National Electrical Contractor, January, 1906.*

6. *National Electrical Contractor, April, 1909.*

7. *"The Fire Risks of Electric Lighting," by Killingsworth Hedges, Electrical World, October 5, 1889*

8. *"The Protection of Secondary Circuits from Fire Risk" by Dr. Cary T. Hutchinson, Trans. AIEE, Vol. XVI, 1899.*

9. *National Electrical Contractor, January, 1903.*

10. *Trans. AIEE, 1911, Part 3, P. 2591.*

11. *Electrical World and Engineer, December 15, 1900*

12. *National Electrical Contractor, April, 1911.*

13. *Electrical World, March 26, 1892.*

14. *Trans. AIEE 1909, Part 2, p. 1513.*

15. *"Means of Attaining Safety in Electrical Distribution" by W. L. R. Emmet, Electrical Review, June 7, 1899.*

Specifications to assure safety

Certain conclusions can thus be established to assist in writing specifications to cover Service and Equipment Grounding to assure safety in installations. They are as follows:

1. All ungrounded systems should have a ground detector and proper maintenance applied to avoid, as far as practicable, the occurrence of grounds on opposite phases at the same time. If proper maintenance cannot be assured for an ungrounded system then a grounded system should be installed.

2. The use of grounding electrodes at a transformer bank and at the building service provide by themselves virtually no protection in the event of a ground fault, as far as overcurrent protection is concerned. However, the lowest practical resistance for such grounds is desirable to serve the second purpose of holding the equipment as near to ground potential as possible.

3. The service ground and the equipment ground must have their common point of connection within the service equipment enclosure. If the service equipment is in the switchboard, then the common point must be within the switchboard section which houses the service equipment. A common grounding conductor to the grounding electrode must be run from that common point of connection.

4. When a transformer bank is grounded, that is, when a service is run to a building from a system which has any point grounded, the grounded conductor must be run to the service equipment whether that grounded conductor is to be used for voltage requirements or not, and that grounded conductor must be connected to the equipment grounding conductor within the service equipment enclosure.

5. When a fault occurs on any grounded system we must not assume it began as a short circuit. (phase/to/phase fault) until we have clearly indisputable evidence to the contrary, if we are to analyze the fault correctly. It is almost self-evident, when the overcurrent devices do not operate promptly that the trouble was owing to a ground fault.

Unless "effective grounding" is followed and carefully observed, we cannot expect a ground fault to clear unless the circuit is manually opened or the ground fault continues and develops into a phase to phase fault. The latter condition may require many minutes during which time the equipment grounding conductor will develop a potential to ground, which may become high enough to be dangerous. Further, we may have a disastrous fire loss. Some method of automatically clearing a ground fault in limited time is highly desirable.

Engineering specifications must be carefully written to cover service and equipment grounding and not left to chance interpretation of how to accomplish safety. It is not an uncommon practice for specifications to read: "Ground in accordance with the provisions of Article 250 of the National Electrical Code®." The engineer is not assuming his full responsibility for the safety of the installation when he writes such specifications. It is the engineer's responsibility to outline in his specifications how the job should be installed to fully comply with all three of the objectives of "effective grounding" and not leave it to chance, as is now often the case.

Ufer Ground

H.G. Ufer, in an IEEE Conference Paper, CP-61-978, describes an installation of made ground electrodes on 24 buildings in 1942, in Arizona, to meet a 5-ohm maximum value. The resistance values were checked bimonthly over an 18-year period, during which time no servicing was required.

In 1960, the maximum reading was 4.8 ohms and the minimum 2.1 ohms. The average value of the 24 installations was 3.57 ohms.

The installations used ½-inch steel reinforcing rods set in a concrete footing. They were at two locations in Arizona. The first was near Tucson, Arizona, which is normally hot and dry during most of the year and has an average annual rainfall of 10.91 inches. The soil is sand and gravel. The second location was near Flagstaff, Arizona, where the soil is clay, shale gumbo and loam with small area stratas of soft limestone. The made electrodes were used as no water piping system was available.

As a result of these installations and the 18-year test period, Mr. Ufer suggested that a No. 4 or larger copper wire be embedded in the concrete footing of a building and that test data be compiled further to verify the effectiveness. Based on this data, CMP-5 accepted a concrete-encased electrode commonly referred to as a "Ufer Ground." The concrete-encased electrode shall consist of at least 20 feet of bare copper not smaller than No. 4 AWG encased in 2 inches of concrete near the bottom of the footing or foundation.

Answers to Questions Chapter One - Seventeen

Answers to questions in Chapters 1-17.
The references included below can be found on the pages in this textbook or in the 1999 National Electrical Code® under the sections given.

Chapter 1
1. c. In textbook
2. d. Section 250-2(a)
3. d. Section 250-2(d)
4. d. In textbook
5. b. In textbook
6. a. In textbook
7. b. In textbook
8. b. In textbook
9. d. In textbook

Chapter 2
1. d. Section 250-20(a)(1)
2. b. Section 250-20(a)(2)
3. a. Section 250-20(a)(3)
4. b. Section 250-20(b)(1)
5. b. Section 250-20(b)(2)
6. c. Section 250-20(b)(3)
7. d. Section 250-21(1)
8. b. Section 250-21(2)
9. d. Section 250-21(3)
10. b. Section 250-21(4)
11. d. Section 250-21(5)
12. b. Section 250-20(c)
13. a. Section 250-20(c)
14. d. In textbook
15. d. Article 100
16. c. Section 250-20(d)
17. b. In textbook
18. c. In textbook
19. d. In textbook
20. d. In textbook
21. a. In textbook
22. c. In textbook
23. b. In textbook
24. d. In textbook

Chapter 3
1. b. Article 100
2. a. Article 100
3. c. Article 100
4. a. Section 200-6(a)
5. c. Section 200-6(e), Exception No. 1
6. d. Section 200-6(b)
7. d. Section 200-6(a)
8. c. Sections 215-8, 230-56 and 384-3(e)
9. d. In textbook

Chapter 3 *continued*
10. a. In textbook
11. d. In textbook
12. b. In textbook
13. d. In textbook
14. c. In textbook
15. b. In textbook
16. d. In textbook
17. c. In textbook
18. b. In textbook
19. c. Section 240-85
20. d. Section 430-36

Chapter 4
1. c. Section 250-24(b)
2. d. Section 250-24(b)
3. a. Section 250-24(b)
4. b. Section 250-24(b)
5. a. Section 250-24(b)
6. c. Section 250-24(b)
7. c. Section 250-24(b)
8. a. In textbook
9. d. In textbook
10. a. Section 250-24(a)(2)
11. c. Table 250-66
12. c. In textbook
13. d. In textbook
14. d. Section 250-36
15. a. Section 250-170
16. b. Section 250-170, Exception
17. d. Section 250-172, Exception

Chapter 5
1. b. Article 100
2. a. Article 100
3. d. Section 250-28(a)
4. b. Section 250-28b)
5. d. Table 250-66
6. d. Section 250-28(d)
7. a. Section 250-28(d)
8. d. Section 250-71
9. b. Section 250-94(i)
10. c. In textbook
11. a. Table 250-102(c)
12. d. Section 250-94

Chapter 6
1. b. In textbook
2. a. Article 100
3. d. Section 250-50(3)
4. c. Section 250-50(c)
5. b. Section 250-50(d)

Chapter 6 *continued*

6. a. Section 250-50(a)
7. d. Section 250-50(a)(2)
8. c. Section 250-50
9. d. Section 250-52
10. a. Section 250-58
11. d. Section 250-6
12. b. Section 250-6(d)
13. d. Section 250-56 (In text)
14. a. In textbook
15. d. In textbook
16. b. Section 250-60

Chapter 7

1. b. Article 100
2. c. Table 250-66
3. c. Table 250-66
4. a. In textbook
5. c. Table 250-66
6. c. Table 250-66
7. d. In textbook
8. d. Table 250-66
9. c. Section 250-64(d)
10. b. Section 250-66(a)
11. b. Section 250-66(b)
12. a. Section 250-70
13. b. Section 250-70
14. a. Section 250-64(b)
15. b. Section 250-64(b)
16. b. Section 250-64(a)

Chapter 8

1. d. Article 100
2. d. Section 250-113
3. d. Section 370-3, Exception Nos. 1 and 2
4. c. Section 250-114
5. b. Section 250-74, Exception No. 4
6. c. Section 250-79(f)
7. d. Section 250-80(b)
8. d. In textbook
9. c. Section 250-80(b)
10. b. Section 250-77
11. a. Section 250-78
12. c. Section 250-95
13. c. Section 250-80(a)
14. c. In textbook
15. b. Section 250-80(a)
16. d. Section 250-80(c)
17. d. Section 250-104(a)(3)
18. a. Section 250-104(a)(2)

Chapter 9

1. a. Article 100
2. d. In textbook
3. c. In textbook
4. a. Table 250-122
5. d. Section 250-118
6. c. Section 250-118
7. c. Section 250-118(7)
8. c. Section 250-120(a)
9. b. Sections 250-122(e)
10. b. Section 250-134(b)
11. c. Section 250-122(f)(1)
12. a. Section 250-122(f)(2)

Chapter 10

1. c. Article 100
2. a. Article 100
3. b. Section 250-110
4. d. Section 250-110
5. d. Section 250-86 Exception No. 1
6. c. Section 250-110, Exception No. 1
7. b. Section 250-110, Exception No. 2
8. d. Section 250-110, Exception No. 3
9. b. Section 250-86, Exception No. 1
10. a. Section 250-112(l)
11. d. Section 250-8
12. b. Section 250-8
13. c. Section 250-80
14. b. Section 250-140
15. b. Sections 250-24(a), 215-6 and 384-20

Chapter 11

1. c. In textbook
2. b. In textbook
3. c. In textbook
4. b. In textbook
5. a. In textbook
6. b. In textbook
7. d. In textbook
8. b. In textbook
9. a. In textbook
10. b. In textbook
11. a. No. 6, In textbook
12. b. No. 2, In textbook
13. c. No. 10, In textbook
14. d. No. 2, In textbook

Chapter 12

1. c. Article 100
2. b. Section 250-30(a)(2) and Table 250-66
3. a. Section 250-30(b)
4. c. Section 250-30(a)(3)

Chapter 12 *continued*
5. b. Section 250-30(a)(3)
6. c. Section 250-30(a)(3)
7. a. Section 250-58
8. d. In textbook
9. a. In textbook
10. b. In textbook
11. d. In textbook

Chapter 13
1. b. Article 100
2. c. Section 250-32(a)
3. c. In textbook
4. a. Section 250-32(a)
5. c. Section 250-32(b)(1)
6. a. Section 250-52
7. c. In textbook
8. c. Section 250-32(a)
9. a. Section 250-142(a)
10. a. Section 250-6(d)
11. d. Section 250-32(e)

Chapter 14
1. b. Section 210-7(d)
2. a. Section 210-7(d), Exception
3. b. Section 210-8(a)(2), Exception No. 2
4. c. Section 210-8(a)(4)
5. c. Section 210-8(a)(5)
6. c. Section 210-63
7. d. Section 210-8(b)(1)
8. d. Section 210-8(a)(3)
9. b. In textbook
10. a. In textbook
11. a. In textbook
12. b. In textbook
13. a. In textbook

Chapter 15
1. b. Article 100
2. b. Section 230-95
3. c. Section 230-95, Exception Nos. 1 and 2
4. b. Section 230-95(a)
5. a. In textbook, Section 215-10
6. c. In textbook
7. a. In textbook
8. d. Section 230-95(c)
9. b. Section 517-17(a)
10. c. Section 215-10

Chapter 16
1. c. In textbook
2. a. Section 250-106
3. c. Sections 501-16(a), 502-16(a) and 503-16(a)
4. d. In textbook
5. a. Section 250-102(e)
6. b. Section 547-4(f)
7. a. Section 547-9(b)
8. c. Section 547-9(d)
9. d. In textbook
10. d. Section 517-13(b)
11. d. Section 517-19(d)
12. c. Section 680-22(a)(5)
13. a. Section 680-22(b)(3)
14. a. Section 680-20(c)
15. d. Section 680-6(a)(3)
16. c. Section 680-5(c)
17. a. Section 680-7
18. c. Section 680-5(b)
19. d. Section 680-20(a)(1)
20. d. Section 680-20(a)(1)
21. b. Section 680-20(b)(3)
22. a. Section 680-20(b)(3)
23. b. Section 680-5(a)
24. b. Section 680-5(a)
25. c. Section 600-7
26. c. Section 600-7
27. d. Section 600-7

Chapter 17
1. b. In textbook
2. a. In textbook
3. b. In textbook
4. c. In textbook
5. c. In textbook
6. a. In textbook
7. c. In textbook
8. b. In textbook
9. c. In textbook
10. d. In textbook
11. a. In textbook
12. b. In textbook

Index

—T—

—U—

—V—

—W—

—Z—

IAEI Soares Book on Grounding

Designed by: Brady Davis

Cover Design: Frank Kripaitis
                       Kirk Massey

Contributor: Travis Lindsey

Technical Review: J. Philip Simmons
                       Paul Dobrowsky
                       Chuck Mello

Illustrations &
  Reprints: AEMC Instruments, Chauvin Arnoux, Inc., Boston, MA
                       AFC Cable Systems, New Bedford, MA
                       AVO International, Biddle Instruments, Dallas, TX
                       Cooper Industries, Bussmann Division, St. Louis, MO
                       Eagle Electric Company, Long Island City, NY
                       Insulated Cable Engineers Association, South Yarmouth, MA
                       International Electronic and Electrical Engineers,
                           Piscataway, PA
                       National Electrical Manufacturers Association, Rosslyn, VA
                       National Fire Protection Association, Quincy, MA
                       NFPA Research Foundation, Quincy, MA
                       Post Glover Resistors, Inc., Erlanger, KY
                       Square D Company, Palatine, IL
                       Steel Conduit Section, National Electrical Manufacturers
                           Association, Rosslyn, VA
                       Thomas & Betts, Memphis, TN
                       Underwriters Laboratories, Northbrook, IL

Graphic Artists: Brady Davis
                       Elaine Flynn
                       Joanne Beverly
                       Richard Church

Project Editor: Kathryn P. Ingley

Project Manager: Elaine Flynn

Composed at
International Association of Electrical Inspectors
in New Century Schoolbook and Helvetica-Narrow
Printed by The Odee Company on Luna Matte text stock
Bound by The Odee Company in Kalima White Coated Cover